BEHAVIORAL GENETICS

A Primer

A Series of Books in Psychology

Editors:

Gardner Lindzey
Jonathan Freedman
Richard F. Thompson

BEHAVIORAL GENETICS

A Primer

Robert Plomin
J. C. DeFries
G. E. McClearn
University of Colorado

W. H. FREEMAN AND COMPANY
San Francisco

Sponsoring Editor: W. Hayward Rogers
Project Editor: Pearl C. Vapnek
Copyeditor: John Hamburger
Designer: Marie Carluccio
Production Coordinator: Linda Jupiter
Illustration Coordinator: Cheryl Nufer
Compositor: Bi-Comp, Inc.
Printer and Binder: The Maple-Vail Book Manufacturing Group

Library of Congress Cataloging in Publication Data

Plomin, Robert, 1948–
Behavioral genetics, a primer.

(A Series of books in psychology)
"Follows an earlier text by G. E. McClearn and J. C. DeFries, Introduction to behavioral genetics, W. H. Freeman and Company, 1973."
Bibliography: p.
Includes index.
1. Behavior genetics. I. DeFries, J. C., 1934– joint author. II. McClearn, G. E., 1927– joint author. III. McClearn, G. E., 1927– Introduction to behavioral genetics. IV. Title.
QH457.P56 591.1′5 79-24456
ISBN 0-7167-1127-3
ISBN 0-7167-1128-1 pbk.

Printed in the United States of America

9 8 7 6 5 4 3 2 1

Contents

Preface xi

1 Overview 1

Plan of the Book 3
Nature and Nurture 5
Interactionism 6
Genes and Behavior 7
Do Genes Determine One's Destiny? 9
Discerning "What Is" from "What Could Be" 9
Individual Differences and Equality 10
Summary 12

2 Historical Perspective 13

The Era of Darwin 15
Darwin's Theory of Evolution 20
The Principle of Natural Selection 21
Reaction to the Theory of Evolution 22
Galton's Contributions 25
Hereditary Genius 26
Pioneering Research in Psychology and Statistics 27

Twins and the Nature–Nurture Problem 28
Galton's Work in Perspective 32
Pre-Mendelian Concepts of Heredity and Variation 33
Heredity 33
Variation 33
The Work of Gregor Mendel 34
Summary 37

3 Evolution 38

The Evolutionary Record: Genetic Variability over Time 38
The Replicators 38
Beginnings of Life 39
Primates and Hominids 42
Behavior and Evolution 47
Genetic Variability Between Species 51
Evolution and Variability 56
Sociobiology 58
Between-Species Examples of Sociobiology 59
Sociobiology and Genetic Differences Within Species 61
Sociobiology and Behavioral Genetics 62
Genetic Variability Within Species 64
Summary 64

4 Single Genes 66

Mendel's Experiments and Laws 66
Monohybrid Experiments 67
Dihybrid Experiments 72
Distributions 75
Mendelian Crosses in Mice 77
The Genetics of Waltzer Mice 77
The Genetics of Twirler Mice 78
The Genetics of Audiogenic Seizures in Mice 79
The Genetics of Squeaking in Mice 81
Behavioral Pleiotropism 83
Recombinant Inbred Strains 85
Mendelian Analysis in Humans 88
Dominant Transmission 89
Recessive Transmission 93
Sex-Linked Transmission 97
Scope of Single-Gene Effects in Human Behavior 101
Summary 101

5 Mechanisms of Heredity and Behavior 102

Mechanisms of Gene Action 102
The Chemical Nature of Genes 104
Genes and Protein Synthesis 105
A New View of Mendel's "Elements" 110

Phenylketonuria 110
Other Recessive Metabolic Errors 113
Regulator Genes 114
Mutation 117
Using Mutations to Dissect Behavior 118
Bacteria *119*
Paramecia *121*
Round Worms *121*
Drosophila *122*
Genetic Mosaics *124*
Genetic Dissection and Development (Fate Mapping) *127*
Genes and Individual Differences 130
Summary 130

6 Chromosomes 132

Mendel and Chromosomes 133
Cell Division 134
Mitosis *135*
Meiosis *135*
Crossing Over 137
Human Chromosomes 142
Karyotypes *142*
Banding *143*
Chromosomal Abnormalities 149
Sex Chromosomes and the Lyon Hypothesis 150
Chromosomes and Genetic Variability 152
Summary 153

7 Chromosomal Abnormalities and Human Behavior 154

Down's Syndrome 157
15/21 Translocation Down's Syndrome *158*
Relation to Maternal Age *160*
Turner's Syndrome(XO) 162
Females with Extra X Chromosomes 165
Klinefelter's Syndrome 167
Males with Extra Y Chromosomes 169
Summary 174

8 Population Genetics 175

Allelic and Genotypic Frequencies 175
Hardy–Weinberg–Castle Equilibrium 177
Uses of the Hardy–Weinberg–Castle Equilibrium *178*
Tasting PTC *179*
Forces That Change Allelic Frequency 181
Migration *181*
Random Genetic Drift *182*

Mutation *183*
Selection *183*
Balanced Polymorphisms 187
Heterozygote Advantage *188*
Frequency-Dependent Selection *188*
Frequency-Dependent Sexual Selection *188*
Forces That Change Genotypic Frequencies 193
Inbreeding *193*
Assortative Mating *197*
Genetic Variability 198
Summary 199

9 Quantitative Genetic Theory 200

Brief Overview of Statistics 202
Statistics Describing Distributions *202*
Statistics Describing the Relationship Between Two Variables *204*
Historical Note 210
The Single-Gene Model 211
Genotypic Value *212*
Additive Genetic Value *212*
Dominance Deviation *213*
The Polygenic Model 214
Epistatic Interaction Deviation *215*
Variance *215*
An Example of the Polygenic Model *217*
Covariance of Relatives 219
Covariance *220*
Shared and Independent Influences *220*
Genotype-Environment Correlation and Interaction *222*
Genetic Covariance Among Relatives *222*
Heritability 224
What Heritability Is Not *225*
What Heritability Is *226*
Path Analysis *227*
Multivariate Analysis *228*
Resemblance of Relatives Revisited 231
Univariate Analysis *231*
Multivariate Analysis *232*
Summary 233

10 Quantitative Genetic Methods: Animal Behavior 234

Family Studies 234
Open-Field Behavior in Mice *237*
Multivariate Analysis *238*
Strain Studies 240
Genetic Effects on Learning *243*
Environmental Effects on Behavior *247*
Classical Analysis *253*
Diallel Design *255*

Selective Breeding Studies 256
Dogs *256*
Heritability *258*
Maze Running in Rats *262*
Open-Field Behavior *265*
Other Behavioral Selection Studies *270*
Summary 273

11 Family Studies of Human Behavior 274

Components of Covariance in Family Studies 275
Cognition 276
Intelligence Quotient (IQ) *277*
Specific Cognitive Abilities *280*
Multivariate Analysis *281*
Psychopathology 282
Schizophrenia *284*
Manic-Depressive Psychosis *285*
Heterogeneity *287*
Summary 288

12 Twin Studies 289

Zygosity Determination 292
Equal Environments Assumption 295
Heritability 299
Environmentality 302
Other Twin Designs 303
Families of Identical Twins *304*
Identical Co-twin Control Studies *305*
Studies of Genetic Similarity Within Pairs of Fraternal Twins *306*
Twin Studies and Behavior 306
Cognition 307
Intelligence Quotient (IQ) *307*
Developmental Differences *309*
Specific Cognitive Abilities *312*
Multivariate Analysis *314*
Psychopathology 318
Schizophrenia *319*
Identical Twins Discordant for Schizophrenia *322*
Manic-Depressive Psychosis *323*
Summary 323

13 Adoption Studies 325

Issues in Adoption Studies 326
Representativeness *326*
Selective Placement *327*
Other Issues *328*
Heritability and Environmentality 329

Intelligence Quotient (IQ) 331
Parent-Offspring Adoption Studies *331*
Sibling Studies *339*
Identical Twins Reared Separately *341*
Specific Cognitive Abilities 341
Psychopathology 342
Schizophrenia *343*
Summary of Adoption Studies on Schizophrenia *350*
Psychopathology Other Than Schizophrenia *351*
Summary 352

14 Directions in Behavioral Genetics 354

Recapitulation and Synthesis 354
Genetics of General Cognitive Ability *355*
Fitting Models to IQ Data *357*
Controversies 358
Nature–Nurture Arguments *358*
Group Differences *366*
Science and Politics *370*
The Importance of Behavioral Genetics 374
Behavior *375*
Genetic Influences on Behavior *375*
Environmental Influences on Behavior *375*
The Interface Between Genes and Environment *376*
Summary 377

References 379

Index of Names 405

Index of Topics 411

Preface

This textbook follows an earlier text by G. E. McClearn and J. C. DeFries (*Introduction to Behavioral Genetics*, W. H. Freeman and Company, 1973). Rather than a revision of the earlier text, it is more of a sequel. The earlier text was written primarily for advanced undergraduate and graduate students taking their first course in behavioral genetics. However, it quickly became apparent that the "little green book" was not truly an introductory text. Therefore, as more undergraduates became attracted to the course, we felt a need for a text that assumed less previous exposure to genetics and statistics. As a new member of the faculty of the Institute for Behavioral Genetics, Robert Plomin undertook this task, in collaboration with McClearn and DeFries. The result is the present text, which includes some sections from the earlier book, but is largely new.

Although behavioral genetics is a complex field, we have written a book that is as simple as possible without sacrificing honesty of presentation. Most importantly, we have written a book that is fundamentally for students, not for our colleagues. Although our coverage is representative, it is by no means exhaustive or encyclopedic; for those who seek more comprehensive coverage, there are other books (for example, Dworkin and Haber, 1980; Fuller and Thompson, 1978; Ehrman and Parsons, 1976; Van Abeelen, 1974). On the basis of comments made by the students who used an earlier draft of this text in an undergraduate class at the University of Colorado, we expect that you will find it challenging but readable. We hope that it will stimulate you to learn more about behavioral genetics and to begin thinking about behavior from the behavioral genetics perspective.

We could not have written a book of this kind if we were not closely associated with an interdisciplinary group of researchers from many academic disciplines. Much of our thinking is shaped by daily interactions with our colleagues at the Institute for Behavioral Genetics. We cannot even begin to enumerate their influences, let alone express proper thanks. Rebecca G. Miles, with her excellent editorial skills, transformed our scrawls and scraps into readable prose. Kimberly A. Myers took the transformed scraps and turned them into a beautifully typed manuscript. Agnes A. Conley, administrative officer of the institute, helped to oversee the project, read the manuscript, and assisted in the bibliographic work.

In addition to acknowledging our associates at the Institute for Behavioral Genetics, we thank Glayde Whitney of Florida State University and Joseph Horn of the University of Texas at Austin for reviewing an earlier draft of the manuscript and for making many useful suggestions. Finally, we appreciate the professionalism of the staff at W. H. Freeman and Company.

November 1979

Robert Plomin
J. C. DeFries
G. E. McClearn

BEHAVIORAL GENETICS

A Primer

1

Overview

In his introduction to *Hereditary Genius: An Inquiry into Its Laws and Consequences,* Francis Galton begins:

> I propose to show in this book that a man's natural abilities are derived by inheritance, under exactly the same limitations as are the form and physical features of the whole organic world. (Galton, 1869, p. 45)

However, more than a century later, in his introduction to *The Science and Politics of I.Q.,* Leon J. Kamin states:

> The present work arrives at two conclusions. The first stems from a detailed examination of the empirical evidence which has been adduced in support of the idea of heritability, and it can be stated simply. There exist no data which should lead a prudent man to accept the hypothesis that I.Q. test scores are in any degree heritable. That conclusion is so much at odds with prevailing wisdom that it is necessary to ask, how can so many psychologists believe the opposite?
>
> The answer, I believe, is related to the second major conclusion of this work. The I.Q. test in America, and the way in which we think about it, has been fostered by men committed to a particular social view. (Kamin, 1974, pp. 1–2)

Have the hundred years of research since the publication of *Hereditary Genius* yielded incontrovertible evidence that mental ability is *not* heritable? We recently reviewed adoption studies concerning the genetics of mental ability and concluded that "a prudent person has no alternative but to reject the hypothesis of zero heritability" (DeFries and Plomin, 1978, p. 501). Later in this text we shall review this evidence, and you may draw your own conclusions. We ask only that you consider the evidence on its own merits and avoid being unduly influenced by your own particular social view. We have no ax to grind on this or any other controversial issue in behavioral genetics and will strive to remain apolitical. As John L. Fuller and W. Robert Thompson have warned:

> . . . human behavior genetics, of all disciplines in science, is always in danger of edging into the political domain. To the extent this happens, it becomes less of a science. Consequently, there is a real onus on all workers in the area to avoid, as far as possible, any political biases that may erode their impartiality as scientists and to follow closely the dictates of facts and logic. (Fuller and Thompson, 1978, p. 225)

We shall make every effort to heed this cogent warning.

Although studies related to IQ are probably most familiar to students, behavioral genetics research has extended into many other areas. For example, in a review of the genetics of specific cognitive abilities, J. C. DeFries, S. G. Vandenberg, and G. E. McClearn stated that "cognitive ability is far too complex to be assessed by a univariate number such as IQ" (1976, p. 180). Psychopathology and personality have also been studied extensively from a behavioral genetics perspective. Furthermore, it often surprises students to learn that more research in behavioral genetics has been devoted to understanding the behavior of nonhuman animals than the behavior of our own species. The primary reason for this emphasis on nonhuman behavior is the experimental control possible in the laboratory.

Behavioral genetics is most controversial when it considers human behavior, particularly when it focuses on socially relevant aspects of human behavior, such as IQ. We shall discuss these controversies. However, our major objective is to communicate the principle that both heredity and environment can be responsible for individual behavioral differences. We shall demonstrate that neither heredity nor environment alone is sufficient for understanding behavior.

This unique perspective, combining genetics and the behavioral sciences, makes behavioral genetics an exciting interdisciplinary area. (See Figure 1.1.) Although research in behavioral genetics has been conducted for many years, the field has only recently emerged as a distinct discipline. The field-defining monograph *Behavior Genetics,* by John L. Fuller and W. Robert Thompson, was published in 1960. Since that date, research in behavioral genetics has undergone exponential growth.

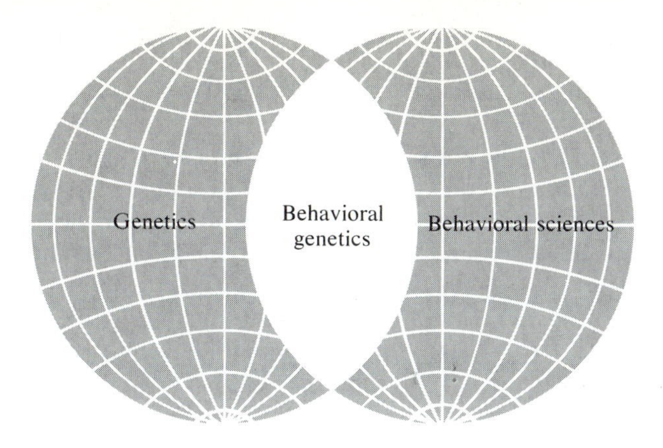

FIGURE 1.1
Behavioral genetics as the intersection between genetics and the behavioral sciences. (From *Introduction to Behavioral Genetics* by G. E. McClearn and J. C. DeFries, W. H. Freeman and Company. Copyright © 1973.)

PLAN OF THE BOOK

We begin with a historical perspective on evolution and genetics, followed by three chapters on basic principles. Behavior is a *phenotype*—that is, an observable characteristic that we can measure. The basic principles of heredity are the same, regardless of the phenotypes we choose to study. Thus, much of this book is devoted to genetic concepts that are not unique to behavioral characters. On the other hand, this book is not simply another genetics text; its focus is on the complexity of behavioral phenotypes. Although behavioral genetics is relevant to the perspectives of anthropologists, economists, educators, political scientists, sociologists, and others, the behavioral phenotypes that we discuss necessarily reflect the fact that most behavioral genetics research currently centers on topics traditionally defined within the domain of psychology.

The historical perspective of Chapter 2 describes the work of Charles Darwin and his cousin Francis Galton. Chapter 3 continues the discussion of Darwin's theory of evolution and relates it to behavioral genetics and sociobiology. The next two chapters provide the basic principles of genetics. Chapter 4 focuses on Gregor Mendel's classic experiments and on single-gene influences on behavior. In Chapter 5 we present a basic description of genes and how they work, with particular reference to the heredity of behavior. Chromosomes and their influence on behavior are dealt with in Chapters 6 and 7. The next two chapters focus on population genetics and quantitative genetics, two topics that tend to be difficult for students who are not statistically inclined. For this reason, we have attempted to discuss these concepts rather than to present them algebraically. Population genetics (Chapter 8) considers the transmission of genes from the population perspective rather than from the individual perspective. Quantitative genetics (Chapter 9) generalizes the single-gene, Mendelian model to a *multifactorial model* that considers the effects of many genes and many environmental factors. The multifactorial model is most relevant to the study of behaviors that interest behavioral scientists.

In Chapters 10 to 13, we look at the methods and findings of behavioral genetics in greater detail. In Chapter 10, we shall consider methods specific to the analysis of nonhuman behavior. Chapters 11, 12, and 13 are devoted to the three major human behavioral genetics methods—family studies, twin studies, and adoption studies. In writing a text on behavioral genetics, it is possible to choose a content-oriented or a method-oriented approach. We have chosen a method-oriented presentation. That is because our goal is to present the methods of behavioral genetics, rather than to provide an exhaustive review of the applications of these methods to various behavioral domains. (For such reviews of the literature, see Fuller and Thompson, 1978.) In Chapter 10, we shall refer to various behaviors in providing examples of nonhuman quantitative genetics methods. In Chapters 11 through 13, we will focus on two major domains of behavior, cognition (IQ and specific cognitive abilities) and psychopathology (primarily the psychoses), although we will provide references to other behavioral domains.

In the last chapter, we recapitulate some of the major issues raised in the preceding chapters, and then address some of the controversial issues in behavioral genetics. These include the origin of group differences (ethnic, class, and sex differences) and the societal implications of behavioral genetics studies. It is likely that you are familiar with some of the controversies surrounding behavioral genetics and you may have some preconceptions about them. Thus, here are a few words about the origin of some of the controversial issues.

NATURE AND NURTURE

Some controversy can be expected surrounding any new approach that questions prevailing views, and behavioral genetics is certainly no exception. Much of the controversy concerning behavioral genetics stems from a misunderstanding of what it means to say that genes influence behavior. For example, behavioral genetics is often mistakenly viewed as pitting nature (genes) against nurture (environment), as if behavior were influenced solely by one or the other. The story behind this view is an interesting bit of history, and it helps to explain the frequent appearance of this view even today.

Historically, the controversy began at the turn of the century with the development of the *behavioristic* point of view, which assumed a dominant role in the developing discipline of psychology, particularly in the United

States. Behaviorism arose as a protest against all forms of "introspective psychology," which was concerned with mental states such as consciousness and will. The term *behaviorism* refers to a strict focus on observable behavioral responses. Because its emphasis on observable responses led to an emphasis on observable environmental stimuli, behaviorism came to imply *environmentalism.* Stimulus-response chains eventually became the only acceptable explanation of behavior. Behaviorism also moved in the direction of environmentalism with its rejection of the instinct doctrine of W. McDougall (1908). At that time, instincts were thought of as inherited patterns of behavior, and the behaviorists attacked this position as redundant and circular. However, in rejecting this naive view of instincts, the behaviorists also discarded the notion that heredity can influence behavior. Thus, the behaviorists explained individual differences completely by environmental factors.

J. B. Watson, the founder of behaviorism, said:

> So let us hasten to admit—yes, there are heritable differences in form, in structure.... These differences are in the germ plasm and are handed down from parent to child.... But do not let these undoubted facts of inheritance lead us astray as they have some of the biologists. The mere presence of these structures tells us not a thing about function.... Our hereditary structure lies ready to be shaped in a thousand different ways—the same structure—depending on the way in which this child is brought up....
>
> Objectors will probably say that the behaviorist is flying in the face of the known facts of eugenics and experimental evolution—that the geneticists have proven that many of the behavior characteristics of the parents are handed down to the offspring.... Our reply is that the geneticists are working under the banner of the old "faculty" psychology. One need not give very much weight to any of their present conclusions. We no longer believe in faculties nor in any stereotyped patterns of behavior which go under the names of "talent" and inherited capacities....
>
> Our conclusion, then, is that we have no real evidence of the inheritance of traits. I would feel perfectly confident in the ultimately favorable outcome of careful upbringing of a *healthy, well-formed* baby born of a long line of crooks, murderers and thieves, and prostitutes. Who has any evidence to the contrary? (Watson, 1930, pp. 97–103)

Watson followed this with the familiar and frequently quoted challenge:

> I should like to go one step further now and say, "Give me a dozen healthy infants, well-formed, and my own specified world to bring them up in and I'll guarantee to take any one at random and train him to become any type of specialist I might select—doctor, lawyer, artist, merchant-chief and, yes, even beggar-man and thief, regardless of his talents, penchants, tendencies, abilities, vocations, and race of his ancestors." I am going beyond my facts and I admit it, but so have the advocates of the contrary and they have been doing it for many thousands of years. (Watson, 1930, p. 104)

R. S. Woodworth (1948) pointed out that this extreme environmentalism was not a necessary consequence of the behavioristic position. He suggested that Watson's stand was taken, in part, "to shake people out of their complacent acceptance of traditional views" (p. 92). For whatever reason, Watson sought to exorcise genetics from psychology, and he succeeded to a remarkable degree. His position in his book *Behaviorism* soon became the traditional view that was complacently accepted by the majority of psychologists.

But this majority view was not without opposition. In fact, since Watson's pronouncement, not a single year has passed without publication of some evidence showing it to be wrong. Collectively, these researches have demonstrated the important role of heredity in many varieties of behavior and in many kinds of organisms.

INTERACTIONISM

During the last decade, there have been clear signs that the behavioral sciences are beginning to accept the theory and methodology of genetics. Rarely does anyone concur with Watson's conclusion that genes are unimportant in behavior. Instead, it is now generally agreed that both nurture and nature play a role in determining behavior. However, the mistaken notions of the nature-nurture argument have too often been replaced with the equally mistaken view that the effects of heredity and environment cannot be analyzed separately, a view called "interactionism" (Plomin, DeFries, and Loehlin, 1977). This topic will be discussed in detail in the last chapter, but we shall say a few words about it here.

Obviously, there can be no behavior without both an organism and an environment. The scientifically useful question is: For a particular behavior, what causes differences among individuals? For example, what causes individual differences in cognitive ability? Various environmental hypotheses leap to mind: Families differ in the stimulation they offer for cognitive growth; environments differ in motivating people toward intellectual goals; educational experiences differ. However, genetic hypotheses should also be considered.

Research in behavioral genetics is directed toward understanding differences in behavior. Methods are employed that consider both genetic and environmental influences, rather than assuming that one or the other is solely important. As a first step, behavioral genetics research studies whether individual behavioral differences are influenced by hereditary differences, and estimates the relative influences of genetic and environmental factors. Although this first step includes most human behavioral genetics research to date, it is only a first step. If there are genetic effects, we can then ask how many genes are involved and on what chromosomes they are located. We can also begin to study the evolutionary significance of these genes, their developmental course, their interaction with the environment, and the pathways between the primary effect of genes on protein production and regulation and their ultimate effect on behavior.

GENES AND BEHAVIOR

Some people are disturbed by the idea that genes can influence behavior. They don't understand the workings of genes and probably picture them as master puppeteers within us, pulling our strings. To the contrary, genes are merely chemical structures. However, encoded in these structures are the messages that enable genes to do their marvelous job of reliably replicating themselves and controlling development. In Chapter 5, we shall see that there is no such thing as a gene for behavior; nor is there a gene for the length of one's nose. Genes are blueprints for the assembly and regulation of proteins, which are the building blocks of our bodies, including the nervous system. Each gene codes for a specific sequence of amino acids that the body assembles to form a protein. If even the smallest part of this chain is altered, the entire protein can malfunction. We shall discuss this in much greater detail in Chapter 5.

Here, we want to emphasize that genes are not mystical entities. Genes do not magically blossom into behavior or anything else. They are stretches of chemical bonds that code for protein production or regulate the activity of other genes. In this sense, all aspects of human beings—behavior as well as our bones—are part of this process. As illustrated in Figure 1.2, proteins do not directly cause behavior. For example, one gene (G_2) codes for a particu-

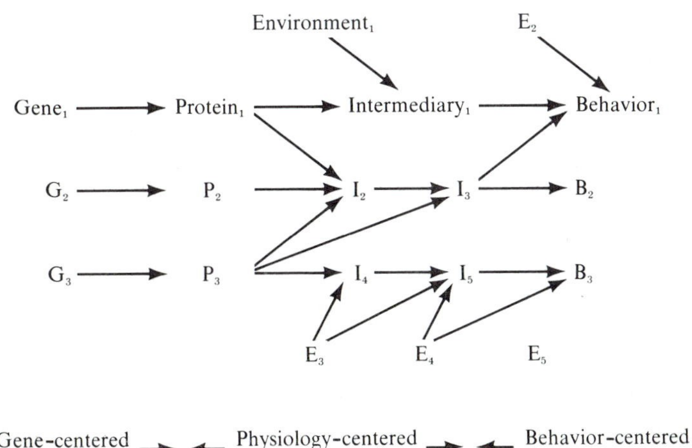

FIGURE 1.2

Genes do not directly cause behavior; they work indirectly via physiological systems.

lar protein (P_2). However, that protein does not cause a particular type of behavior. There is no gene or protein, for example, that repeatedly causes a person to lift shot glasses and perhaps become an alcoholic. Proteins interact with other physiological intermediaries (such as I_2), which may be other proteins, such as hormones or neurotransmitters, or may be structural properties of the nervous system. Environmental factors (such as E_2, which might represent nutrition) may also be involved. These influences can ultimately and indirectly influence behavior in a certain direction. For example, differences in neural sensitivity to ethanol may tip the scale in the direction of alcoholism for an individual who imbibes frequently. Various genes, chemical and structural brain differences, and environmental factors may be at the root of such differences in neural sensitivity.

So, when we talk about genetic influences on behavior, we do not mean robotlike, hard-wired circuits. We are referring to indirect and complex paths between genes and behavior via proteins and physiological systems. Ernst Mayr (1974) distinguished *closed behavioral programs* from *open behavioral systems* to replace the hopelessly entangled arguments concerning instinct and learning. Closed behavioral programs are relatively impervious to individual experiences. Even so, there is no implication that a gene directly produces a particular behavior; genes always work by controlling the production of proteins. Open behavioral systems are more susceptible to individual experience. Closed and open systems differ only in the degree of flexibility. Behavioral systems of higher organisms tend toward the "open" end of this continuum.

Figure 1.2 also points out different approaches to the study of behavioral genetics. Researchers using these different approaches tend to have different goals. The *gene-centered approach* starts with a simple genetic effect and studies its effect on behavior. For example, we can study a single-gene mutation and observe its behavioral effect. Another approach, which characterizes the field of biopsychology, can be called *physiology-centered.* This focuses on the physiological intermediaries between genes and behavior. The physiology-centered approach ideally works back to genes and forward to behavior. (See Figure 1.2.) In both the gene-centered and physiology-centered approaches, behavior is really only a tool for understanding the workings of both genes and various physiological systems. The third approach begins with behavior. The behaviors are not selected for their genetic or physiological simplicity, but rather because of their intrinsic interest or social relevance. A first step in understanding the etiology of such behavior is to ask the extent to which genetic and environmental influences make a difference for a particular behavior. Once in a great while, we find relatively simple genetic and physiological systems. This is what happened for one type of mental retardation, phenylketonuria, which is caused by a single-gene defect. More often, particularly for the complex behaviors that interest behavioral scientists, we find that both genetic and environmental differences are important. Usually, many genes are involved, and we cannot identify the physiological systems involved. However, it is an important first step to know that genetic factors influence a particular behavior, just as it is important to narrow the search for salient environmental influences.

DO GENES DETERMINE ONE'S DESTINY?

This text will demonstrate that genetic differences can account for a substantial portion of individual variation in many important behaviors. Some people do not look favorably on such findings. These people believe that if genes are shown to influence behavior, there is nothing that can be done to alter that behavior short of eugenic (breeding) intervention or genetic engineering. Thus, if a certain form of psychopathology was shown to be caused primarily by genes, it might be mistakenly assumed that psychotherapy and other environmental intervention would be useless.

This pessimistic view is simply wrong. Genes do not determine one's destiny. A genetically determined behavioral problem may be bypassed, ameliorated, or remediated by environmental interventions. The best example is phenylketonuria (PKU), a single-gene defect, which formerly resulted in severe retardation and was responsible for about 1 percent of institutionalized retarded individuals. Biochemical studies of the gene-behavior pathway indicated that the ultimate cause of the retardation was the inability to break down a particular chemical, phenylalanine, which led to its accumulation at high levels in the blood. This caused severe damage to the developing brain. As we will discuss in Chapter 5, PKU individuals do not suffer retardation if a diet low in phenylalanine is provided during the developing years. Thus, an environmental intervention was successful in bypassing a genetic problem. This important discovery was made possible by recognition of the genetic basis for this particular type of retardation.

DISCERNING "WHAT IS" FROM "WHAT COULD BE"

Most research in human behavioral genetics has involved analyzing behavioral differences among individuals and estimating the relative extent to which these differences are due to heredity and environment, given the mix of genetic and environmental influences at a particular time. If either the genetic or environmental influences change, the relative impact of genes and environment will change. Thus, if a new environmental treatment were introduced, a behavior strongly influenced by heredity could nonetheless be altered.

Confusion between *what is* and *what could be* caused some of the heat in reaction to a well-known monograph by Arthur Jensen (1969), entitled "How much can we boost I.Q. and scholastic achievement?" Behavioral genetics research, as reviewed in Chapters 11 to 13, suggests that genes have a substantial influence on IQ scores. In other words, genetic differences among people account for a substantial portion of variation in IQ, given the genetic and environmental differences in the samples studied. However, this does not mean that we cannot boost IQ. The same behavioral genetics research also provides the best evidence for environmental influence on IQ scores. However, even if the environmental differences that currently exist do not significantly affect IQ scores, we should not conclude that untried environmental interventions will not make a difference.

Behavioral genetics data concerning *what is* should not dampen research efforts aimed at determining *what could be*. The two should be viewed as complementary. Knowledge about what is can help to guide research concerning what could be. For example, if currently existing environmental differences did not make much of a difference for a particular behavior, then it would not seem reasonable to tinker with these environmental factors if one wanted to change this behavior. It would make more sense to go outside of the traditional environmental factors to find novel environmental interventions.

INDIVIDUAL DIFFERENCES AND EQUALITY

Another reason some people have difficulty in accepting the role of genes in influencing behavior is that they do not recognize the wide range of individual differences for most behaviors. One mouse is not like every other mouse; nor is every human like every other. In fact, as we shall see, the principles of genetics demonstrate that there has never been nor will ever be another human being who is genetically exactly like you. Variability is the key to understanding evolution and genetics, and it provides a needed perspective for behavioral scientists.

J. Hirsch (1963) outlined three different approaches that have been used to study behavior. The three approaches differ in the extent to which they recognize variability. The first view recognizes no important differences between or within species. This is a position suggested by much of the older research in learning, which assumed that the laws of learning for one species would generalize to all other species. This approach is typified by titles such as *The Behavior of Organisms* (Skinner, 1938). This view has largely given way to a form of *typological thinking* (Mayr, 1965), which recognizes differences *between*, but not *within*, species. It characterizes comparative psychology and, indirectly, much current research on human behavior. Researchers interested in human behavior may study humans rather than other animals because they assume that important species differences exist, but they do not study differences within the human species. The third approach, which considers variability between *and* within species, is characteristic of behavioral genetics.

To some extent, these three views are like the different powers of a telescope. To a visitor from another planet, peering at us from their orbiting space capsule, we humans might not appear all that different from squirrels. Humans are bigger and have somewhat less fur, but both humans and squirrels have two eyes, a nose, and four limbs. If the two species were examined with a more powerful lens, the differences between humans and squirrels would readily emerge. At this power setting, however, all humans would seem essentially identical. Some scientists choose this power setting to study behavior—clearly, there are some common features among all humans in the way we perceive stimuli and learn. However, it is our belief that a powerful science of behavior will not emerge until we switch to a higher resolving power that reveals differences within, as well as between, species. The questions that most often confront scientists studying human behavior are those dealing with differences among people. And genetics, the study of variation of organisms, is uniquely qualified to aid us in analyzing these individual differences.

This focus on individual differences may be difficult for some of you to accept philosophically. Are not all men created equal? This was a self-evident truth to the signers of the Declaration of Independence. Arguments that some men are inherently more able to learn than others may appear to be at odds with the democratic ideal and to imply the principle of rule by the elite (even though it is obvious that there are inherent differences in height, strength, and other characteristics). A more thorough understanding of behavioral genetics suggests a very different philosophical conclusion.

If there is a central message from behavioral genetics, it involves individual differences. With the trivial exception of members of identical multiple births, each one of us is a unique genetic experiment, never to be repeated again. Here is the conceptualization on which to build a philosophy of the dignity of the individual! Human variability is not simply imprecision in a process that, if perfect, would generate unvarying representatives of a Platonic ideal. Individuality is the quintessence of life; it is the product and the agency of the grand sweep of evolution.

It is important not to confuse biological identity with the political concept of equality. The two concepts are very different, though the problem is to an extent a semantic one. The Greeks had not one, but many words for equality, distinguishing, for example (Hutchins and Adler, 1968, p. 305):

isonomia: equality before the law
isotimia: equality of honor
isopoliteia: equality of political rights
isokratia: equality of political power
isopsephia: equality of votes or suffrage
isegoria: equality in right to speak
isoteleia: equality of tax or tribute

isomoiria: equality of shares or partnership
isokleria: equality of property
isodaimonia: equality of fortune

Perhaps we confuse ourselves by using only the word *equality*.

SUMMARY

Behavioral genetics lies at the interface between genetics and the behavioral sciences. In this book, we present basic genetic concepts, including evolution, transmission genetics, chromosomes, population genetics, and quantitative genetic theory and methods, but always with an eye toward behavior. The last chapter focuses on controversial issues in behavioral genetics. But because some of you have heard about these controversies and have wondered about them, we have discussed some origins of these controversies in this introductory chapter:

1. Controversy arises when nature (genes) is pitted against nurture (environment). One of the oldest schools of thought in psychology is behaviorism. It led to an environmentalism that rejected the possibility of genetic influences on behavior.
2. Although genetic and environmental influences interact, this does not imply that the separate effects of genes and environment cannot be untangled.
3. Genes do not act as master puppeteers within us. They are chemical structures that control the production of proteins, thereby indirectly affecting behavior.
4. Genes do not determine one's destiny. A genetically determined behavioral problem can sometimes be alleviated environmentally.
5. Behavioral genetics research usually describes *what is*—the relative impact of genes and environment at a particular time for a particular population—although it may provide us with knowledge about *what could be*.
6. Finally, recognition of individual differences—regardless of their environmental or genetic etiology—is independent of acknowledgment of the values of political equality.

2

Historical Perspective

To illustrate a point concerning the inheritance of gestures, Darwin quoted an interesting case that had been brought to his attention by Galton.

> A gentleman of considerable position was found by his wife to have the curious trick, when he lay fast asleep on his back in bed, of raising his right arm slowly in front of his face, up to his forehead, and then dropping it with a jerk so that the wrist fell heavily on the bridge of his nose. The trick did not occur every night, but occasionally.

Nevertheless, the gentleman's nose suffered considerable damage, and it was necessary to remove the buttons from his nightgown cuff in order to minimize the hazard.

> Many years after his death, his son married a lady who had never heard of the family incident. She, however, observed precisely the same peculiarity in her husband; but his nose, from not being particularly prominent, has never as yet suffered from the blows.... One of his children, a girl, has inherited the same trick. (Darwin, 1872, p. 34)

Probably everyone could cite some examples, perhaps less quaint than Galton's, in which some peculiarity of gait, quality of temper, degree of talent, or other trait is characteristic of a family. Such phrases as "a chip off the old block," "like father, like son," and "it runs in the family" give ample evidence of the notion that behavioral traits, like physiological ones, can be inherited. The aim of this chapter is to consider the history of scientific inquiry into these matters, with major emphasis on developments during the latter half of the nineteenth and the early twentieth centuries.

It is extremely difficult to pinpoint the earliest expression of a view concerning any subject. While the present topic is no exception, we should note that its origins must be very remote indeed. The concept that "like begets like" has had great practical importance in the development of domesticated animals, which have been bred for behavioral, as well as morphological, characteristics. We can postulate that the notion of inheritance, including inheritance of behavioral traits, may have appeared in human thought as early as 8000 B.C., when the domestication of the dog began.

The workings of inheritance have been of great interest to men throughout recorded history, and many interesting conjectures have been made (Zirkle, 1951). One of the most familiar of the early statements is that of Theognis, who, in the sixth century B.C., commented on contemporary mores:

> We seek well-bred rams and sheep and horses and one wishes to breed from these. Yet a good man is willing to marry an evil wife, if she bring him wealth; nor does a woman refuse to marry an evil husband who is rich. For men reverence money, and the good marry the evil, and the evil the good. Wealth has confounded the race. (in Roper, 1913, p. 32)

About the same time, the Spartans practiced infanticide to eliminate both those of unsound soul and those of defective body. As Roper points out, "To the Greeks, believing only in the beauty of the spirit when reflected in the beauty of the flesh, the good body was the necessary correlation of the good soul" (1913, p. 19).

In *The Republic,* Plato suggested a course of action whereby the principles of inheritance of behavior could be used to develop an ideal society:

> It necessarily follows ... from what has been acknowledged, that the best men should as often as possible form alliances with the best women, and the most depraved men, on the contrary, with the most depraved women; and the offspring of the former is to be educated, but not of the latter, if the flock is to be of the most perfect kind. (in Davis, 1849, p. 144)

Biological thought during the ensuing centuries was dominated by Aristotle's pronouncements on natural history, and by the teachings of Galen, a Roman, concerning anatomy. Progress in understanding biological phenomena was virtually halted during the general stagnation of secular pursuits that typified the Middle Ages. Then came the Renaissance. Leonardo da Vinci's study of anatomy in connection with art characterizes the far-ranging inquisitiveness of the Renaissance scholars. Less well known is a family incident that reveals a deep conviction about the workings of heredity. Leonardo was an illegitimate child resulting from a liaison between Piero, a notary from the village of Vinci, and a peasant girl named Caterina. As a modern biographer of Leonardo puts it:

> There was an interesting and deliberate attempt to repeat the experiment. Leonardo had a step-brother, Bartolommeo, by his father's third wife. The step-brother was forty-five years younger than Leonardo, who was already a legend when the boy was growing up and was dead when the following experiment took place. Bartolommeo examined every detail of his father's association with Caterina and he, a notary in the family tradition, went back to Vinci. He sought out another peasant wench who corresponded to what he knew of Caterina and, in this case, married her. She bore him a son but so great was his veneration for his brother that he regarded it as profanity to use his name. He called the child Piero. Bartolommeo had scarcely known his brother whose spiritual heir he had wanted thus to produce and, by all accounts, he almost did. The boy looked like Leonardo, and was brought up with all the encouragement to follow his footsteps. Pierino da Vinci, this experiment in heredity, became an artist and, especially, a sculptor of some talent. He died young. (Ritchie-Calder, 1970, pp. 39–40)

THE ERA OF DARWIN

The applied husbandry of the da Vinci family cannot be said to have had a pivotal effect on the subsequent development of biological thought. However, there were other concurrent developments that did. Andreas Vesalius's exhaustive work on human anatomy, published in 1543, was based on detailed and painstaking dissection of human bodies. In 1628, William Harvey made his momentous discovery of the circulation of the blood. These findings were of far-reaching importance, for they opened the way to experimentation on the phenomena of life.

After Harvey's discovery, the pace of biological research quickened, and many fundamental developments in technique and theory ensued in the following century. One of the cornerstones of biology was laid by the Swede Karl von Linné (better known as Linnaeus), who, in 1735, published *Systema Naturae,* in which he established a system of taxonomic classification of all known living things. In so doing, Linnaeus emphasized the separateness and distinctness of species. As a result, the view that species were fixed and unchanging became the prevailing one. This notion, of course, fit the biblical account of creation. However, not everyone was persuaded that species are unchangeable. For example, the Englishman Erasmus Darwin suggested, in the latter part of the eighteenth century, that plant and animal species appear capable of improvement, as well as degeneration. A more influential view on this subject was promoted by the Frenchman Jean Baptiste Lamarck, who argued that the deliberate efforts of an animal could result in modifications of the body parts involved, and that the modification so acquired could be transmitted to the animal's offspring. For example:

> We perceive that the shore bird, which does not care to swim, but which, however, is obliged ... to approach the water to obtain its prey, will be continually in danger of sinking in the mud, but wishing to act so that its body shall not fall into the liquid, it will contract the habit of extending and lengthening its feet. Hence it will result in the generations of these birds which continue to live in this manner, that the individuals will find themselves raised as if on stilts, on long naked feet: namely, denuded of feathers up to and often above the thighs. (in Packard, 1901, p. 234)

Changes of this sort were presumed to accumulate, so that eventually the characteristics of the species would change. Lamarck was not the first to assume that changes acquired in this manner could be transmitted to the next generation, but he crystallized and popularized the notion. Thus, it has come to be called Lamarckism, or the law of use and disuse. As we shall see, this is an incorrect view of evolution. But it was significant, in that it questioned the prevailing view that species do not change.

The strict and literal interpretation of the account in Genesis of the creation of the earth and its inhabitants was being challenged most seriously on the basis of geological evidence. The discovery of fossilized animal bones deep in strata beneath the earth's surface proved difficult to accommodate to Bishop Ussher's calculations that the earth had been created in 4004 B.C. A theory of "catastrophism" was put forward to account for these fossils. The Deity was regarded to have created and extinguished life on many successive occasions, with catastrophes such as floods and violent upheavals, which caused the bones to be buried at various depths. Many geologists questioned, however, whether catastrophic events were responsible for the geological record. A school of "uniformitarians" argued that the processes at work in the past were the same as those of the present, and thus that the accumulation of strata required millions of years rather than the six thousand-odd years attributable to Bishop Ussher's postulated date of creation. A leader of this uniformitarian school of thought, and one of the dominant intellects of the time, was Charles Lyell (see Eiseley, 1959). Lyell published the first volume of his *Principles of Geology* in 1830, and one of the early copies found its way into the baggage of a young man about to embark on what was probably the most important voyage in the history of biological thought.

Erasmus Darwin's grandson Charles (Figure 2.1) had been a student of medicine at Edinburgh, but was so unnerved by the sight of blood during surgery that he gave up further medical study. He then went to Cambridge, where, although a student of mediocre accomplishment, he received a degree in 1831. Darwin appeared to be destined for a career as a clergyman, when

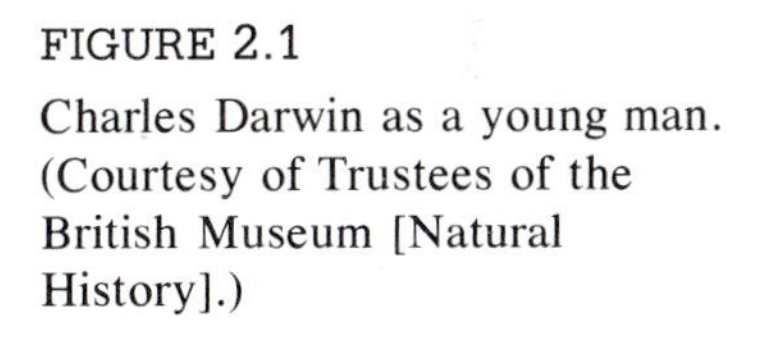

FIGURE 2.1
Charles Darwin as a young man. (Courtesy of Trustees of the British Museum [Natural History].)

suddenly and unexpectedly, through the recommendation of one of his old professors, he was nominated for the unpaid post of naturalist aboard the H.M.S. *Beagle,* a survey ship of the Royal Navy about to embark on a long voyage. It was not uncommon for a naturalist to be taken on trips of this kind, and the young and devout captain of the *Beagle,* Captain Robert Fitz-Roy, was pleased at the prospect, for he expected a naturalist to be able to produce yet more data in support of "natural theology." A central theme of natural theology was the so-called argument from design, which viewed the adaptation of animals and plants to the circumstances of their lives as evidence of the Creator's wisdom. Such exquisite design, so the argument went, implied a "Designer." As exploration opened up hitherto unexplored parts of the world, new evidence of the Designer's works was uncovered. It was with this end in mind that Captain Fitz-Roy welcomed the young Darwin.

During the next five years, Darwin experienced chronic seasickness, tropical fever, volcanic eruptions, earthquakes, tidal waves, and the high adventure of encounters with rebels and life with Argentine gauchos. He filled many notebooks with observations on fossils, primitive men, and various species of animals—and their remarkable and specific adaptation to their environments. He made particularly compelling observations about the 14 species of finch found in a small area on the Galápagos Islands. The principal differences among these finches were in their beaks, and each was exactly appropriate for the particular eating habits of the species (see Figure 2.2). Somehow, thought Darwin, these birds derived from a common ancestral

FIGURE 2.2

The 14 species of Galápagos and Cocos Island finches. (a) A woodpeckerlike finch that uses a twig or cactus spine instead of its tongue to dislodge insects from tree-bark crevices. (c, d, e) Insect-eaters. (f, g) Vegetarians. (h) The Cocos Island finch. The birds on the ground eat mostly seeds. Note the powerful beak of (i), which lives on hard seeds. (From "Darwin's finches" by D. Lack. Copyright © 1953 by Scientific American, Inc. All rights reserved.)

group. "Seeing this gradation and diversity of structure in one small, intimately related group of birds, one might really fancy that from an original paucity of birds in this archipelago, one species had been taken and modified for different ends" (Darwin, 1896, p. 380). However, influenced by Lyell's book, and his own observations on geology and biology, Darwin was not inclined to make the argument in favor of the "argument from design." On his return to England, Darwin began work on several reports summarizing his observations on coral reefs, barnacles, and other matters. Meanwhile, he gradually and systematically marshaled evidence that species evolve one from another, and pondered the possible mechanisms through which this evolution could occur. He shared his developing theory with a few friends, including Lyell himself, an eminent botanist named J. D. Hooker, and T. H. Huxley. He gradually convinced some but not all of them of the merit of his theory.

Realizing the kind of opposition that a theory contradicting the biblical account of creation would encounter, Darwin hesitated. He planned a monumental work in which he would present an overwhelming mass of evidence. Though his friends warned him that he should publish a brief version immediately, lest someone anticipate him, he continued to work slowly and carefully on amassing the evidence and anticipating the objections. He did, however, take time to write out short sketches in correspondence with his friends and confidants. Finally, in 1858, a blow fell. A young man named Alfred Wallace sent Darwin a manuscript for his comments. In it, with much less evidence in hand than Darwin had, Wallace arrived at essentially the same theory that Darwin had been developing for more than two decades. Darwin was greatly concerned over the course of action he should take. As he said in a letter to Lyell:

> I should be extremely glad now to publish a sketch of my general views in about a dozen pages or so; but I cannot persuade myself that I can do so honourably. Wallace says nothing about publication, and I enclose his letter. But as I had not intended to publish any sketch, can I do so honourably, because Wallace has sent me an outline of his doctrine? I would far rather burn my whole book, than that he or any other man should think that I had behaved in a paltry spirit. (Darwin, 1858, p. 117)

Lyell and Hooker took the initiative and resolved the issue by arranging for the simultaneous presentation, at a meeting of the Linnean Society in 1858, of a sketch Darwin had prepared in 1844 and Wallace's paper.

With the theory now out in the open, Darwin began work on what he called an abstract. This abstract, published in 1859 under the title *On the Origin of Species by Means of Natural Selection, or the Preservation of Favoured Races in the Struggle for Life,* proved to be one of the most influential books ever written. A contemporaneous *London Times* review of his book, which sold out on the first day, is excerpted in Box 2.1.

Box 2.1
1859 Review of Darwin's *The Origin of Species*

The Origin of Species made an immediate impact on the scientific world. The following is a short excerpt from a long review (over 5,000 words) of Darwin's book that appeared on December 26, 1859, in the *Times* (London).

> There is a growing immensity in the speculations of science to which no human thing or thought at this day is comparable.... Hence it is that from time to time we are startled and perplexed by theories which have no parallel in the contracted moral world.... The hypothesis to which we point, and of which the present work of Mr. Darwin is but the preliminary outline, may be stated in his own language as follows: "Species originated by means of natural selection, or through the preservation of the favored races in the struggle for life." ... When we know that living things are formed of the same elements as the inorganic world, that they act and react upon it, bound by a thousand ties of natural piety, is it probable, nay is it possible, that they, and they alone, should have no order in their seeming disorder, no unity in their seeming multiplicity, should suffer no explanation by the discovery of some central and sublime law of mutual connexion?
>
> Questions of this kind have assuredly often arisen, but it might have been long before they received such expression as would have commanded the respect and attention of the scientific world, had it not been for the publication of the work which prompted this article. Its author, Mr. Darwin, inheritor of a once celebrated name, won his spurs in science when most of those now distinguished were young men, and has for the last 20 years held a place in the front ranks of British philosophers. After a circumnavigatory voyage, undertaken solely for the love of his science, Mr. Darwin published a series of researches which at once arrested the attention of naturalists and geologists; his generalizations have since received ample confirmation, and now command universal assent, nor is it questionable that they have had the most important influence on the progress of science. More recently Mr. Darwin, with a versatility which is among the rarest of gifts, turned his attention to a most difficult question of zoology and minute anatomy; and no living naturalist and anatomist has published a better monograph than that which resulted from his labours. Such a man, at all events, has not entered the sanctuary with unwashed hands, and when he lays before us the results of 20 years' investigation and reflection we must listen even though we be disposed to strike. But, in reading his work it must be confessed that the attention which might at first be dutifully, soon becomes willingly given, so clear is the author's thought, so outspoken his conviction, so honest and fair the candid expression of his doubts.

Darwin's Theory of Evolution

Darwin was honored by being buried near Sir Issac Newton in Westminster Abbey primarily because he convinced the world of the reasonableness of evolution. As we shall see, however, his theory of the evolution of species had some serious gaps, mainly because the mechanism for heredity, the gene, was not yet understood.

The elements of Darwin's theory can be stated as follows. Within any species, many more individuals are born each generation than survive to maturity. Great variation exists among the individuals of a population. These individual differences are due, at least in part, to heredity. If the likelihood of surviving to maturity and reproducing is influenced, even to a slight extent, by a particular trait, offspring of the survivors and reproducers should manifest slightly more of the trait than their parents' generation. Thus, bit by bit, the characteristics of a population can change. Over a sufficiently long period, the cumulative changes are so great that in retrospect the latter and the earlier populations are, in effect, different.

Darwin's theory of evolution thus has three components, as indicated in Table 2.1. Some mechanism causes and maintains variation (as discussed later in this chapter), and natural selection uses this variability to shape a species. We shall see that Darwin and his contemporaries knew little about the source of variability or the process by which it is maintained and transmitted from one generation to another. However, Darwin did discover the process of natural selection, and we shall discuss this first.

The Principle of Natural Selection

Darwin's most notable contribution to the theory of evolution was his principle of natural selection. He used the phrase "survival of the fittest" to characterize the principle, but it could more appropriately be called "differential reproduction of the fittest"—or, more simply, "reproductive fitness." Mere survival is necessary, but this is not sufficient. As Darwin himself put it:

> Owing to this struggle [for life], variations, however slight and from whatever cause proceeding, if they be in any degree profitable to the individuals of a species, in their infinitely complex relations to other organic beings and to their

TABLE 2.1
Model of evolution: what Darwin knew and what we know now

	Darwin	Now
Induction of variation	Environmental modification and "use and disuse"	Mutation of DNA
Maintenance of variation	Pangenesis	Segregation and genetic transmission
Selection of variation	Natural selection—"survival of the fittest" and "reproductive fitness"	Natural selection—"reproductive fitness"

physical conditions of life, will tend to the preservation of such individuals, and will generally be inherited by the offspring. The offspring, also, will thus have a better chance of surviving, for, of the many individuals of any species which are periodically born, but a small number can survive. (Darwin, 1859, pp. 51–52)

It is clear that Darwin considered behavioral characteristics to be just as subject to natural selection as physical traits. In *The Origin of Species* an entire chapter is devoted to discussion of instinctive behavior patterns. A later book, *The Descent of Man and Selection in Relation to Sex,* gave detailed consideration to comparisons of mental powers and moral senses of animals and man, as well as to the development of intellectual and moral faculties in man. In these discussions Darwin was satisfied that he had demonstrated that the difference between the mind of man and the mind of animals "is certainly one of degree and not of kind" (1871, p. 101). This is an essential point, since one of the strongest objections to the theory of evolution was the qualitative gulf that was supposed to exist between the mental capacities of man and of lower animals.

In an explicit summary statement, based largely on observations of "family resemblance," Darwin said:

So in regard to mental qualities, their transmission is manifest in our dogs, horses, and other domestic animals. Besides special tastes and habits, general intelligence, courage, bad and good temper, etc., are certainly transmitted. With man we see similar facts in almost every family; and we now know, through the admirable labours of Mr. Galton, that genius which implies a wonderfully complex combination of high faculties, tends to be inherited; and, on the other hand, it is too certain that insanity and deteriorated mental powers likewise run in families. (Darwin, 1871, p. 414)

Darwin's writings focused on selection of individuals and their characteristics. A different perspective has been emphasized by researchers in an area called sociobiology, which applies evolutionary theory to the study of social behavior. Sociobiologists look at evolution from the standpoint of the gene rather than the individual. Just as the chicken may only be the egg's way of producing other eggs, sociobiology suggests that the individual may only be the gene's way of producing more genes. This way of thinking suggests an interesting analysis of certain social behaviors, such as altruism, as well as the importance of selection for genetically similar individuals, a theory called *kinship selection*. This view of evolution will be discussed in the next chapter.

Reaction to the Theory of Evolution

The Origin of Species caused a violent reaction. Fierce denunciation came from those whose sensibilities were shocked by this contradiction of the biblical account of creation. There was opposition, too, from other scientists, whose theories were challenged by the new conceptions. At Harvard University, the eminent Louis Agassiz, a confirmed antievolutionist, was opposed by Asa Gray, a recipient of one of Darwin's prepublication sketches of his theory. A lively controversy followed at Harvard and elsewhere.

Darwin's health had failed, perhaps because of long-term consequences of a tropical infection he contracted while on the *Beagle*. This, combined with his great timidity concerning public encounters, left the actual evolutionary debate in Britain largely up to his friends. This distinguished group, which included Wallace, Lyell, Hooker, and Huxley, was certainly equal to the task. In what must be the most famous confrontation in the history of biology, Huxley found himself debating Bishop Wilberforce at a meeting of the British Association for the Advancement of Science in 1860. Wilberforce, who was an extremely effective public speaker, had been carefully coached by an anti-Darwinian named Richard Owen. After some stirring oratory on the matter, Wilberforce turned to Huxley and inquired whether it was through his grandfather or grandmother that he claimed to be descended from the apes. A letter of Huxley's describes his response:

> When I got up I spoke pretty much to the effect—that I had listened with great attention to the Lord Bishop's speech but had been unable to discover either a new fact or a new argument in it—except indeed the question raised as to my personal predilections in the matter of ancestry—That it would not have occurred to me to bring forward such a topic as that for discussion myself, but that I was quite ready to meet the Right Rev. prelate even on that ground. If then, said I, the question is put to me would I rather have a miserable ape for a grandfather or a man highly endowed by nature and possessing great means and influence and yet who employs those faculties and that influence for the mere purpose of introducing ridicule into a grave scientific discussion—I unhesitatingly affirm my preference for the ape. (in Montagu, 1959, pp. 2–3)

Huxley's rejoinder caused quite an uproar. In the hubbub one particularly plaintive voice was heard. Robert Fitz-Roy, now an admiral and the contributor of a paper on meteorology to an earlier session of the association meetings, waved a Bible and denounced the views of his old friend and shipmate. After order was restored, Hooker was called on to speak, and he too spoke persuasively in favor of the Darwinist view. In many ways this meeting represented a turning point, for after two of the most eminent scientists in Britain had publicly defended the evolutionary theory, thinking men everywhere had to take it seriously.

After 120 years, some resistance to the theory of evolution still exists today. For example, the conflicts underlying the famous Scopes trial, dramatized in the play *Inherit the Wind,* can be seen today—and not just on the stage. In 1925, John T. Scopes taught evolution in a Tennessee public school in violation of a statute prohibiting it. He was correctly found guilty of violating the statute and planned to appeal the decision in order to have the statute found unconstitutional. But his attempted appeal was not considered. The trial provided a grandstand confrontation between William Jennings Bryan, as prosecutor, and Clarence Darrow, Scopes's defense attorney. This debate served to popularize the concept of evolution. The Tennessee statute was eventually repealed in 1967. But in 1973 the Tennessee Senate (by a vote of 28 to 1) and the House (54 to 14) passed a new antievolution measure requiring that equal time in teaching and equal space in texts be given to biblical and Darwinian accounts of evolution. Although this Tennessee statute was ruled unconstitutional in 1975, several states continue to pressure publishers to impede the teaching of evolution (Grabiner and Miller, 1974). For example, in 1974, the Texas State Board of Education adopted a statement requiring the presentation of evolutionary theory as only one of several theories and that it be "clearly presented as theory rather than verified." In 1975, evolution was mentioned in less than 20 percent of the biology texts in Texas (Fox, 1975).

Darwin anticipated the major reason for reticence to accept his theory of the origin of species: "The chief cause of our natural unwillingness to admit that one species has given birth to clear and distinct species, is that we are always slow in admitting great changes of which we do not see the steps" (1859, p. 368). In some cases, we can see the steps by which populations of the same species drift apart and select different niches. This results in genetic changes that can eventually lead to reproductive infertility when animals from the different populations (now different species) are mated. Sometimes such evolution occurs rapidly and can be documented (Dobzhansky and Pavlovsky, 1971). However, because of the great gulfs of time involved in the evolution of higher animals, it is difficult to demonstrate conclusively that one species evolves from another. Nonetheless, the logical appeal of the theory of evolution is strong, as Darwin noted:

> If then, animals and plants do vary, let it be ever so slightly or slowly, why should not variations or individual differences, which are in any way beneficial, be preserved and accumulated through natural selection, or the survival of the fittest? If man can by patience select variations useful to him, why, under changing and complex conditions of life, should not variations useful to nature's living products often arise, and be preserved or selected? What limit can be put to this power, acting during long ages and rigidly scrutinising the whole constitution, structure, and habits of each creature,—favouring the good and rejecting the bad? I can see no limit to this power, in slowly and beautifully adapting each form to the most complex relations of life. The theory of natural selection, even if we look no farther than this, seems to be in the highest degree probable. (Darwin, 1859, p. 360)

GALTON'S CONTRIBUTIONS

Among the supporters and admirers of Darwin at this time was another one of Erasmus Darwin's grandsons, Francis Galton (Figure 2.3). Galton had already established something of a reputation as a geographer, explorer, and inventor. By the time *The Origin of Species* was published, he had invented a printing electric telegraph, a type of periscope, and a nautical signaling device. The effect on him of Darwin's work is revealed in a letter he later wrote to Darwin:

> I always think of you in the same way as converts from barbarism think of the teacher who first relieved them from the intolerable burden of their superstition. I used to be wretched under the weight of the old-fashioned arguments from design; of which I felt though I was unable to prove to myself, the worthlessness. Consequently the appearance of your *Origin of Species* formed a real crisis in my life; your book drove away the constraint of my old superstition as if it had been a nightmare and was the first to give me freedom of thought. (in Pearson, 1924, vol. I, plate II)

The Origin of Species directed Galton's immense curiosity and talents to biological phenomena, and he soon developed what was to be a central and abiding interest for the rest of his life: the inheritance of mental characteristics.

FIGURE 2.3

Francis Galton, in 1840, from a portrait by O. Oakley. (Courtesy of the Galton Laboratory.)

Hereditary Genius

In 1865, two articles by Galton, jointly entitled "Hereditary Talent and Character," were published in *Macmillan's Magazine.* Four years later a greatly expanded discussion was published under the title, *Hereditary Genius: An Inquiry into Its Laws and Consequences.* The general argument presented in this work is that a greater number of extremely able individuals is found among the relatives of persons endowed with high mental ability than would be expected by chance. Furthermore, Galton discovered that the closer the family relationship, the higher the incidence of such superior individuals. Galton applied Quetelet's "law of deviation from an average," which at the time was a recent development, but later became familiar as the normal curve. Galton distinguished 14 levels of human ability, ranging from idiocy through mediocrity to genius.

Since there was no satisfactory way of quantifying natural ability, Galton had to rely on reputation as an index. By "reputation," he did not mean notoriety for a single act, nor mere social or official position, but "the reputation of a leader of opinion, of an originator, of a man to whom the world deliberately acknowledges itself largely indebted" (1869, p. 37). The designation "eminent" was applied to those individuals who constituted the upper 250-millionths of the population (i.e., 1 in 4,000 persons would attain such a rank), and the discussion focused on such men. Indeed, the majority of individuals Galton presented in evidence were, in his estimation, the cream of this elite group, and were termed "illustrious." These were men whose talents ranked them one in a million.

On the basis of biographies, published accounts, and direct inquiry, Galton evaluated the accomplishments of well-known judges, statesmen, peers, military commanders, literary men, scientists, poets, musicians, painters, Protestant religious leaders, and Cambridge scholars. (Oarsmen and wrestlers of note were also examined to extend the range of inquiry from brain to brawn.) The approximately 1,000 men who were designated as "eminent" were found to belong to 300 families. With the overall incidence of eminence only 1 in 4,000, this result clearly illustrated the tendency for eminence to be a familial trait.

Taking the most eminent man in each family as a reference point, the other individuals who attained eminence (in the same or in some other field of endeavor) were tabulated with respect to closeness of family relationship, as indicated in Figure 2.4. Briefly stated, the results showed that eminent status was more likely to appear in close relatives, with the likelihood of eminence decreasing as the degree of relationship became more remote. Eminence was attained by 26 percent of the fathers of the 100 most distinguished men, 23 percent of their brothers, and 36 percent of their sons. Second-degree relatives such as grandfathers, uncles, nephews, and grandsons achieved eminence to a much lower degree (7.5 percent, 4.5 percent, 4.8 percent, and 9.5 percent, respectively). However, these percentages are still

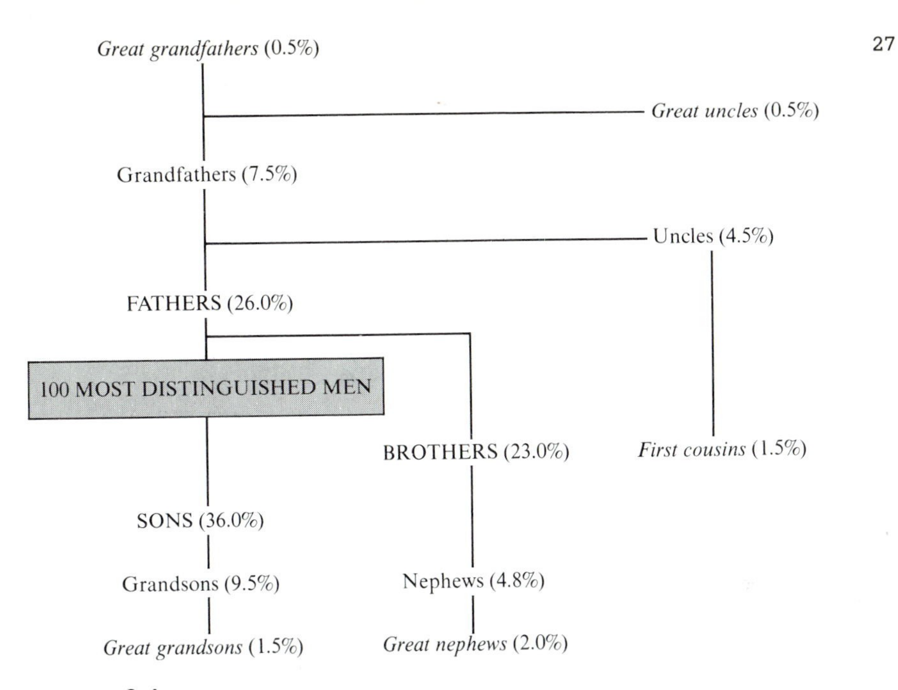

FIGURE 2.4

Percentage of eminent men among relatives of the 100 most distinguished men in Galton's study. (From *Hereditary Genius* by Francis Galton, 1869.)

high when compared to the overall incidence of only 1 in 4,000 (0.025 percent).

Galton was aware of the possible objection that relatives of eminent men share social, educational, and financial advantages, and that the results of his investigation might be interpreted as showing the effectiveness of such environmental factors. To demonstrate that reputation is an indication of *natural* ability, and not the product of environmental advantages, three arguments were presented. First, Galton stressed the fact that many men had risen to high rank from humble family backgrounds. Second, it was noted that the proportion of eminent writers, philosophers, and artists in England was not less than that in the United States, where education of the middle and lower socioeconomic classes was more advanced. He felt that the educational advantages in America had spread culture more widely without producing more persons of eminence. Finally, Galton compared the success of adopted kinsmen of Roman Catholic popes, who were given great social advantages, with the sons of eminent men, and the latter were judged to be more distinguished.

Pioneering Research in Psychology and Statistics

In order to further his researches, it was necessary for Galton to find ways of assessing mental characteristics. In a prodigious program of research, he developed apparatus and procedures for measuring auditory thresholds, visual acuity, color vision, touch, smell, judgment of the vertical, judgment of length, weight discrimination, reaction time, and memory span. In addition, he employed a questionnaire technique to investigate mental imagery.

One particularly intriguing, although not especially successful, investigation involved the use of composite portraiture. Photographs of a number of individuals were superimposed to yield their common features. Figure 2.5, which shows front and profile composite photographs of three sisters, taken by Galton, illustrates his notion that a composite photograph is usually better-looking than the individual photographs. Galton developed this technique in an effort to determine what relationship, if any, existed between the facial characteristics of certain groups and various attributes of their personality, morality, and health. Since there did not seem to be a good way of measuring facial configurations, Galton used composite portraiture to produce an archetype of certain groups. For example, Galton's contemporaries were interested in facial features as they might relate to criminality. When Galton used composite portraiture to investigate this possibility, he found no distinctive facial features in composite photographs of three groups of criminals: those convicted of murder and manslaughter, felons such as forgerers, and sexual offenders.

The problems of properly expressing and evaluating the data obtained from such researches were formidable, and Galton also turned his remarkable energies to statistics. He pioneered the development of the concepts of the median, percentiles, and correlation.

Since it was, of course, desirable to have data from large numbers of individuals, Galton employed various stratagems to this end. For example, he arranged for an "Anthropometric Laboratory for the measurement in various ways of Human Form and Faculty" to be located at an International Health Exhibition during 1884 and 1885. Some 9,337 people paid threepence or fourpence each for the privilege of being measured for various bodily and sensory characteristics. (See Figure 2.6.) After the International Health Exhibition closed, a permanent Anthropometric Laboratory was established. During the next seven years, another 7,500 individuals were tested (at the bargain rate of one penny each), including Galton himself. His data are reproduced in Figure 2.7. On another occasion a contest was sponsored in which awards of £7 were given to those submitting the most careful and complete "Extracts from their own Family Records." In this way, Galton was able to obtain a large number of pedigrees that he could examine for evidence of human inheritance.

Twins and the Nature–Nurture Problem

Galton introduced (1883) the twin-study method to assess the roles of nature (inheritance) and nurture (environment). The essential question in his

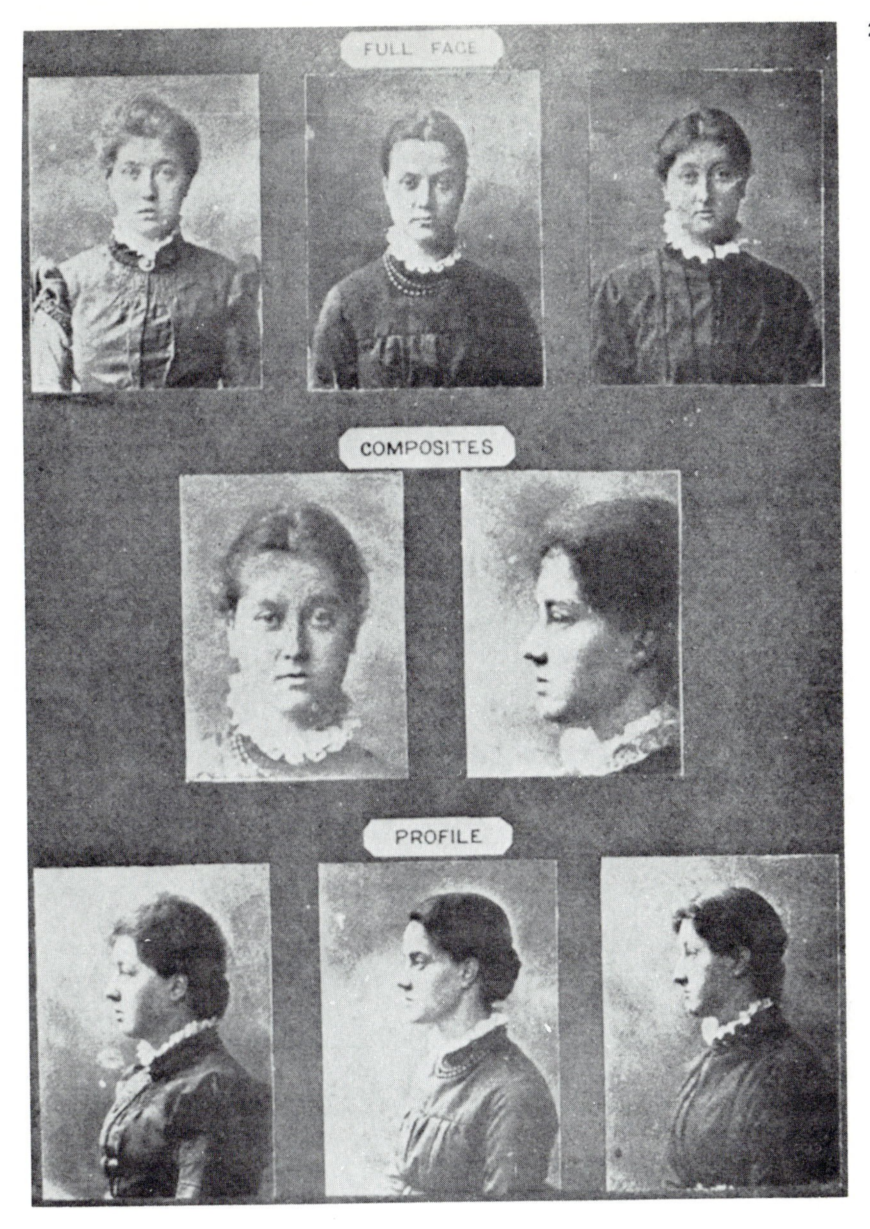

FIGURE 2.5

Galton's composite portraiture of three sisters. (Courtesy of the Galton Laboratory.)

examination of twins was whether twins who were alike at birth became more dissimilar as a consequence of any dissimilarities in their nurture. Conversely, did twins who were unlike at birth become more similar as a consequence of similar nurture? Galton acknowledged two types of twins—those arising from separate eggs, and those arising from the same egg. Yet he

ANTHROPOMETRIC
LABORATORY

For the measurement in various ways of Human Form and Faculty.

Entered from the Science Collection of the S. Kensington Museum.

This laboratory is established by Mr. Francis Galton for the following purposes:—

1. For the use of those who desire to be accurately measured in many ways, either to obtain timely warning of remediable faults in development, or to learn their powers.

2. For keeping a methodical register of the principal measurements of each person, of which he may at any future time obtain a copy under reasonable restrictions. His initials and date of birth will be entered in the register, but not his name. The names are indexed in a separate book.

3. For supplying information on the methods, practice, and uses of human measurement.

4. For anthropometric experiment and research, and for obtaining data for statistical discussion.

Charges for making the principal measurements:
THREEPENCE each, to those who are already on the Register.
FOURPENCE each, to those who are not:— one page of the Register will thenceforward be assigned to them, and a few extra measurements will be made, chiefly for future identification.

The Superintendent is charged with the control of the laboratory and with determining in each case, which, if any, of the extra measurements may be made, and under what conditions.

H & W. Brown, Printers, 20 Fulham Road, S.W.

FIGURE 2.6
Handbill of 1884 announcing Galton's Anthropometric Laboratory. (Courtesy of Cedric A. B. Smith, the Galton Laboratory, University College, London.)

did not distinguish between the two types in his discussion, except as they fell into his categories, "alike at birth" or "unlike at birth."

Galton gathered his evidence from answers to questionnaires and biographical and autobiographical material. He observed that, among 35 pairs who had been alike at birth, and who had been reared under highly similar conditions, the similarities within the twinship persisted after the members had grown to adulthood and gone more-or-less separate ways. From 20 pairs of originally dissimilar twins, there was no compelling evidence that any had

F Galton 12

Initials of			Father and mother first cousins?	Date of birth	Eye color	Sex	Married Single Widowed	Color sense	Occupation
Self	Father	Mother							
FG	G	D	No	16 2 22	Light Blue	M	M	Normal	Private Gen.

Date of measurement			Head length maximum from root of nose		Head breadth maximum		Height standing less heels of shoes		Span of arms from opposite fingertips		Weight in ordinary clothing	Strength of squeeze		Breathing capacity	Keenness of sight distance of reading diamond numerals		Snellen's type read at 90 feet
												Right hand	Left hand		Right eye	Left eye	
Day	Month	Year	Inches	Tenths	Inches	Tenths	Inches	Tenths	Inches	Tenths	Pounds	Pounds	Pounds	Cubic inches	Inches	Inches	Number of type
28	2	88	7	1	6	1½	68	7	71	8	171	80	71	159	17	19	D9

Height sitting above seat of chair		Height of top of knee when sitting less heels left arm		Length of elbow to finger of left hand		Length of middle finger of left hand		HEARING		REACTION TIME	
								Keenness of hearing	Highest audible note	To sight	To sound
Inches	Tenths	Inches	Tenths	Inches	Tenths	Inches	Tenths		Vibrations per second	Thousandths of second	Thousandths of second
36	5	21	6	18	4	4	4				

FIGURE 2.7
Galton's own data from the records of his Anthropometric Laboratory. For some reason, hearing and reaction-time data were not recorded for Galton. (Courtesy of Cedric A. B. Smith, the Galton Laboratory, University College, London.)

become more alike through exposure to similar environments. According to Galton, "There is no escape from the conclusion that nature prevails enormously over nurture when the differences of nurture do not exceed what is commonly to be found among persons of the same rank of society and in the same country. My fear is, that my evidence may seem to prove too much, and be discredited on that account, as it appears contrary to all experience that nurture should go for so little" (1883, p. 241).

Galton's Work in Perspective

The ten years between *The Origin of Species* and *Hereditary Genius* had not been sufficient for the idea of man as an animal to be completely accepted. For many of those who accepted Darwin's theory, of course, Galton's work was a natural and logical extension: human beings differ from animals most strikingly in mental powers; humans, like other animals, have evolved; evolution works by inheritance; mental traits are heritable. For those whose faith in the special creation of man remained firm, Galton's views were unacceptable, atheistic, and reprehensible.

But there were scholars whose inquiries stemmed from a general desire to understand the mind. This philosophical approach was dominated by the British philosophers, whose emphasis was clearly on experience (nurture). They based their views on John Locke's seventeenth-century tabula rasa dictum that ideas are not inborn, but come from experience. The role of experience was also emphasized by experimental psychology, which is usually dated from Wilhelm Wundt's establishment in 1879 of the Psychologisches Institut at Leipzig, Germany. In spite of the fact that Wundt had come to psychology from physiology, his approach was not biological in the same sense as Galton's. The goal of Wundt's institute was the identification, through introspection, of the components of consciousness. Individual differences, which formed the heart of Galton's investigations, were nuisances in this search for principles that could be generally applied to all. One notable exception to this general trend was provided by the American, J. McK. Cattell, who, as a student of Wundt, insisted on studying individual differences. After Cattell left Leipzig, he worked for a while with Galton, with whom he strengthened and confirmed his belief in the importance of individual differences. Cattell had an important influence on the development of American psychology, and inspired some of the earliest experimental work in behavioral genetics.

From the foregoing, it is clear that Galton's work was neither completely in step, nor completely out of step, with his times. Galton lived during a period of great intellectual turmoil in biology. His work was both a product and a cause of the advances that were made. Galton was not the first to insist on the importance of heredity in traits of behavior. At the beginning of this chapter, we read explicit statements on this matter by the ancient Greeks. Nor was Galton the first to place his conclusions in an evolutionary context. Herbert Spencer introduced an "evolutionary associationism" in

1855 (Boring, 1950, p. 240). But it was Galton who championed the idea of inheritance of behavior, and vigorously consolidated and extended it. In effect, we may regard Galton's inspired efforts as the beginning of behavioral genetics.

PRE-MENDELIAN CONCEPTS OF HEREDITY AND VARIATION

Neither Darwin nor Galton understood the mechanism by which heredity works or how heritable variation is maintained. The answers were being worked out by a contemporary scientist working in what is now Czechoslovakia. But since this research was not known to Darwin or to Galton, they had to work within the prevailing views of heredity and heritable variation.

Heredity

Long before Darwin and Galton, there had been substantial evidence of the importance of heredity, although its laws had proved extremely resistant to analysis. In particular, a vast amount of data had been accumulated from plant and animal breeding. Many offspring bore a closer resemblance to one parent than to the other. It was also common for the appearance of offspring to be intermediate between the two parents. But two offspring from the same parents could be quite unlike. As J. L. Lush described the situation considerably later, the first rule of breeding is that "like produces like," while the second rule is that "like does not always produce like" (1951, p. 496).

In Darwin's time, the theory of heredity that seemed to explain most adequately the confusion of facts at the time was the "provisional hypothesis of pangenesis." In this view, the cells of the body, "besides having the power, as is generally admitted, of growing by self-division, throw off free and minute atoms of their contents, that is gemmules. These multiply and aggregate themselves into buds and the sexual elements" (Darwin, 1868, p. 481). Gemmules were presumably thrown off by each cell throughout its course of development. In embryogenesis and later development, gemmules from the parents, originally thrown off during various developmental periods, would come into play at the proper times, thus directing the development of a new organ like that of the parents. The theory of pangenesis was quite reasonable (although it was wrong). It was particularly compelling because it was compatible with Lamarck's notion of "use and disuse" as the source of variability in evolution (see Table 2.1).

Variation

The source of heritable variation was the most difficult component of the model of evolution for Darwin to explain. Without heritable variation in each generation, evolution could not continue. Because children often exhibit some of the characteristics of each of their parents, it was commonly accepted that characteristics of parents merged or blended in their offspring. The troublesome implication of such a "blending" hypothesis is that variability would be greatly reduced (in fact, roughly halved) each generation. For example, if one parent were tall and the other short, the offspring would be of average height. Thus, the blending hypothesis implies that variability would rapidly diminish to a trivial level if it were not replenished in some manner. Although Darwin worried about the problem, he never resolved it. He suggested two ways in which variability might be induced, but both of them assumed that environmental factors altered the stuff of heredity. The theory of pangenesis suggested that gemmules (miniature replicas of the parents' cells) could reflect changes in environment. Darwin vaguely concluded that changes in the conditions of life in some way altered gemmules in the reproductive systems of animals so that their offspring were more variable than they would have been under stable conditions. Ordinarily, this increased variability would be random. Natural selection would then preserve those deviants that by chance happened to be better adapted as a consequence of their deviation.

Sometimes, however, an environmental condition might induce *systematic* change. Darwin hesitatingly accepted the Lamarckian theory of use and disuse to suggest that acquired characteristics can be inherited. In *The Descent of Man,* Darwin speculated about the alleged longer legs and shorter arms of sailors as compared to soldiers: "Whether the several foregoing modifications would become hereditary, if the same habits of life were followed during many generations, is not known, but it is probable" (1871, p. 418). In some of his writings, Darwin seemed sure that variations in life experiences can increase genetic variability: "there can be no doubt that use in our domestic animals has strengthened and enlarged certain parts, and disuse diminished them; and that such modifications are inherited" (1859, p. 102). Likewise, he stated, with respect to behavioral characteristics, that "some intelligent actions, after being performed during several generations, become converted into instincts and are inherited" (1871, p. 447). However, for the most part, Darwin was unsure of the source of variability: "Our ignorance of the laws of variation is profound. Not in one case out of a hundred can we pretend to assign any reason why this or that part has varied.... Habit in producing constitutional peculiarities and use in strengthening and disuse in weakening and diminishing organs, appear in many cases to have been potent in their effects" (1859, p. 122).

THE WORK OF GREGOR MENDEL

Although Darwin struggled with these issues, in his files was an unopened manuscript by an Augustinian monk, Gregor Mendel (Allen, 1975). Through

FIGURE 2.8
Gregor Johann Mendel, 1822–1884. A photograph taken at the time of his research. (Courtesy of V. Orel, Mendel Museum, Brno, Czechoslovakia.)

his research on pea plants in the garden of a monastery at Brunn, Moravia, Mendel had provided the answer to the riddle of inheritance and variability.

As we will discuss in detail in Chapter 4, Mendel summarized his many experiments with two laws. The law of segregation states that there are two elements of heredity for a single character. These two elements segregate, or separate, "cleanly" during inheritance so that offspring receive one of the two elements from each parent. The important implication of this law is that the hereditary elements are discrete and do not blend. Because the elements do not blend, the reduction in inherited variation that concerned Darwin does not occur. Mendel's second law, the law of independent assortment, concerns the inheritance of two traits, each with two elements. When we look at the inheritance of two traits at the same time, the elements for each trait assort independently of the elements for the other trait. In other words, the inheritance of one trait does not affect the inheritance of the other.

We can translate Mendel's theory into more modern terms. As we will discuss in Chapter 5, a *gene* is a coded stretch of DNA (deoxyribonucleic acid). Genes are located on chromosomes in the nucleus of a cell, and chromosomes occur in pairs—one from the mother, and one from the father. Alternate forms of genes (*alleles*) are at the same place (*locus*) on the matching chromosomes. Most humans have 23 pairs of chromosomes.

Now, let's look at a specific example. At the tip of each member of one medium-sized pair (designated chromosome number 9) is an allele (either *A*, *B*, or *O*) of the gene determining the *ABO* blood group. At this locus on one of the chromosomes is the allele (e.g., *A*) from the mother; on the matching chromosome is the allele (e.g., *O*) from the father. In this example, the offspring would have blood group *AO*. Genes do not blend in inheritance, and genetic variation is maintained. But what initially causes the genetic variation? The major source of genetic variability is *mutation,* or changes in the genetic code of DNA.

Mendel's results and his theory were read to the Brunn Society of Natural Science in 1865, and were later published in the proceedings of the society. The crucial experiments had, thus, been done and reported prior to Darwin's statement of pangenesis. But Darwin was not alone in overlooking Mendel's ideas. For 34 years, the "Versuche Ueber Pflanzenhybriden" (Mendel, 1866) remained almost completely unacknowledged.

In 1900, three investigators—C. Correns, Hugo de Vries, and Erich von Tschermak—almost simultaneously "rediscovered" Mendel's work. Thus, a period of intensive research was inaugurated, in which the Mendelian theory was confirmed and extended. The vigorously developing area of research was given the name *genetics* by William Bateson in 1905, and in 1909 the name *gene* was proposed for the Mendelian elements by Wilhelm Johannsen. At the same time, Johannsen made a fundamental distinction between *genotype,* which is the genetic composition of the individual, and *phenotype*, which is the apparent, visible, measurable characteristic. The importance of this distinction is that it makes clear that the observable trait is not a perfect index of the individual's genetic properties. Given a number of individuals of the same genotype, we might nonetheless expect differences among them—those caused by environmental agents. Thus, two beans might be from the same "pure line," and have identical genotypes for size; yet one might be larger than the other because of differences in nurture, such as soil conditions. Nevertheless, their genotypes would remain unaffected, and the beans of the plants grown from these two beans would be of the same average size. The inheritance of acquired characteristics obviously has no place in this scheme.

The fundamental message of Mendelian genetics is variability. The law of segregation means that genes are discrete units that do not blend in inheritance and thus maintain genetic variability. The law of independent assortment means that considerable variability can be expected because traits vary independently of each other. By considering the variability caused by Mendelian segregation and assortment, it is obvious that each individual is truly unique. Mendel happened to study characters in the pea plant that involved only two forms (alleles) of the gene. When only two alleles are considered for a particular gene, an individual must be one of three possible genotypes: A_1A_1, A_1A_2, A_2A_2. However, if more than one gene is involved, the number of possible genotypes increases exponentially (3 raised to the power of the number of loci). Moreover, many genes have more than two alleles. With only 4 alleles for each of 10 genes, the number of different possible genotypes is 10 billion!

Thus, even when relatively few alleles for only 10 genes are considered, the number of possible genotypes is considerably larger than the present human population of the earth. However, complex genes have been studied in which the number of alleles is greater than 200. Research has begun to suggest that multiple alleles—perhaps as many as a dozen to three dozen—tend to be the rule rather than the exception (Coyne, 1976; Singh, Lewontin, and Felton, 1976). Although the exact number of genes in man is not known, it is conservatively estimated (Stern, 1973) to be between 10,000 and 100,000, and one-third or more of these loci may be segregating for two or more alleles (Hopkinson and Harris, 1971). Thus, segregation and independent assortment provide mechanisms for generating astronomical numbers of different genotypes.

Before considering Mendel's experiments in detail in Chapter 4, we shall in the next chapter discuss modern evolutionary theory.

SUMMARY

Although ancient concepts of heredity are interesting, the history of behavioral genetics really began with Darwin, Galton, and Mendel. Darwin's theory of natural selection as an explanation for the origin of species made a major impact on scientific thinking. Galton was the first to study the inheritance of mental characteristics and to suggest using twins to study nature-nurture problems. Mendel solved the riddle of inheritance with his experiments on garden peas. He demonstrated that heredity involves discrete elements, now called genes, and he formulated two laws of inheritance—segregation and independent assortment. We now know that heritable variability is caused by mutations of DNA, the genetic material, and is maintained and transmitted according to the laws discovered by Mendel.

3

Evolution

Darwin's formulation of evolutionary theory caused great furor and vigorous contention. During the ensuing years, however, his principal themes became well substantiated for behavioral, as well as physical, characteristics. Because evolution is the primary shaper of genetic variability, it is appropriate to begin the story of genetic influences on behavior by discussing evolutionary origins.

It is clear from Darwin's writings that he viewed evolution as a generally slow process, involving minute changes over many generations. However, it is difficult to grasp the expanses of evolutionary time without an analogy. A comparison with a football field will be useful. If we take the beginning of the earth some 4.5 billion years ago to be one goal line (the zero-yard line) and take the other goal line to be the present, life began at the 20-yard line. *Homo sapiens* has been around for the last 0.024 inch. We shall begin the story at the 20-yard line.

THE EVOLUTIONARY RECORD: GENETIC VARIABILITY OVER TIME

The Replicators

Life began some 3.5 billion years ago, about a billion years after the formation of the earth, according to our current understanding (Dickerson,

1978). It began in the oceans, when a particular set of atmospheric ingredients was exposed to high temperatures and lightning, producing the forerunners of amino acids. Amino acids, which are the building blocks of protein, accumulated in the oceans. In this "organic soup," combinations of amino acids and other organic material formed and yielded complex molecules. The most important step in evolution was the formation of a particular molecule that could replicate itself. Richard Dawkins (1976) has called this molecule the *replicator:*

> At some point a particularly remarkable molecule was formed by accident. We will call it the *Replicator*. It may not necessarily have been the biggest or the most complex molecule around, but it had the extraordinary property of being able to create copies of itself. This may seem a very unlikely sort of accident to happen....
>
> Actually a molecule which makes copies of itself is not as difficult to imagine as it seems at first, and it only had to arise once. Think of the replicator as a mould or template. Imagine it as a large molecule consisting of a complex chain of various sorts of building block molecules. The small building blocks were abundantly available in the soup surrounding the replicator. Now suppose that each building block has an affinity for its own kind. Then whenever a building block from out in the soup lands up next to a part of the replicator for which it has an affinity, it will tend to stick there. The building blocks which attach themselves in this way will automatically be arranged in a sequence which mimics that of the replicator itself. It is easy then to think of them joining up to form a stable chain just as in the formation of the original replicator. This process could continue as a progressive stacking up, layer upon layer. This is how crystals are formed. On the other hand, the two chains might split apart, in which case we have two replicators, each of which can go on to make further copies.* (Dawkins, 1976, pp. 16–17)

The replicator was the ancestor of DNA, the double coils of genetic information in the nucleus of each of the cells in all living things.

Beginnings of Life

Once genetic material could be replicated, life was off and running. Figure 3.1 provides an overview of the unfolding of life. For 2 billion years, life consisted of single-celled organisms (Schopf, 1978). The early cells, called prokaryotic cells, resembled modern bacteria, which have no nucleus (Dickerson, 1978). The step from single-celled to multicelled organisms, whose cells are eukaryotic (have a nucleus), was apparently more complex, and there were many false starts (Valentine, 1978). Photosynthesis was a

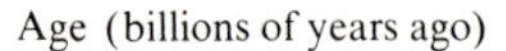

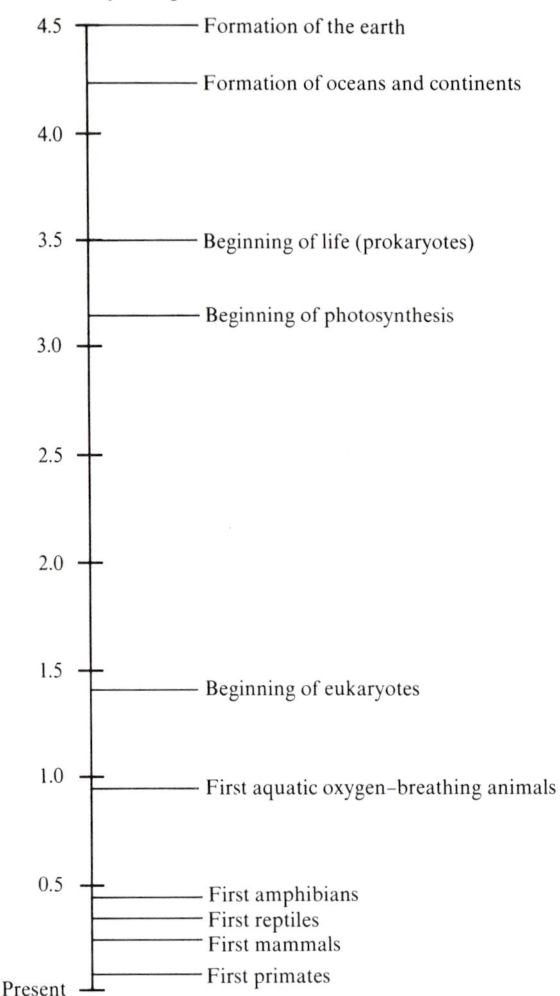

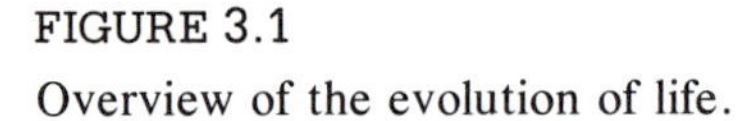

FIGURE 3.1
Overview of the evolution of life.

fundamental new development during this early period of single-celled organisms. This enabled these primitive life forms to acquire a new set of attributes that prepared them to survive in a particular ecological niche. Photosynthesis permitted the building up of food (carbohydrates) from the carbon dioxide that was accumulating in the atmosphere. The energy of light plus carbon dioxide and chlorophyll not only generated carbohydrates, but also produced oxygen as a waste product.

A long time passed before the first aquatic oxygen-breathing animals appeared, a little less than a billion years ago (at the 80-yard line in our football field analogy). The ability to breathe oxygen was a successful development because it permitted mobility and led to an explosion of life forms occupying a wide variety of niches.

Green plants and invertebrates (animals without a backbone or spinal column) led the invasion of the land. In this virgin territory, the variety of plants increased tremendously and came to blanket the earth. At the same time, invertebrates, with their living tissues encased in exoskeletons, were doing well. The earliest insects evolved from some of the invertebrates that first went ashore. Amphibians arose from the invertebrate lobe-finned fish, which had a much fleshier and more muscular padding on their fins than the ray-finned fish we see today. Over the course of millions of years, some of these lobe-finned fish developed the capacity to survive out of water for short periods of time, perhaps moving from one dried-out mud hole to another. This basic plan allowed amphibians to swarm into new and unoccupied ecological niches. The ability to live on land, however, was tempered by the necessity of returning to the water for reproduction and the early development of the next generation.

The next major evolutionary event was the development of animals with larger eggs (to hold more food) and with leathery skins to keep them from drying up on land. A hundred million years after the first amphibians, these animals evolved into a class we call *reptiles*. This was about 400 million years ago (at the 91-yard line). A diversity of reptiles, including the dinosaurs, dominated the scene until about 100 million years ago. Some evolved away from the typical limb structure of reptiles (with fore and hind limbs protruding at right angles to the body axis) and developed limbs parallel to the body, permitting a more efficient stride. Others evolved mechanisms for controlling body temperature, thus freeing the organism from complete dependence on temperature fluxes in the environment. Other developments included a hairy coat, internal fertilization, and the development of the young inside the mother. All of these adaptations also characterize the early mammals.

Adaptations may be specialized to greater or lesser degrees. *Specialized adaptations* are those that precisely fit an organism to its environment. For example, the different species of finch that Darwin observed on the Galápagos Islands (as shown in Figure 2.2) evolved beaks to fit very specialized niches. One species, with an especially powerful beak, survives by cracking open hard seeds. Although such specialized adaptations may at first seem ideal in terms of evolutionary progress, there is a price to pay. Organisms with adaptations specialized for a certain environment suffer when environmental conditions change. *Generalized adaptations* may not fit a specific environment particularly well, but they fit a range of environments. The human hand, for example, is not as specialized as the paws of most other animals. But the hand's generalized adaptations make it useful for a large number of tasks, such as digging, hitting, climbing, and grasping. Although behavioral adaptations in mammals are sometimes specialized, mammals show a predominance of generalized behavioral adaptations. Human beings are part of this trend, with few behavioral patterns that are fixed and many that are generalized.

The rapid divergence of our mammalian ancestors testifies to the apparent success of their adaptations. The class Mammalia includes 19 orders as diverse as monotremes (such as anteaters and platypuses), marsupials (opossums and kangaroos), insectivores (shrews and moles), chiroptera (bats), endentata (sloths and armadillos), rodents, cetacea (whales), carnivores, elephants, and ungulates. Eventually a mammal developed with features that characterize the primate order: mobile digits on hands and feet, a shortened snout, frontally placed eyes, a tendency toward an upright posture, usually singleton offspring, and a brain (the cerebral cortex, in particular) that is large relative to body size.

Primates and Hominids

There are now approximately 193 primate species, of which 53 are little animals (such as the tree shrews, lemurs, and tarsiers) that do not look much like primates. The Pongidae family of gibbons, orangutans, chimpanzees, and gorillas began to diverge about 50 million years ago from the Cercopithecidae family, which includes Old World monkeys and baboons. About 9 million years ago (at the 99.8-yard line), *Australopithecus* roamed Africa. *Australopithecus* was only about 4 feet tall, was a hunter, and lived in small groups. Then *Homo erectus* appeared—a primate that was a member of *Homo,* our genus (the taxonomic category between family and species), although he was of a different species. *Homo erectus* includes the Java and Peking archeological finds.

There is much uncertainty concerning the evolutionary record of the human species. The basic question has to do with intraspecific variability. That is, are the observed differences among the various skulls merely differences among members of the same species or do they represent different species? It has been estimated that the representativeness of the current skeleton samples from archeological digs may be comparable to picking two individuals at random from the population of the United States (Walker and Leakey, 1978). These two individuals, of course, may be quite different in skull shape and would by no means represent the entire population of the country. Thus, archeological data for the human species is quite limited.

Most of the skulls that have been found may now be classified into one of two *genera* (plural of *genus*): *Australopithecus* and *Homo.* Until the 1970s, the standard story was quite simple. *Australopithecus robustus* was the name given to the most common fossil representatives found decades ago. These fossils were characterized by a large face and massive jaws. Although the face was large, the brain case was small, about 500 cubic centimeters (as compared to the contemporary human average of 1,360 cc.). Another type of skull, found by Louis Leakey in the Olduvai Gorge of East Africa (see Figure 3.2), was given the name *Zinjanthropus boisei,* but was later changed

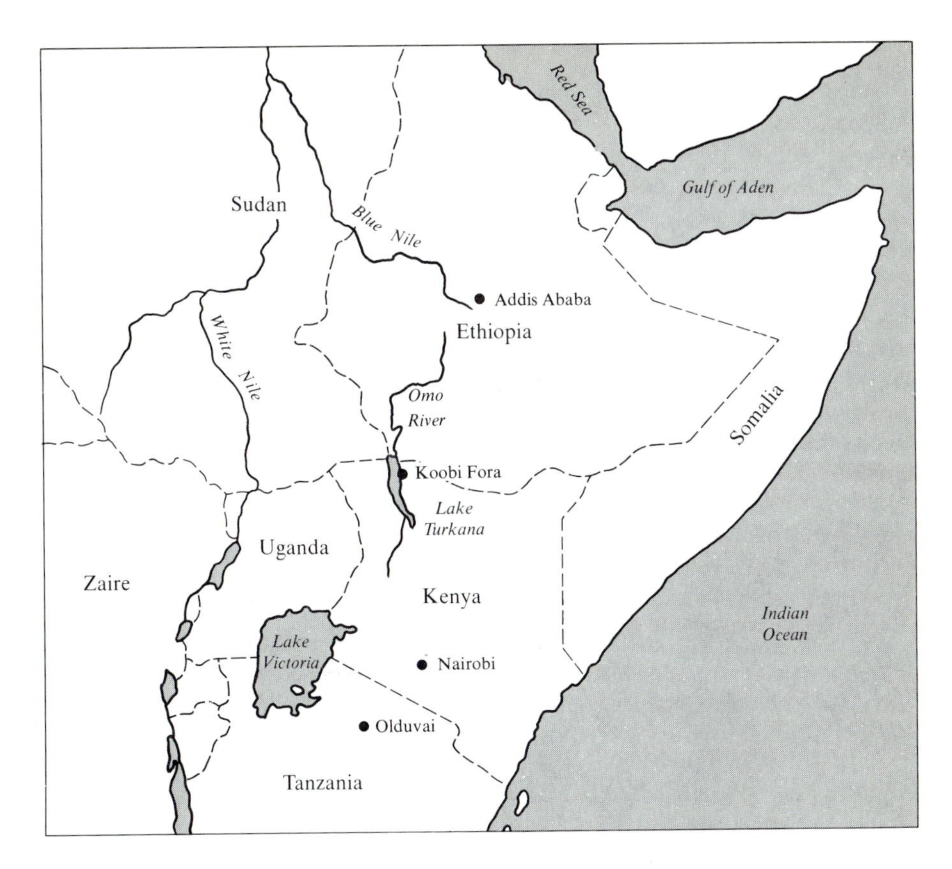

FIGURE 3.2
The Olduvai and Koobi Fora (Lake Turkana) fossil areas in East Africa. (From "The hominids of East Turkana" by A. Walker and R. E. F. Leakey. Copyright © 1978 by Scientific American, Inc. All rights reserved.)

to *Australopithecus boisei.* (See Box 3.1 for a discussion of names of hominid skulls.) Then another type of skull (*Australopithecus africanus*) was found that was smaller and less "robust." This type was thought by many to be the female version of *A. robustus.* However, in 1972, this simple picture changed considerably with the discovery of several specimens at Lake Turkana in Kenya (see Figure 3.2). One very old skull had a large but lightly built brain case with a capacity of about 775 cc. There is no general agreement as to whether this specimen and others like it belong to the genus *Homo* or *Australopithecus;* Richard Leakey prefers to call it *Homo habilis.* Another type of skull with a lightly built but small brain case was also found. R. Leakey suggests that this type may represent a different species—perhaps the same as the *A. africanus* specimens formerly considered to be the female version of *A. robustus.* Figure 3.3 indicates the vast differences among these three types of skulls.

Various hypotheses have been made concerning these three types of skulls, from those who argue that all three forms are diverse members of the

Box 3.1
What's in a Name?

One of the advantages of having studied Latin or Greek is that it helps in understanding the origin of some of the names given to the hominids. The genus *Australopithecus* has nothing to do with Australia. The prefix, *Australo,* comes from the Latin *australis,* meaning "south," and the suffix, *pithecus,* is from the Greek *pithek,* which means "ape." *A. robustus,* of course, refers to the heavy features of this type of Australopithecine. *A. africanus* also has an obvious interpretation, although Richard Leakey prefers to call it *A. gracilis* in order to contrast its smaller size and lighter build with *A. robustus*. *Boisei* refers to "forest," so *A. boisei* means 'the southern ape from the forests." Louis Leakey originally called the *A. boisei* skulls *Zinjanthropus* because *Zinj* is an Arabic word for "eastern Africa" and *anthrop* is a Greek word meaning "human."

The names of *Homo erectus* ("erect" man) skulls can be even more easily explained. For the most part, the early skulls were named after the place where they were found. Java man (also called *Pithecanthropus,* which means "ape-man") was found on the Indonesian island of Java. Similarly, there are *Homo erectus* skulls that were said to represent Peking man and Rhodesian man. Neanderthal man was first found in the Neander valley near Dusseldorf, Germany, and is referred to as *Homo sapiens neanderthalensis*. Cro-magnon man was named after a cave in France where the first skulls were found. Cro-magnon, like modern humans, is referred to as *Homo sapiens sapiens,* a dubious title meaning "wise, wise man."

An interesting sidelight is the story of the Piltdown hoax. In 1912, near the small town of Piltdown, England, a skull was found with a humanlike cranium and an apelike jaw. Piltdown man, as the skull came to be identified, was thought to be the missing link between apes and humans. In 1953, the skull was shown to be a fraud consisting of fragments of modern human skulls and modern ape jaws. The bone fragments had been doctored to make them appear very old. No one thought that the prankster would be found. However, in a recent posthumous statement (Halstead, 1978), British geologist J. A. Douglas accused his predecessor at Oxford, W. J. Sollas. Douglas, who worked in Sollas's laboratory, said that Sollas wanted to damage the reputation of a rival by tricking him into accepting a fake skull as authentic. The trick backfired, however. Because the skull was accepted as authentic, not only by Sollas's rival, but by the entire scientific establishment, Sollas could not expose the fraud without injuring many others besides his rival.

same species to those who argue that each represents a different species. A. Walker and R. Leakey (1978), for example, argue that the variability of these three types is greater than that found among living anthropoid apes. They, therefore, suggest that each represents a different species—*A. robustus, A. africanus,* and *H. habilis*.

Most surprising of all, a skull found in East Africa in 1975 is very similar to the *Homo erectus* skulls found near Peking in the 1930s. Like that of "Peking man," this skull is thick-boned with a large brain case (about 850

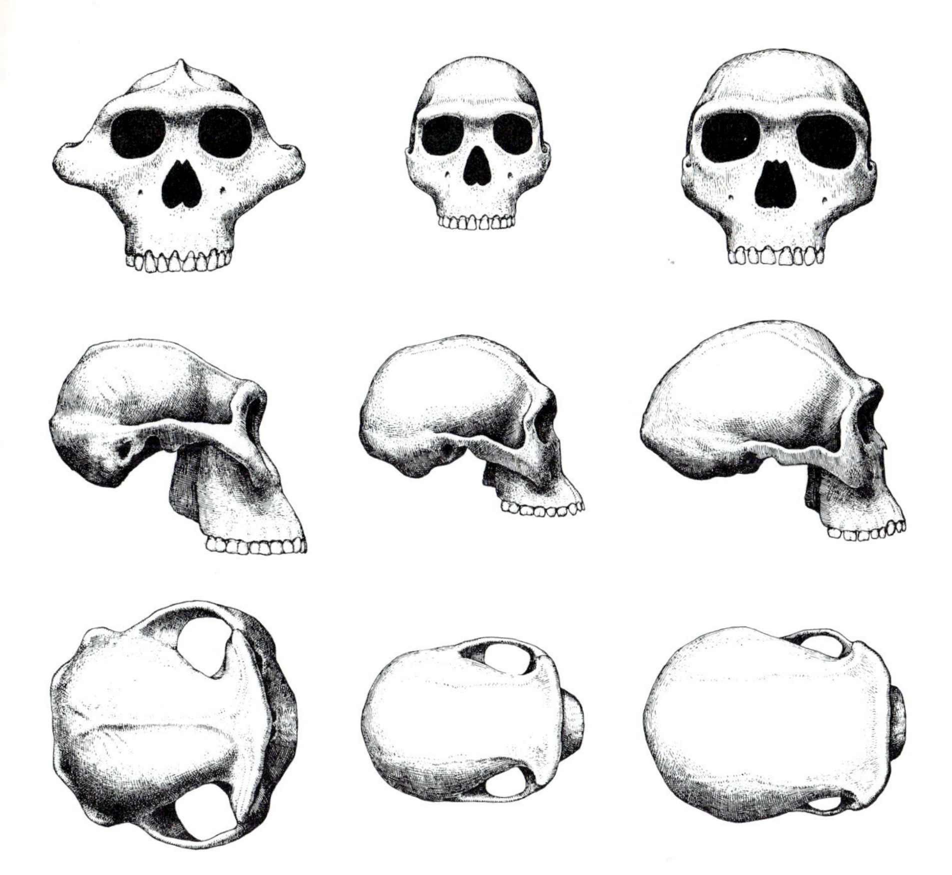

FIGURE 3.3

Three forms of hominid skull from the Turkana fossils. Although there is considerable speculation about these skulls, one possibility is that the skull on the left is *A. robustus*, the one in the middle is *A. africanus*, and the one on the right is *Homo habilis*. (From "The hominids of East Turkana" by A. Walker and R. E. F. Leakey.

cc.) and the characteristic brow ridges. However, the African specimen is about a million years older than the specimens from northern China. This exciting evidence suggests that the genus *Homo* was remarkably stable over a million years of evolution, and that it coexisted with some of the more primitive Australopithecenes. The latter, therefore, may be only collateral ancestors, rather than direct ancestors, of modern mankind. What is the origin of *Homo erectus?* While any of the hominids represented in Figure 3.3 are a possibility, one current theory is that the Australopithecenes (*robustus* and *africanus*), although quite humanlike, eventually became extinct without successors and that *Homo habilis* evolved into *Homo erectus* (Walker and Leakey, 1978).

Figure 3.4 depicts the evolutionary scene from about 10 million years ago to the present. *Australopithecus* was dying out, and *Homo erectus* was

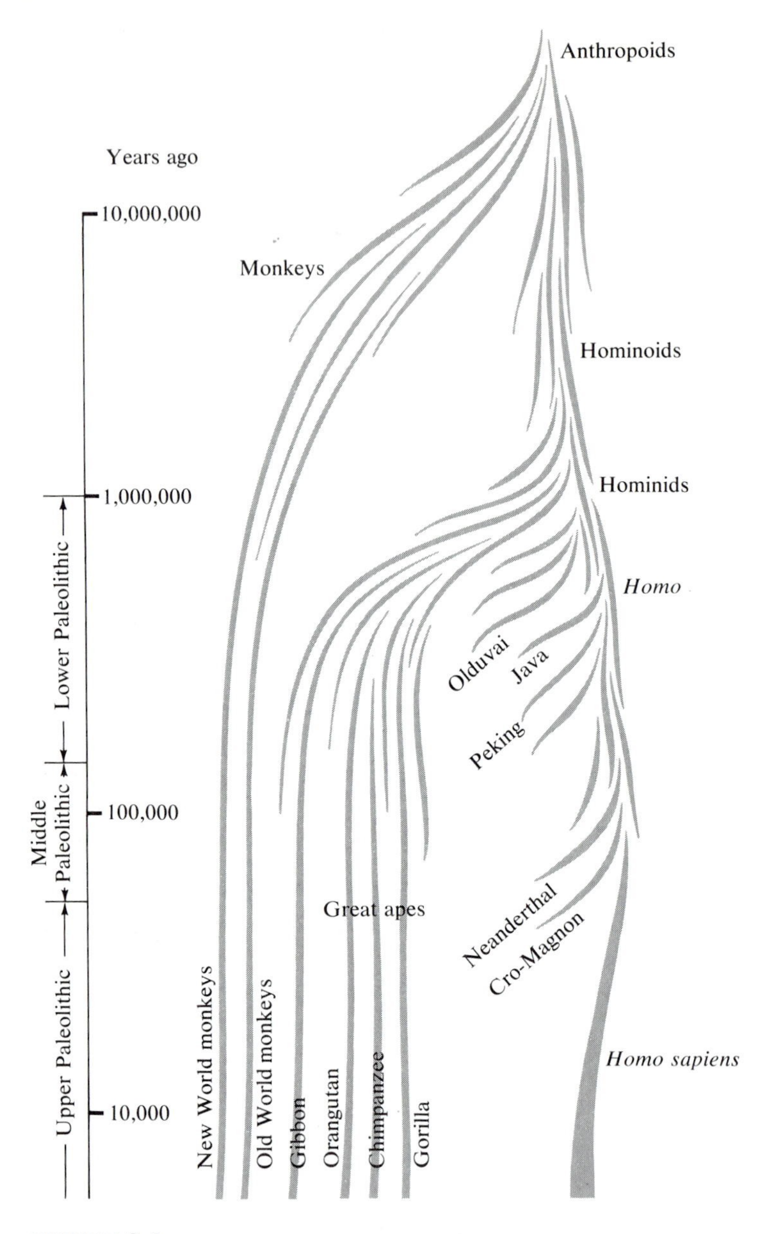

FIGURE 3.4

Pedigree of *Homo sapiens* and his relatives on an exponential time scale. (After "Tools and human evolution" by S. L. Washburn. Copyright © 1960 by Scientific American, Inc. All rights reserved.)

well established; *Homo neanderthalensis,* the Neanderthal man, appeared. The remains of Neanderthal have been found throughout Europe and western Asia. The finds include evidence of religious beliefs and burial ceremonies, and some would like to consider him as belonging to the same species as modern man. Anywhere from 40,000 to 300,000 years ago, *Homo erectus* began to dwindle and the Neanderthal man became dominant. With the eventual appearance of *Homo sapiens,* modern man, Neanderthal man soon became extinct. The skulls of these human progenitors are illustrated in Figure 3.5. The cave paintings, bows and arrows, and pottery of *Homo sapiens* are 30,000 years old.

Behavior and Evolution

Although we are accustomed to thinking about evolution in terms of anatomical records, biologists have come to realize the importance of behavior in the process of speciation. Mayr (1978), for example, suggests that behavior may be the critical factor in the creation of species:

> . . . behavior often—perhaps invariably—serves as a pacemaker in evolution. A change in behavior, such as the selection of a new habitat or food source, sets up new selective pressures and may lead to important adaptive shifts. There is little doubt that some of the most important events in the history of life, such as the conquest of land or of the air, were initiated by shifts in behavior. (Mayr, 1978, pp. 54–55)

The role of behavior as a mechanism for isolating incipient species can be clearly seen in the fruit fly, *Drosophila. Drosophila* are often used in genetic research because their generation interval is only two weeks from egg to adult, they are easily reared, and they provide several unique advantages for genetic analysis, as we shall see in Chapter 5. *Drosophila* have also been studied extensively to understand the evolutionary significance of behavior. There are thousands of species of *Drosophila,* which rarely mate with each other in nature. Some species (called *allopatric*) are isolated simply by geographic separation. However, many species (called *sympatric*) live in overlapping regions; mating behavior that restricts intercrossing serves as the primary isolating mechanism (Ehrman and Parsons, 1976). In Chapter 8, we shall discuss other evolutionary implications of *Drosophila* mating behavior.

Since behavior does not fossilize, the behavior of our evolutionary forebears cannot be studied directly. Fortunately, some products of behavior can be studied from the fossil record. Behavior of human progenitors can be inferred from burial remains, tools, and other cultural artifacts. For example, many of the known *Homo erectus* skulls have been found to have large holes

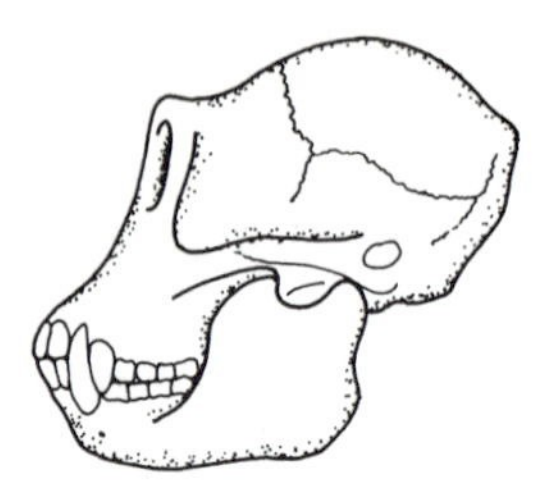

Chimpanzee

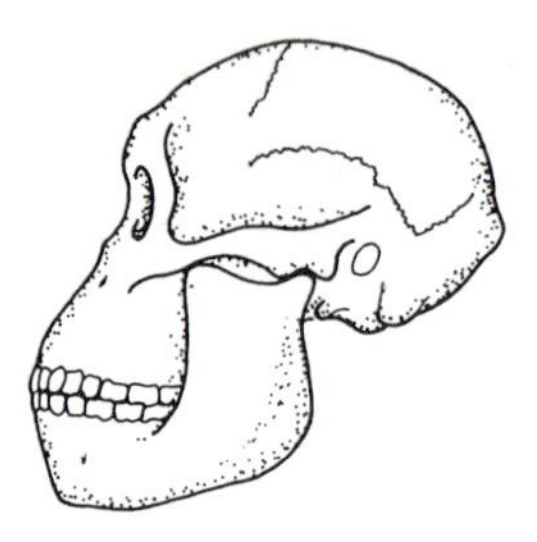

Australopithecus

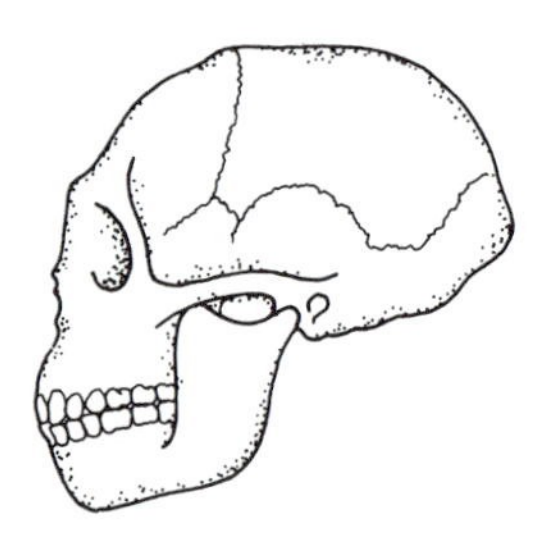

Homo erectus (Java Man)

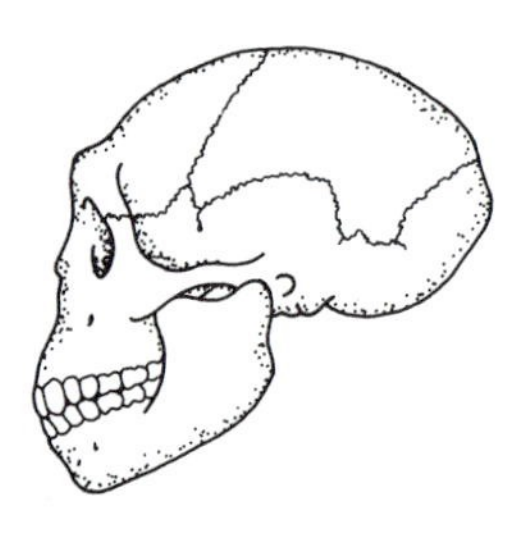

Homo erectus (Peking man)

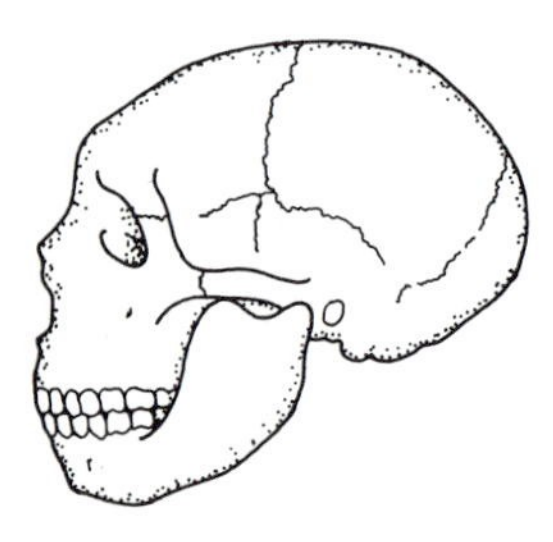

Homo neanderthalensis

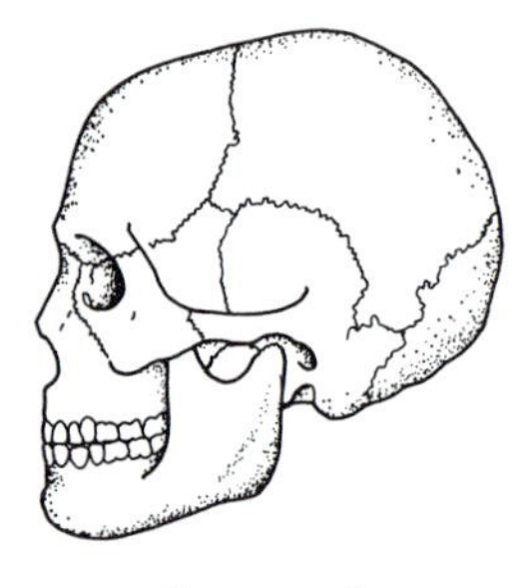

Homo sapiens

FIGURE 3.5

Five hominid skulls, as well as an anthropoid ape skull for comparison. (From *Evolution and Genetics: The Modern Theory of Evolution* by D. J. Merrell. Copyright © 1962 by Holt, Rinehart and Winston, Inc. Reprinted by permission of Holt, Rinehart and Winston, Inc.)

bashed in, suggesting a violent demise. And it is clear from the remains of animal bones around *Homo neanderthalensis* sites that the Neanderthals were superb hunters. The behavioral consequences of hunting are extensive. It is possible that "our intellect, interest, emotions, and basic social life ... are evolutionary products of the success of the hunting adaptation" (Washburn and Lancaster, 1968, p. 213). The hunting of large animals requires not only the efficient use of tools, but also a high degree of coordination and cooperation. Men capable of cooperating to kill large animals were also able to wage effective war on other men. A more comprehensive treatment of the behavioral consequences that followed from the hunting-gathering way of life can be found in R. B. Lee and Irven DeVore (1968).

In the case of a few species, there are some behavioral records that are less inferential. Carl J. Berg (1978) has provided an interesting example of a fossilized record of behavior in predatory marine snails. Some snails prey on clams and other snails by grabbing them with their large foot, moving their prey to a certain position, and drilling a hole through the shell (Figure 3.6). Berg found that each species of predatory snail bores into its prey in a distinct area near the center of the shell. In addition to the position of the hole, Berg was interested in the consistency of hole-drilling behavior within species. Because the hard shells of the "bored" snails fossilize, they can be used to study the evolutionary history of this predatory behavior. Berg found that snail shells from 100 million years ago show essentially no systematic patterning of boreholes. Beginning about 60 million years ago, some patterning can be observed. Because the holes in the fossils could have been made by different species of predatory snails, Berg focused on more recent ones, in which the holes could be attributed to one particular predatory species. These data, spanning the past 20 million years, show no significant change in positioning and a slight increase in consistency of behavior during that time.

In addition to studying the fossil record, some insight into the evolution of human behavior can be gained by studying contemporary human populations that exist under conditions similar to those in which early man evolved.

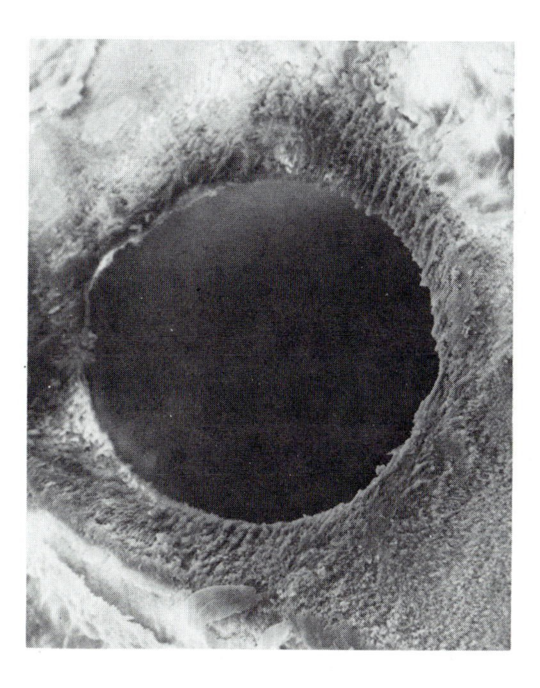

FIGURE 3.6

Scanning electron photomicrograph of a predatory borehole made by a young snail. The diameter of the borehole is approximately 125μ. (Courtesy of Carl J. Berg.)

Most human evolution occurred prior to the appearance of agriculture about 10,000 years ago—at the 99.9998-yard line of our football field analogy. The life style of these early humans revolved around hunting and gathering. Of several populations in the world today that are hunter-gatherers, the best known is the San (formerly called Bushmen) of the Kalahari Desert in southern Africa. They do not cultivate crops; most of their foodstuffs come from gathering nuts, occasionally supplemented by meat obtained on hunting expeditions. Studies of the San have filled in some of the lines in our evolutionary story (Konner, 1977). For example, children are spaced four to five years apart. Infants are completely indulged by their mothers. A baby is carried in a sling in a vertical position so that the head, arms, and legs are free and the baby has continual access to the breast and to a string of beads hanging around the mother's neck (see Figure 3.7). As in most primate species, San fathers clearly have a secondary role in child rearing. Although such studies are informative, you should keep in mind that these populations

FIGURE 3.7

Hunter-gatherer San mother with child in the Kalahari Desert. (Courtesy of Melvin J. Konner.)

have continued to evolve, and may not represent the hunter-gathering way of life of our progenitors.

Another approach is to compare contemporary species to find behavioral similarities, as well as differences. As in the previous method, there is the problem that living representatives of a species have been subject to recent evolutionary forces and may have changed substantially from ancestral forms. Behavior of the most primitive contemporary birds, for example, may not be at all representative of the archaeopteryx (the first bird to evolve from reptiles). Nonetheless, comparisons between species, like studies of contemporary human populations living in hunter-gathering environments, can be used to sketch some broad strokes in the picture of our evolutionary past.

For example, there are two general reproductive strategies that tend to be used by different species. One emphasizes quantity, while the other emphasizes quality, but each works well in various species. The quantity strategy, which is wasteful but effective, is to produce as many offspring as possible. Because of the large numbers of offspring, minimal parental care can be given to each. This strategy works as long as the death rate is lower than the birth rate. The quality strategy, more characteristic of mammals, is to produce few offspring, but to provide for and protect them. Delayed maturation and increased parental care coevolved with this strategy (Wilson, 1975). Because humans, like other primates, have relatively few offspring, we expect them to provide substantial parental care. Other evolutionary features that characterize most primates are greater maternal than paternal care of the young, mechanisms to promote attachment between mothers and their young, and an important role for peer play.

There are also important behavioral differences between humans and other primates. Cognitive development of Old World and New World monkeys closely parallels that of human children until the third year of life, when human cognitive development skyrockets and development in other primates slows down (Scarr, 1975). Human beings also appear to be the only natural users of language. The famous chimpanzee Washoe needed extensive training in order to learn American Sign Language at a level comparable to that shown by 3-year-old human children. This suggests that one unique evolutionary trend in the human line was toward the symbolic capabilities required for language use. Such behavioral and anatomical differences between humans and chimps raise interesting questions, given the considerable genetic similarity between these species. (See Box 3.2 and Box 3.3.)

GENETIC VARIABILITY BETWEEN SPECIES

The purpose of the previous section was to present a temporal scale on which to examine human origins from the invertebrate past through the development of the amphibians, reptiles, mammals, primates, and modern

Box 3.2
Methods for Determining Genetic Differences Among Species

Four major methods have been used to investigate genetic differences between species, or between populations within a species: (1) amino acid sequencing, (2) immunological techniques, (3) electrophoresis, and (4) DNA hybridization.

AMINO ACID SEQUENCING

Amino acids are the building blocks of proteins, and the sequence in which amino acids are linked to form proteins is coded by genes, as described in Chapter 5. We can thus compare the sequence of amino acids in similar proteins in different species in order to determine how similar they are genetically.

IMMUNOLOGICAL TECHNIQUES

Before biochemists discovered a technique for determining the sequence of amino acids, various immunological techniques were used to compare proteins in different species. *Immunology* is the study of antibodies formed in the serum of blood when a foreign substance, called an *antigen* (a protein or carbohydrate), is introduced. The ABO blood-typing system is an example of an immunological system. When antibodies against a particular protein are formed, they link with the protein, creating a clump, or *precipitate*. In 1904, G. H. Nuttall used the immune response to study differences between species. Consider three species: chimps, rabbits, and human beings. When chimp serum is injected into rabbits, the rabbits will form antibodies to proteins in the chimp serum. Then serum from the rabbit (with the "anti-chimp" antibodies) is mixed with human blood. The anti-chimp antibodies in the rabbit serum will link with human proteins that are similar to the chimp proteins. The great similarity between the chimp and human proteins will create a strong precipitation, or clumping reaction. This method has been shown to agree with the more recent amino acid sequencing results (Goodman, et al., 1971), and has been used much more extensively in primate studies.

ELECTROPHORESIS

Another technique, called electrophoresis (literally, "carried by electricity"), compares the overall charges of amino acids in an electrical field rather than actually analyzing their sequences in proteins. This technique, which has been used extensively in population genetics (Chapter 8), is shown in the figure on the facing page.

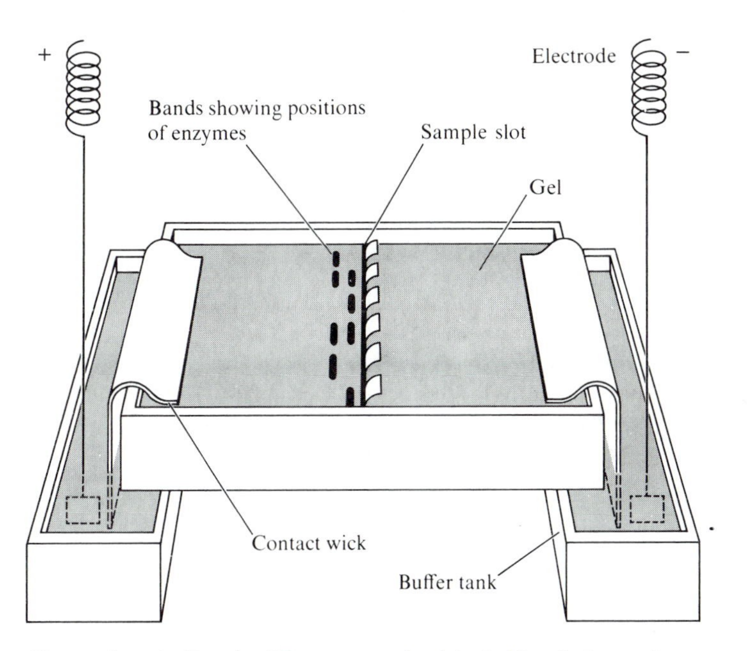

Electrophoresis. Protein differences can be detected by placing various proteins on a gel made of starch, agar, or polyacrylamide. The gel with the protein samples is subjected to an electric current, usually for a few hours. Each protein in the sample migrates through the gel in a direction and at a rate determined by its net electric charge and molecular size. A chemical solution is applied, which reacts with the protein to produce a colored band at those spots to which the protein has migrated. Differences between proteins can be assessed by the number and position of the electrophoretic bands. (Adapted from "The mechanism of evolution" by F. J. Ayala. Copyright © 1978 by Scientific American, Inc. All rights reserved.)

Amino acid sequencing data suggest that those proteins that show electrophoretic differences are likely to differ in only one amino acid out of the hundreds in each protein.

DNA HYBRIDIZATION

Another method combines DNA from different species. It compares the temperature at which the combined DNA for two species separates to the temperature at which recombined DNA from the same species separates.

Box 3.3
Genetic Differences Between Humans and Chimps

The methods described in Box 3.2 have been applied extensively to primates. Together, the data suggest that *Homo sapiens* is closer genetically to the chimpanzee and the other African apes than to any other extant primate species.

Amino acid sequence differences between humans and chimps have been studied for 8 proteins, with a total of more than 1,000 amino acid sites. For these proteins, only 2 amino acid differences were found between humans and chimps (King and Wilson, 1975). The figure opposite shows immunological differences (in arbitrary units) between selected mammals. Human beings are clearly more similar to the African apes than to other primates, and differ relatively little from chimpanzees and gorillas (Bruce and Ayala, 1978). When electrophoresis was used to compare 44 different proteins in chimps and humans, about half were identical (King and Wilson, 1975). We know that most of the proteins that differ are likely to differ in only one amino acid out of the hundreds in each protein. DNA hybridization techniques show similar results: chimp-human DNA combinations lead to an estimate of only a 1 percent difference in DNA between the two species.

All of these methods demonstrate considerable genetic similarity between man and chimp. In fact, the genetic distance between them is about that found for closely related sibling species of other mammals. However, these data are commonly misinterpreted to mean that chimps and humans are about 99 percent genetically similar. Some people (e.g., Washburn, 1978b) even go on to argue that humans must, therefore, be more than 99 percent genetically similar to each other. Note that we have said that human and chimp DNA differs by only 1 percent. This is not the same as saying that only 1 percent of their genes differ. As we shall see in Chapter 5, about 300 amino acids, on the average, are linked to form a protein. If a single one of these amino acids is missing or in the wrong sequence, the resulting protein can be completely different. For example, sickle-cell anemia is caused by a single amino acid substitution in a polypeptide, called the beta chain, of hemoglobin. There are 146 amino acids in the beta chain, but a single substitution causes the protein to malfunction. Thus, the difference between normal hemoglobin and the mutant form is less than the 1 percent difference in DNA between chimps and humans (a difference in only 1 out of 146 amino acids). However, the proteins are functionally very different. Most mutations are of this type.

In other words, even though chimp DNA and human DNA are very similar, their protein products can be quite different. If the average protein has 300 amino acids, and if 1 in every 100 amino acids differs between humans and chimps, every protein in the human body could conceivably be different from every protein in the chimp body. In fact, the probability that humans and chimps are identical for a particular gene product is only 52 percent, as computed from the electrophoretic data cited above (Plomin and Kuse, 1979).

In summary, although humans are certainly more similar genetically to the African apes than to other primates, considerable caution is required in answering questions about absolute genetic differences between the species. One must be clear about the meaning of the word *genetic*, and one must face the difficult issue, "How different is different?"

Immunological differences. Immunological distances between selected mammals are indicated by the separation, as measured along the horizontal axis, between the branches of this "divergence tree." For example, the monotremes (primitive egg-laying mammals) are removed

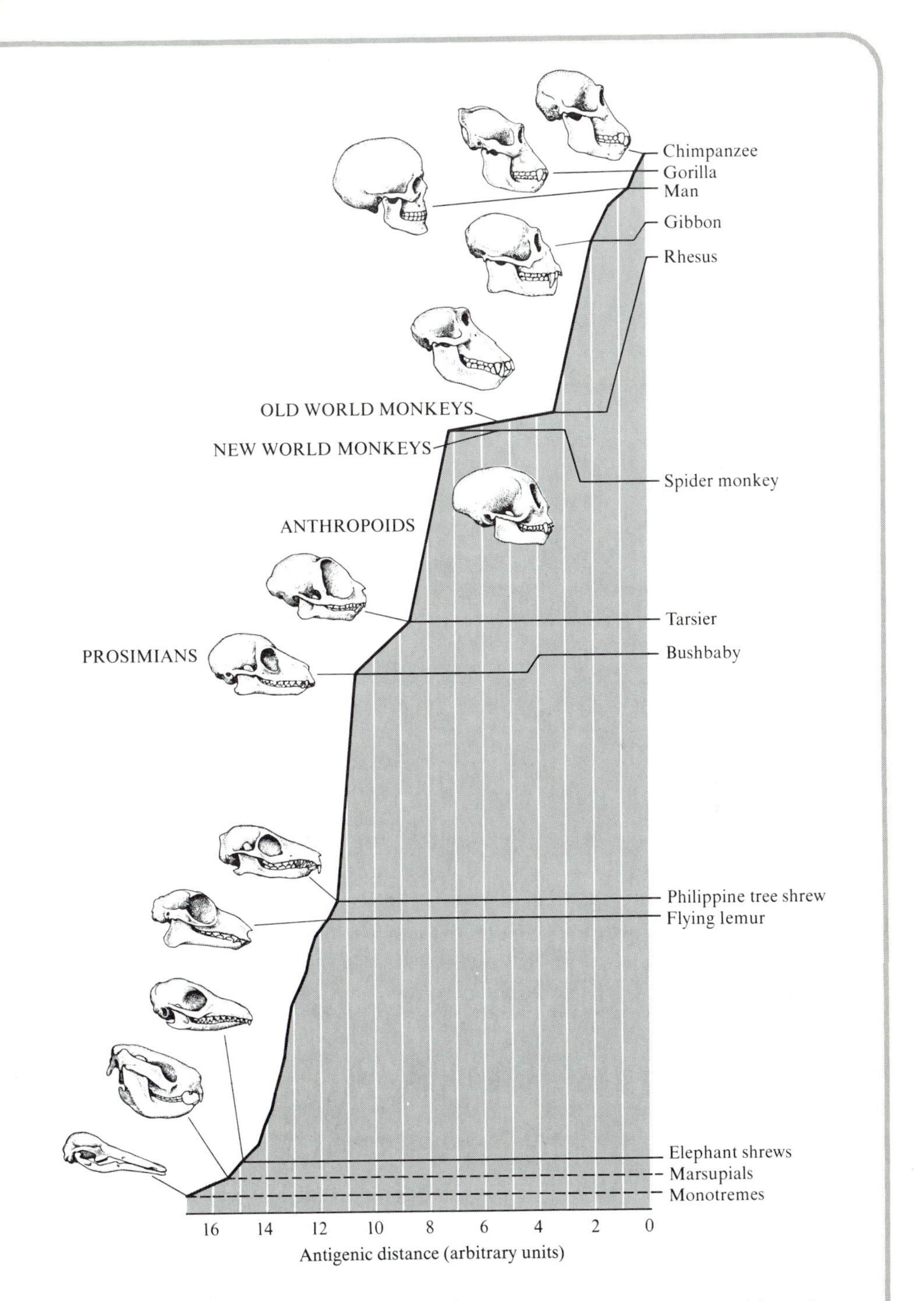

from the marsupials by a distance (in arbitrary units) of only 1.5, but are removed from the chimpanzee by a distance of nearly 17. The distance between man and the Old World monkeys is a little more than 3, between man and the Asiatic gibbons 2, and between man and the gorillas and the chimpanzees of Africa less than 1. The data are from Morris Goodman of Wayne State University. (Adapted from "The evolution of man" by S. L. Washburn. Copyright © 1978 by Scientific American, Inc. All rights reserved.)

man. The examples show genetic diversity growing over a long period of time. In this section, we shall consider contemporary genetic variability between species.

When Linnaeus defined the word *species* in 1735, he thought that each species was an immutable unit specially created by God. We now know that no system of taxonomy can perfectly capture genetic variability in unchanging discrete units (species), because there is sizable genetic variability within each species. Before discussing such intraspecies variability, we shall first consider genetic variability between species.

Although the genetic variability of organisms closest to man in an evolutionary sense is most interesting to us, it needs to be placed in a larger context. The primate order belongs to the mammalian class, which includes 18 other orders, some of which were listed earlier. There are about 4,300 species in the mammalian class alone, but this great variability represents only a small part of the phylum Chordata, which consists primarily of vertebrates. There are about 3,000 amphibian species, 6,000 reptile species, 11,000 species of birds, and 28,000 species of bony fish. All told, there are about 55,000 species of vertebrates. This genetic diversity, although impressive, is minute when compared to the invertebrates. There are an estimated 1,055,000 species in the phylum Chordata.

A microscopic view of single-celled organisms also adds to our perspective on variability. It has been estimated that there are about 100 octillion living cells in the world today—that is, 100,000,000,000,000,000,000,000,000,000 (Hockett, 1973). Of these 100 octillion living cells, perhaps as many as 99 octillion (at least 90 octillion) are tied up in single-celled organisms such as bacteria. We, the *metazoan* (multicellular) organisms, are in a decided minority. It should also be pointed out that this variability includes only the survivors of the process of natural selection. Many more species are extinct than extant.

EVOLUTION AND VARIABILITY

As discussed in Chapter 1, comparisons between species tend to be typological, assuming invariant behavior among members of each species. Although comparisons among contemporary species can provide clues concerning evolution, we must also study variation within species. Until recently, evolutionary thinking has focused on differences between species and has scarcely considered intraspecies differences. There is a common, but mistaken, tendency to consider selection as a force that creates an ideal match between a species and its environment, and to assume that all members of the species conform to this ideal and show no important variations. Because of the prevalence of this view, we shall discuss it briefly at this time, and in more detail in Chapter 8.

As indicated in the previous chapter, Darwin recognized that variability is the key to natural selection. He was uncomfortable with the prevalent blending notion of inheritance because it would halve variability each generation. We now know that inheritance does not involve blending, but rather the discrete hereditary elements discovered by Mendel.

Natural selection is often equated with directional selection—selection for an extreme of a character. Perhaps we are accustomed to thinking this way because artificial selection is most often directional. We select for hens that lay the most eggs, cows that produce the most or richest milk, and horses that run the fastest. If selection is directional, genetic variability is diminished because alleles favoring the selected characteristic tend to become "locked into" the species. (See Chapter 8 for details.) Because directional selection squeezes out genetic variability, it leads to typological thinking. It has even been argued that traits that show genetic variability within species are "genetic junk," that is, unimportant evolutionarily.

However, *stabilizing selection* is another important selection strategy, and it can maintain genetic variability. Even with artificial selection, stabilizing selection operates to keep the organism in balance. For example, as hens are selected for extreme egg production, their fertility and general viability usually diminish, thus counterbalancing the effects of extreme artificial selection. Stabilizing selection is not nearly as dramatic as directional selection, but it is very common in nature. Edward O. Wilson notes that "examples of counteracting (stabilizing) selection forces are easy to find in nature" and cites aggressive behavior, dominance systems, and sexual behavior as examples (1975, p. 132).

Francis Galton, who was responsible for so many firsts, was one of the first to consider stabilizing selection. During his travels in Africa, Galton noticed the strong herding behavior of his pack-oxen. He considered the survival value of this behavior in a paper published in 1871, and concluded that stabilizing selection, rather than directional selection, was the best evolutionary bet. A moderate amount of gregariousness appeared optimal for both grazing and protection. Too little gregariousness would leave the oxen in a vulnerable position for predators, and too much gregariousness would be inefficient for grazing.

In their book on population genetics, J. F. Crow and M. Kimura conclude: "Several times in this book we have noted that for almost any metrical trait (except fitness itself) the most fit individuals are likely to have an intermediate value for the trait. If size is the trait, the species typically has a characteristic size and individuals that are too large or too small are less viable or fertile, and similarly for other quantitative traits" (1970, p. 293).

Thus, stabilizing selection maintains the mean of a population. It also can maintain genetic variability—as, for example, when heterozygotes have a selective advantage. When many genes are involved, there may be some reduction in variability, but this is a complex issue (Crow and Kimura, 1970). These issues are discussed at greater length in Chapter 8.

During the last decade, population geneticists have come to consider stabilizing selection in detail because it provides a possible evolutionary answer to a dramatic finding: at least a third of all genes are polymorphic (that is, have two or more alleles—alternate forms of genes) in most species. Although population geneticists agree that genetic variability within species is ubiquitous, they disagree as to its evolutionary cause. "Neutralists" argue that most of the variability is selectively neutral: it remains because it doesn't make a difference in terms of reproductive fitness. "Selectionists," however, argue that variability is actively maintained by some sort of selectional balance. While both positions are probably correct to some extent for certain characteristics, several interesting examples of stabilizing selection have been discovered in recent years (see Chapter 8).

For completeness, we should mention a third type of selection. *Disruptive selection* favors both extremes of a characteristic, thus leaving the mean unchanged but greatly increasing genetic variation.

In summary, because natural selection may actively maintain genetic variability, we must consider genetic differences *within* species as well as average differences *between* species. These issues concerning evolution and genetic variability help in understanding the new field of sociobiology, and in recognizing its relationship to behavioral genetics.

SOCIOBIOLOGY

Although behavioral scientists have long been interested in evolutionary differences between species, one evolutionary theory has recently received considerable attention (Campbell, 1975; Gregory, Silvers, and Sutch, 1978; Wispé and Thompson, 1976). *Sociobiology: The New Synthesis* is the title of a book by E. O. Wilson (1975) that capped a new wave of research on an old problem. The old problem that sociobiologists have addressed is "altruism," self-sacrificing behavior. How can such behavior be explained in Darwinian terms? Darwin himself worried about this problem:

> I will not here enter on these several cases, but will confine myself to one special difficulty, which at first appeared to me insuperable, and actually fatal to the whole theory. I allude to the neuters or sterile females in insect-communities; for these neuters often differ widely in instinct and in structure from both the males and fertile females, and yet, from being sterile, they cannot propagate their kind. (Darwin, 1859, pp. 203–204)

Darwin's answer to the problem involved group selection:

> . . . if such insects had been social, and it had been profitable to the community that a number should have been annually born capable of work, but incapable of procreation, I can see no especial difficulty in this having been affected through natural selection. (Darwin, 1859, p. 204)

In other words, if individual members of a group sacrifice themselves for the good of the group, then that group is more likely to survive. The group selection hypothesis was brought into prominence by a book in which V. C. Wynne-Edwards (1962) used group selection to explain why many animals seem to reduce their reproduction altruistically.

However, group selection is difficult to reconcile with individual selection. In an altruistic group, one might expect selfish individuals to have a selective edge over the altruists. If the altruists are selected against because some do not reproduce, and if the selfish individuals are favored in selection, then the offspring of the selfish individuals might be expected to take over.

In *Sociobiology*, Wilson suggests a different answer to the old problem of altruism: "The answer is kinship: if the genes causing the altruism are shared by two organisms because of common descent, and if the altruistic act by one organism increases the joint contribution of these genes to the next generation, the propensity to altruism will spread through the gene pool" (1975, pp. 3–4). Sociobiologists argue that the notion of individual fitness should be extended to inclusive fitness. *Inclusive fitness* is the fitness of an individual plus that part of the fitness of kin that is genetically shared by the individual (Hamilton, 1964). J. B. S. Haldane reportedly anticipated this view when he jokingly announced that he would lay down his life for two full siblings or eight first cousins. Either two siblings (each sharing approximately ½ of his genes) or eight cousins (each sharing ⅛) would be his genetic equivalent.

Inclusive fitness and kinship selection suggest that the unit of selection is not the group or the individual, but the gene. This is the premise of Dawkins's (1976) book, *The Selfish Gene*. The title of the book implies that genetic selfishness can explain seemingly altruistic acts of an individual if the net result of the altruistic act helps more of that individual's genes survive and helps transmit them to future generations. This is the point of considering inclusive fitness rather than individual fitness. Wilson (1975) summarizes the sociobiological view of altruism: "A genetically based act of altruism, selfishness, or spite will evolve if the average inclusive fitness of individuals within networks displaying it is greater than the inclusive fitness of individuals in otherwise comparable networks that do not display it" (p. 118).

Between-Species Examples of Sociobiology

Kinship selection theory makes the prediction that altruism should occur more frequently as genetic similarity increases. Nearly all of the tests of this hypothesis have come from comparisons between species. That is, why do some species show a certain kind of altruism while others do not? For example, in one of the papers forming the foundation for sociobiology, W. D. Hamilton (1964) used kinship selection theory to explain why some species, such as honey bees, ants, and wasps, have a caste of female workers who do not reproduce. In evolutionary terms, this is the ultimate act of altruism. Nonreproductive female workers share three-quarters of their genes with their sisters (future queens as well as other female workers) because of a special genetic arrangement in which the females, like most organisms, have two sets of genes (diploid), but the males have only one set (haploid). If these female workers reproduced, they would share only half of their genes with their offspring. Thus, their inclusive fitness is better served by working for the rest of the community than by rearing their own offspring (Smith, 1978).

There are other interspecies illustrations of these principles. For example, why do zebras, but not wildebeests, defend their calves? Zebras live in family groups, while wildebeest herds are mixed family groupings. The answer may be that adult zebras are genetically related to the calves they defend, whereas adult wildebeests may be risking their lives for an unrelated calf (West-Eberhard, 1975). Occasionally, sociobiological research has focused on populations within a species—for example, why paternal behavior of hoary marmots in isolated family units differs from paternal behavior in populous colonies (Barash, 1975). Nonetheless, the perspective is typological—assuming invariant behavior for each group or species studied. (See Chapter 1.)

With regard to the human species, it has been argued that "sociobiology deals with biological universals that may underlie human social behavior" (Barash, 1977, p. 278). This is also a typological approach. One example of sociobiological speculation concerning human behavior is the prediction of conflicts between parents and their offspring. If you don't think too deeply about it, you might expect sociobiology to predict harmony between parents and offspring because they are, after all, kin. However, sociobiology reinterprets the relationship between parents and their offspring in terms of the genetic selfishness of inclusive fitness, which leads to conflict (Trivers, 1974). For parents, maximal inclusive fitness comes from having many offspring, which reduces their relative genetic investment in each offspring. For the offspring, however, it is a different story. The inclusive fitness of each offspring may be better served by attempting to maximize his parents' investment in himself rather than in his siblings, thus creating conflict between parents and offspring, as well as among siblings.

Sociobiology also provides an explanation for greater maternal than paternal care of offspring in the vast majority of mammalian species, including humans. Unless a species is completely monogamous (as are eagles, for example), males have less invested in the young. Males may have offspring by several females, but each female must devote energy to the pregnancy and, in mammals, continue to provide sustenance after birth. Although parental care generally increases with parental investment, the ultimate issue is inclusive fitness. Because of the investment females have made in their young, their fitness is better served by increased care of their present offspring. In many cases, however, the males' investment is little more than copulation, and they may maximize their inclusive fitness by having more offspring by different females. A related reason for greater maternal than paternal care may be that females can always be sure that they share half of their genes with their young. Males, however, cannot be so sure. According to sociobiology, the greater altruistic attention of mothers to their offspring is no less selfish from a genetic point of view than that of fathers.

Sociobiology has taken the first step in explaining some broad facts about human behavior. However, in order to pass beyond these broad evolutionary comparisons between species, we need to consider differences within species. Why do some mothers neglect or even abuse their children? Why is there more parent-offspring conflict in some families than in others?

Sociobiology and Genetic Differences Within Species

Sociobiology has focused on differences between rather than within species. This focus is surprising in view of the fact that the foundation of sociobiology rests on genetic differences within species. Kinship selection theory begins with degrees of genetic relationship: full siblings share roughly half of their segregating genes, half-siblings share one-quarter, cousins share one-eighth, and so on. This important concept will be discussed in Chapter 9, but the word *segregating* needs to be emphasized now. As discussed in Chapter 2, segregation refers to genes that have alternate forms (alleles) that separate according to the laws discovered by Mendel. If all genes had only one allele (that is, if there were no genetic variation within a species), then all members of that species would be identical genetically. In other words, kinship would not matter because all members of the species would be the same genetically. Thus, sociobiological theory implicitly requires genetic variability within a species.

There have been a few attempts to apply kinship selection predictions to the study of behavioral differences within species in order to explain why some members of a species are more altruistic than others. Kinship selection theory predicts that the answer is genetic selfishness; we are more altruistic to those who share more genes with us.

One study (Sherman, 1977) considered alarm calls in response to predators within a species of ground squirrels. (See Figure 3.8.) During more than 3,000 hours of observation, researchers recorded 102 predator encounters that led to the death of 9 squirrels. Because squirrels who call out in warning are more likely to be stalked or chased by predators, alarm calls qualify as altruism. Kinship selection predicts that alarm calls will be more prevalent among individuals with more relatives. This prediction is supported by the observation that females with kin gave more alarm calls than females without kin. Females who were pregnant, lactating, or living with postweaning young gave alarm calls 14 of the 19 times that they were present when a predator appeared. Females with no known kin gave alarm calls when a predator appeared on only 2 of 14 occasions. The data for the males did not support kinship selection theory. For example, frequency of copula-

FIGURE 3.8

Ground squirrel in position for alarm call. (Photograph by George D. Lepp; courtesy of Paul W. Sherman.)

tion by males was not related to the frequency of their alarm calls. However, it is difficult to interpret the data for the males because their total number of alarm calls was low and because ground squirrel society is decidedly matrilineal.

In this study, variation was observed in alarm calls within a species. The kinship selection theory prediction that such altruism will be more common among individuals who have more relatives was confirmed for females. However, this does not prove that altruism has a genetic base, because relatives share environments as well as genes. For example, it is possible that females who rear young learn to protect them as they were themselves protected. If kinship selection theory is to be accurately applied, it is necessary to determine the extent to which observed variability is genetic in origin.

Sociobiology and Behavioral Genetics

Sociobiology and behavioral genetics are both concerned, in part, with the inheritance of social behavior. Part of the difficulty in defining the relationship between sociobiology and any other field of study is that the term *sociobiology* has been defined to include all of evolution, genetics, and behavioral science. Wilson defined sociobiology as the "systematic study of the biological basis of all social behavior" (p. 595). There is no apparent justification for such an all-encompassing definition.

The definitive contribution of sociobiology to evolutionary theory involves only the extension of individual fitness to inclusive fitness. When this definition is used, behavioral genetics differs from sociobiology in several important ways. First, behavioral genetics includes many genetic perspectives on behavior: Mendelian single-gene approaches, molecular genetics, quantitative genetics, population genetics, and evolutionary genetics. Sociobiology is concerned primarily with the last two, although an evolutionary approach to behavior can incorporate the other perspectives (Broadhurst, 1979). Second, behavioral genetics is not limited to studying social behavior. For example, in studies of human beings, individual behavior such as specific cognitive abilities and psychopathology have been the focus of many behavioral genetic investigations. Third, behavioral genetics focuses on differences within species, while sociobiology emphasizes differences between species. Although both perspectives are useful, the distinction between them is important. The causes of average differences between species may be unrelated to the causes of individual differences within species. For example, similarities between humans and chimps may be due to their genetic similarity, but either genetic or environmental factors may be primarily responsible for differences within either species. Similarly, differences within either species could be largely genetic in origin, yet average differences in behavior between the species could be caused by the influence of culture on human behavior. Finally, behavioral genetics has developed methods to assess the extent to which observed behavioral variation can be ascribed to genetic or environmental influences. It is important to note that most human and nonhuman behaviors studied by behavioral geneticists are found to be influenced by both heredity and environment. In contrast, sociobiological discussions of human behavior tend to rely on average differences and similarities between the human species and other contemporaneous species, average differences and similarities among contemporaneous human cultures, and speculations concerning the adaptive value of behaviors in hunter-gatherer societies. E. O. Wilson's (1978) book, *On Human Nature,* exemplifies a sociobiological approach to human behavior.

For example, here is how a behavioral geneticist would approach the issue of parental care in humans. Rather than taking a typological approach by asking how paternal care differs on the average from maternal care, a behavioral geneticist would study the variability in parental care. Parents vary from smothering protectiveness of their children to neglect, rejection, and even abuse. After describing such variations in human parental care, the next question concerns the extent to which genetic and environmental influences are involved in causing such variation. Environmental causes seem likely at the outset. For example, a common hypothesis is that child abusers were themselves abused as children and thus learned to react to the frustrations of rearing a child in this destructive manner. However, it is also possible that genetic factors are involved. For example, child abuse may run in families because of an inherited propensity to be emotionally labile, or short-tempered. However, such armchair speculation will not answer the question. In families in which parents rear their own children, both genes and environment are shared, so we cannot separate their effects. Behavioral geneticists, therefore, use special designs, such as the adoption study, in which either genes are shared or environments are shared, but not both. In this way, a behavioral geneticist can study differences in parental care and determine the extent to which that variation is genetic or environmental in origin.

GENETIC VARIABILITY WITHIN SPECIES

An important lesson to be learned from evolution is the pervasiveness of genetic variability. Most evolutionary theorists focus on genetic variability from one species to another. However, as described in Chapter 2, genetic variation within species is the basis of the evolutionary process. Humans have at least 100,000 genetic loci, and at least a third of these may have two or more alleles. The potential variability within the human species is so great that it is next to impossible to imagine that there have ever been two individuals with the same combination of genes. In fact, each of us has the capacity to generate 10^{3000} different eggs or sperm. The number of sperm of all men who have ever lived is only 10^{24}. If we consider 10^{3000} possible eggs being generated by an individual woman and the same number of sperm being generated by an individual man, the likelihood of anyone else with your genotype in the past or in the future becomes infinitesimal (Bodmer and Cavalli-Sforza, 1976).

It might be said of the behavioral sciences that they have finally become aware of Darwin, but have not yet taken note of Mendel. This exaggerated statement means that behavioral scientists have begun to recognize the evolutionary origins of differences between species, but have not yet understood the variability inherent in the genetic reshuffling of sexual reproduction. It creates a different picture of variability. Our biological system is not merely tolerant of differences. The system generates differences, and depends on them. They are the sine qua non of evolution, the quintessence of life. This subject is the topic of the next chapter.

SUMMARY

Genetic varibility is the key to understanding evolution. We can look at it dynamically as it varies over time in the evolutionary record, or we can take a contemporary cross section of the evolutionary process by considering genetic variability *between* species and genetic variability *within* species. Life began about 3.5 billion years ago with "the replicators." Green plants and invertebrates led the invasion of the land, and reptiles and mammals followed in a relatively short time. Primates and the hominid lines are very recent. Although behavior does not fossilize, there are methods to narrow down the possibilities concerning behavioral evolution. One method is to study current species and attempt to piece together the puzzle of the evolution of behavior. Sociobiology makes such interspecies predictions based on kinship selection theory. Although sociobiologists tend to focus on interspecies genetic variability, behavioral geneticists concentrate on genetic variability within a species.

4

Single Genes

Although genetics teachers can dream up very complicated problems, the basic principles of Mendelian genetics are elegantly simple. This is not to say, of course, that the field of genetics is without its complexities. The search for detailed understanding of the transmission of hereditary factors and their mode of action has led investigators to look into cytology, embryology, physiology, biochemistry, biophysics, and mathematics. However, the fundamentals of Mendelian genetics were established without knowledge of the physical or chemical nature of the hereditary material. It is still convenient to introduce the principles of genetics by treating the hereditary determinants as hypothetical factors.

MENDEL'S EXPERIMENTS AND LAWS

To understand the logic of Mendel's experiments, it is helpful to remember that in the 1800s no one knew about genes, and that the prevailing theory of inheritance was pangenesis (as discussed in Chapter 2). Much of the research on heredity involved crossing plants of different species. A critical drawback to this approach is that the offspring are usually sterile, meaning that succeeding generations cannot be studied. Also, the features of the plants that were investigated were too complex for clear analysis. Mendel's success can be attributed in large part to circumventing these problems. He crossed different varieties of pea plants of the same species, and thus obtained fertile offspring that could be crossed to study subsequent generations. In addition, he picked simple qualitative ("either-or") traits that happened to have a number of fortunate characteristics that we shall consider in a moment. Mendel also counted all the progeny rather than being content, as his predecessors had, with a verbal summary of the typical result.

Mendel's research involved two kinds of experiments. The first, which used *monohybrid* crosses, followed the inheritance of one character at a time. These experiments led to Mendel's first law, the *law of segregation.* This law states that there are two "elements" of heredity for each character, and that these two elements separate, or segregate, during inheritance. Offspring receive one of the two elements from each parent. These elements do not blend in inheritance, as the theory of pangenesis suggested. The second type of experiment utilized *dihybrid* and *trihybrid* crosses, and traced the inheritance of characters considered two or three at a time. The dihybrid and trihybrid experiments led Mendel to conclude that the hereditary elements for one character assort independently of the elements for other characters. In other words, the inheritance of one trait in no way influences the inheritance of the other. This conclusion is now known as Mendel's second law, the *law of independent assortment.*

Monohybrid Experiments

Mendel looked at seven qualitative traits of the pea plant. Three of these included: whether the seed was green or yellow inside (the cotyledon), whether the seed was smooth or wrinkled, and whether the stem was long or short. He obtained 22 varieties of the pea plant that differed in some of these characteristics. For each of the seven traits, he crossed two *truebreeding* varieties. Truebreeding plants are those that always yield the same result when self-pollinated or crossed with the same kind of plant.

In one experiment, Mendel crossed truebreeding plants with round seeds to truebreeding plants with wrinkled seeds. Later in the summer, when he opened the pods containing their offspring (called the F_1, or first filial generation), he found that all of them had round seeds. This result indicated that the traditional view of blending inheritance was not correct. The F_1 did not have seeds that were moderately wrinkled. Because these F_1 offspring were fertile, Mendel was able to take the next step of self-fertilizing plants from the F_1 generation to study their offspring, known as F_2. The results were striking: ¾ of the offspring had round seeds, and ¼ had wrinkled seeds. Of the 7,324 seeds from the F_2, 5,474 were round and 1,850 were wrinkled. This result suggests that the factor responsible for wrinkled seeds had not been lost in the F_1 generation, but had been dominated by the factor causing round ones. Excerpts from Mendel's paper in which he described these results are presented in Box 4.1.

Box 4.1
Mendel's Classic Paper: Monohybrid Results

Mendel presented the results of eight years of research on the pea plant at a meeting of naturalists in Brunn, Moravia, in 1865, and his paper was published in 1866. Because this paper is the cornerstone of genetics, we have excerpted the following sections, which focus on his monohybrid results.* Box 4.2 describes his dihybrid results.

EXPERIMENTS IN PLANT-HYBRIDIZATION

Experience of artificial fertilisation, such as is effected with ornamental plants in order to obtain new variations in colour, has led to the experiments which will here be discussed. The striking regularity with which the same hybrid forms always reappeared whenever fertilisation took place between the same species induced further experiments to be undertaken, the object of which was to follow up the developments of the hybrids in their progeny....

The paper now presented records the results of such a detailed experiment. This experiment was practically confined to a small plant group, and is now, after eight years' pursuit, concluded in all essentials. Whether the plan upon which the separate experiments were conducted and carried out was the best suited to attain the desired end is left to the friendly decision of the reader....

The Forms of the Hybrids [F_1]

. . . in this paper those characters which are transmitted entire, or almost unchanged in the hybridisation, and therefore in themselves constitute the characters of the hybrid, are termed the *dominant,* and those which become latent in the process *recessive*. The expression "recessive" has been chosen because the characters thereby designated withdraw or entirely disappear in the hybrids, but nevertheless reappear unchanged in their progeny, as will be demonstrated later on.... Of the differentiating characters which were used in the experiments the following are dominant:

1. The round or roundish form of the seed with or without shallow depressions.
2. The yellow colouring of the seed albumen [cotyledons].

* Translated by William Bateson and the Royal Horticultural Society of London. The original paper was published in *Verh. naturf. Ver. in Brunn, Abhandlungen,* 1866.

Given these facts, Mendel deduced a simple explanation involving two hypotheses, summarized in Figure 4.1. First, each individual has two hereditary factors, now called alleles (alternate forms of a gene), that determine whether the seed is wrinkled or smooth. Thus, each parent has two alleles, but passes only one of those to its offspring. The second hypothesis was that

3. The grey, grey-brown, or leather-brown colour of the seed-coat, in association with violet-red blossoms and reddish spots in the leaf axils.
4. The simply inflated form of the pod.
5. The green colouring of the unripe pod in association with the same colour in the stems, the leaf-veins and the calyx.
6. The distribution of the flowers along the stem.
7. The greater length of stem....

The Generation From the Hybrids [F_2]

The relative numbers which were obtained for each pair of differentiating characters are as follows:

> Expt. 1. Form of seed.—From 253 hybrids 7,324 seeds were obtained in the second trial year. Among them were 5,474 round or roundish ones and 1,850 angular wrinkled ones. Therefrom the ratio 2.96 to 1 is deduced.
>
> Expt. 2. Colour of albumen.—258 plants yielded 8,023 seeds, 6,022 yellow, and 2,001 green; their ratio, therefore is as 3.01 to 1....
>
> Expt. 3. Colour of the seed-coats.—Among 929 plants 705 bore violet-red flowers and grey-brown seed-coats; 224 had white flowers and white seed-coats, giving the proportion 3.15 to 1.
>
> Expt. 4. Form of pods.—Of 1,181 plants 882 had them simply inflated, and in 299 they were constricted. Resulting ratio, 2.95 to 1.
>
> Expt. 5. Colour of the unripe pods.—The number of trial plants was 580, of which 428 had green pods and 152 yellow ones. Consequently these stand in the ratio 2.82 to 1.
>
> Expt. 6. Position of flowers.—Among 858 cases 651 had inflorescences axial and 207 terminal. Ratio, 3.14 to 1.
>
> Expt. 7. Length of stem.—Out of 1,064 plants, in 787 cases the stem was long, and 277 short. Hence a mutual ratio of 2.84 to 1....

If now the results of the whole of the experiments be brought together, there is found, as between the number of forms with the dominant and recessive characters, an average ratio of 2.98 to 1, or 3 to 1.

one allele could dominate the other. These two hypotheses neatly explain the data. The truebreeding parent plant with round seeds has two alleles for round seeds (A_1A_1). The truebreeding parent plant with wrinkled seeds is A_2A_2. Thus, the F_1 offspring will have one allele from each parent (A_1A_2). If A_1 dominates A_2, the F_1 will have round seeds. The real test is the F_2 population. Mendel's theory would predict that when F_1's are self-fertilized or crossed with other F_1 individuals (A_1A_2 with A_1A_2), ¼ of the F_2's should be A_1A_1, ½

	Observed	Hypothesized
Truebreeding Parents	Truebreeding round × Truebreeding wrinkled	$A_1A_1 \times A_2A_2$
F_1	All round	All A_1A_2 (A_1 dominant)
F_2	¾ round, ¼ wrinkled	¼ A_1A_1, ½ A_1A_2 (round); ¼ A_2A_2 (wrinkled)

FIGURE 4.1

Summary of Mendel's monohybrid experiments.

A_1A_2, and ¼ A_2A_2. However, assuming A_1 dominates A_2, then A_1A_2 should have round seeds like the A_1A_1. Thus, ¾ of the F_2 should have round seeds and ¼ wrinkled, which is exactly what Mendel's data indicated. Mendel completed many experiments with other characteristics and other varieties of pea plants, and they all confirmed this theory.

At this point, we need to introduce several terms in addition to genes and alleles. Individuals with the same alleles for a particular gene are called *homozygotes;* those with different alleles are *heterozygotes.* The truebreeding parental varieties were all homozygotes (A_1A_1 or A_2A_2) and the F_1 were all heterozygotes (A_1A_2). *Genotype* refers to the genetic constitution of an individual. With two alleles for a gene, three possible genotypes exist for that particular gene: A_1A_1, A_1A_2, and A_2A_2. We observe the *phenotype,* not the genotype. For example, in the F_2 population, the genotypes were distributed in a 1 : 2 : 1 ratio (¼ A_1A_1, ½ A_1A_2, and ¼ A_2A_2), while the phenotypes were in a 3 : 1 ratio (¾ round, ¼ wrinkled). In other words, the phenotype does not necessarily reflect the genotype, since the A_1A_2 genotype has the same appearance as the A_1A_1 genotype. It was fortunate for Mendel's theory that dominance of the A_1 allele over the A_2 allele was complete. If there had been no dominance, the A_1A_2 phenotype would have been in between the A_1A_1 and the A_2A_2 phenotypes, and it would have appeared as if blending had occurred in the F_1. Different types of gene expression are represented in Figure 4.2. For Mendel's seven pea plant characteristics, there was complete dominance in the sense that the phenotypic value (appearance) of the heterozygote genotype was the same as that of one of the homozygote genotypes. The other side of the coin is *recessiveness*. Rather than saying that the A_1 allele dominates the A_2 allele, we could say that the A_2 allele is recessive to the A_1 allele. If, on the other hand, A_1 were completely recessive to A_2, then the A_1A_2 genotype would look the same as the A_2A_2 genotype. These examples consider only complete dominance and recessiveness, but both could be partial. That is, the A_1A_2 heterozygote can lie anywhere in between the observed value for one homozygote and that of the other.

Mendel summarized his theory with the *law of segregation:* Inheritance is particulate; that is, there are discrete and inviolable units of inheritance

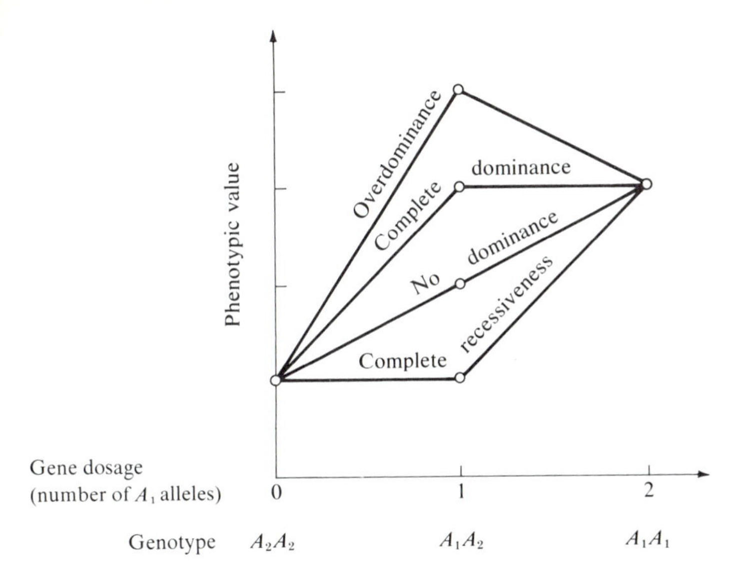

FIGURE 4.2
Graphical representation of four different types of gene expression.

that are transmitted so that offspring have two units, one from each parent. In other words, alleles of a gene pair separate cleanly, with no residual effects on each other. An A_1 allele transmitted from an A_1A_1 parent is no different from one transmitted by an A_1A_2 parent. The ratios in the F_2 population are called *monohybrid segregation ratios:* "mono" because they consider only one trait, such as the roundness of the seed; "hybrid" because they are the result of a cross between two truebreeding populations; and "segregation" because they show that the alleles separate out, or segregate, in the F_2 generation. The 1 : 2 : 1 ratio is the monohybrid *genotypic* segregation ratio, and the 3 : 1 ratio is the monohybrid *phenotypic* segregation ratio.

Another way of looking at the law of segregation is described in Table 4.1. Each gamete (sperm or egg) contains one allele from each gene pair carried by the parent who produced the gamete. F_1 individuals are all heterozygous, A_1A_2. When they are crossed to breed the F_2 generation, half of the sperm will carry the A_1 allele and the other half will carry the A_2 allele.

TABLE 4.1
F_2 offspring from a monohybrid F_1 cross

		Sperm	
		½ A_1	½ A_2
Eggs	½ A_1	¼ A_1A_1	¼ A_1A_2
	½ A_2	¼ A_1A_2	¼ A_2A_2

This is also true for the female gametes. If sperm and eggs unite at random, the offspring produced are those described in Table 4.1. There are three different genotypes in the F_2. Their relative frequencies are: ¼A_1A_1, ½A_1A_2, and ¼ A_2A_2. If dominance is complete, no difference will be observed between the A_1A_1 and A_1A_2 genotypes, thus yielding a phenotypic segregation ratio of 3 : 1.

Dihybrid Experiments

Mendel also experimented with crosses between varieties of plants differing with respect to two traits, resulting in *dihybrid segregation ratios* in the F_2 population. Box 4.2 contains a short excerpt from Mendel's paper detailing some of these results. As described in Box 4.2, Mendel crossed truebreeding plants that had both yellow cotyledons (the inside of the seed, which Mendel calls "albumen") and round seeds with truebreeding plants that had green cotyledons and wrinkled seeds. The F_1 seeds were all round with yellow cotyledons, because roundness and yellowness dominate. The exciting question concerned the F_2. Would yellowness and roundness be transmitted as a package, or would they be inherited independently? When he opened the pods of the F_2, he found that about 3/16 of the seeds were yellow and wrinkled, and a similar number were green and round. Thus, Mendel concluded that the two traits were inherited independently. The observed ratio was 9 : 3 : 3 : 1. (See Box 4.2.) That is, 9/16 were dominant for both traits (yellow and round), 3/16 were dominant for color and recessive for texture, 3/16 were recessive for color and dominant for texture, and 1/16 were recessive for both traits. This ratio is now known as the *dihybrid phenotypic segregation ratio.*

These findings are summarized by Mendel's second law, the *law of independent assortment:* When two traits are inherited, the alleles for each gene assort independently of the other gene. In other words, the alleles of each of the genes segregate as they would have in a monohybrid cross. For example, 12/16 of the seeds were yellow and 4/16 were green, which is the 3 : 1 monohybrid segregation ratio for color of the cotyledon.

If we generalize the dihybrid cross to eggs and sperm, we see in Table 4.2 that the dihybrid individuals (for traits *A* and *B*) of the F_1 generation produce four different kinds of gametes in equal frequencies: ¼A_1B_1, ¼A_1B_2, ¼ A_2B_1, and ¼ A_2B_2. Because the two genes, *A* and *B*, are transmitted independently, we can determine their joint frequency by multiplying their separate probabilities. For example, the probability of having an A_1 allele in the sperm of an A_1A_2 individual of the F_1 generation is ½. (See Table 4.1.) For the *B* gene, the probability of having a B_1 allele is also ½. Thus, the probability that F_1 sex cells contain both A_1 and B_1 is the product of their respective probabilities: $\frac{1}{2} \times \frac{1}{2} = \frac{1}{4}$, as indicated in Table 4.2.

Because the F_2 generation results from crosses or self-pollination of F_1

Box 4.2
Mendel's Classic Paper: Dihybrid Results

Box 4.1 contains quotations from Mendel's presentation of his monohybrid results. The following is an excerpt from his discussion of the results when two traits were investigated simultaneously.* The concluding sentence is his statement of the law of independent assortment:

THE OFFSPRING OF HYBRIDS IN WHICH SEVERAL DIFFERENTIATING CHARACTERS ARE ASSOCIATED

In the experiments above described plants were used which differed only in one essential character. The next task consisted in ascertaining whether the law of development discovered in these applied to each pair of differentiating characters when several diverse characters are united in the hybrid by crossing....

In order to facilitate the study of the data in these experiments, the different characters of the seed plant will be indicated by A, B, C, those of the pollen plant by a, b, c, and the hybrid forms of the characters by Aa, Bb, and Cc.

Expt. 1.—AB, seed parents;
A, form round;
B, albumen yellow.
ab, pollen parents;
a, form wrinkled;
b, albumen green.

The fertilised seeds appeared round and yellow like those of the seed parents. The plants raised therefrom yielded seeds of four sorts, which frequently presented themselves in one pod. In all, 556 seeds were yielded by 15 plants, and of these there were:

315 round and yellow,
101 wrinkled and yellow,
108 round and green,
32 wrinkled and green....

. . . the relation of each pair of different characters in hybrid union is independent of the other differences in the two original parental stocks.

* Translated by William Bateson and the Royal Horticultural Society of London. The original paper was published in *Verh. naturf. Ver. in Brunn, Abhandlungen,* 1866.

TABLE 4.2
F_2 offspring from a dihybrid F_1 cross

		Sperm			
		¼ A_1B_1	¼ A_1B_2	¼ A_2B_1	¼ A_2B_2
Eggs	¼ A_1B_1	$^1/_{16}$ $A_1A_1B_1B_1$	$^1/_{16}$ $A_1A_1B_1B_2$	$^1/_{16}$ $A_1A_2B_1B_1$	$^1/_{16}$ $A_1A_2B_1B_2$
	¼ A_1B_2	$^1/_{16}$ $A_1A_1B_1B_2$	$^1/_{16}$ $A_1A_1B_2B_2$	$^1/_{16}$ $A_1A_2B_1B_2$	$^1/_{16}$ $A_1A_2B_2B_2$
	¼ A_2B_1	$^1/_{16}$ $A_1A_2B_1B_1$	$^1/_{16}$ $A_1A_2B_1B_2$	$^1/_{16}$ $A_2A_2B_1B_1$	$^1/_{16}$ $A_2A_2B_1B_2$
	¼ A_2B_2	$^1/_{16}$ $A_1A_2B_1B_2$	$^1/_{16}$ $A_1A_2B_2B_2$	$^1/_{16}$ $A_2A_2B_1B_2$	$^1/_{16}$ $A_2A_2B_2B_2$

individuals, we can determine the kinds and expected frequencies of F_2 offspring from a consideration of the sex cell combinations illustrated in Table 4.2. This results in a table with 16 cells, with some of the same genotypes appearing in more than one cell. For example, there are two cells that each contain $^1/_{16}$ $A_1A_1B_1B_2$. Notice that the genotypes add up to the 9:3:3:1 dihybrid phenotypic segregation ratio.

Mendel was lucky in his studies of monohybrid ratios to find single-gene, two-allele traits that operate with complete dominance. There are many other traits, such as the size of the seed, that are influenced by more than one gene. And there are many genes with more than two possible alleles. These complications would have made Mendel's search for laws of inheritance much more difficult. His luck held when he considered dihybrid segregation ratios. We know that genes are not just floating around in eggs and sperm. They are carried on *chromosomes,* which literally means "colored bodies" because they stain differently from the rest of the nucleus of the cell. Genes are located at places called *loci* (singular: *locus*) on chromosomes. Eggs contain one chromosome from each pair of the mother's chromosomes and sperm contain one from each pair of the father's. An egg fertilized by a sperm thus has the full chromosome complement—in humans, 23 pairs of chromosomes. If Mendel had studied two genes that were close together on the same chromosome, the results would have surprised him. The two traits would not have been inherited independently. In fact, if they had been very close together on the same chromosome, Mendel would not have found yellow, wrinkled seeds nor round, green seeds. We will discuss this issue in more detail in Chapter 6.

When Mendel read the paper about his theory of inheritance in 1865, reprints were sent to scientists and libraries in Europe and the United States, and one even landed in Darwin's office. However, Mendel's findings on the pea plant were ignored by most biologists, who were more interested in the evolution of higher animals. Mendel died in 1884, without knowing the profound impact that his experiments would have during the twentieth century.

Mendelian crosses and expected segregation ratios are widely used today, primarily to determine whether a particular phenotype is influenced by a single gene. Although complex behaviors such as twirling, squeaking, and audiogenic seizures in mice, or mental retardation in humans, seem far removed from pea seeds, the laws of heredity discovered by Mendel apply to behavior as well as to peas. We shall now consider single-gene influences for a few behaviors in human and nonhuman animals.

DISTRIBUTIONS

If a particular behavior is primarily influenced by a single gene, it can be expected to display an "either-or" expression. For example, Mendel considered his peas to be either round or wrinkled. This sort of distribution is called qualitative, or discontinuous. However, most behaviors are not distributed in this either-or fashion. Instead, they show smooth, continuous distributions. This may be because these behaviors are influenced by many genes, each having small effects, as well as by many environmental factors. For such behaviors, simple Mendelian crosses that assume a single gene are not appropriate. Instead, we need to use the methods of quantitative genetics, as discussed in Chapters 9 to 13. However, some behaviors show an either-or distribution, thus suggesting the influence of a single major gene.

One example of an approximate either-or distribution in humans is the ability to taste a bitter substance called phenylthiocarbamide (PTC). It can be tasted by 70 percent of the Caucasians in the United States. However, it is not quite as simple as saying that someone can either taste PTC or not; it depends on the concentration. Taste sensitivity to PTC and related compounds can be assessed by placing a small drop of a test solution on subjects' tongues and asking if they can taste it. A variety of concentrations is usually administered so that a taste threshold can be established for each subject. While some people can taste highly dilute solutions of PTC, others can distinguish only very strong solutions from plain water. Distributions of taste thresholds for English, Africans, and Chinese individuals are presented in Figure 4.3. Individuals who could taste solution number 13, the most dilute solution, and all stronger solutions were most sensitive, and thus had the lowest taste threshold. Individuals who could not even taste solution number 1, the strongest solution, were least sensitive.

Unlike most behavioral characters, PTC tasting is not normally distributed with most people scoring at the average. Instead, the distributions, particularly for the English, have two humps (called a bimodal distribution)—one for high thresholds, and one for low thresholds. Variation within the taster and nontaster categories could be due to segregation at other loci that have relatively small effects on taste sensitivity to PTC, to environmental variation, or to error of measurement. This bimodality suggests that a single major gene may underlie PTC tasting. Simple Mende-

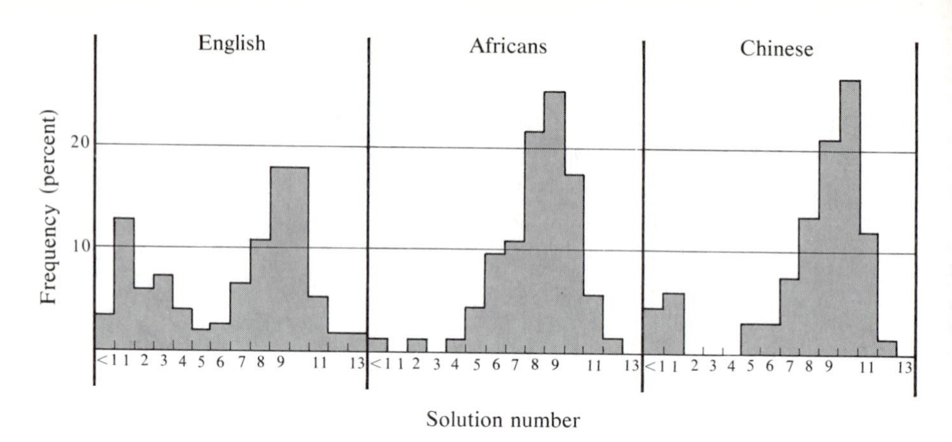

FIGURE 4.3

Distributions of taste thresholds for PTC in 155 English males, 74 Africans, and 66 Chinese. The strongest solution (1) had a concentration of 0.13 percent in water, the next (2) half this strength, and so on through 13. (From "Taste deficiency for phenylthiourea in African Negroes and Chinese" by N. A. Barnicot. *Annals of Eugenics*, 1950, 15, 248–254. Reprinted by permission of Cambridge University Press.)

lian crosses, as described in the next section, support the hypothesis that it is a single-locus, two-allele character, with the allele for tasting PTC dominant. Other examples of discontinuous distributions include genetic diseases with behavioral effects. For example, PKU is a single-gene, recessive condition that causes severe retardation if untreated, as discussed later in this chapter.

Although most behavioral characters show smooth, quantitative variation, rather than qualitative variation, this does not rule out the possibility of single major genes. A major gene may be at work, although its role may be diminished by genetic and environmental effects that cause quantitative variation. A related factor is that genotypes do not always express themselves in the same way in all individuals, due to the complexity of the developmental pathways by which genes are expressed in the phenotypes. Characters that are not expressed in all individuals with the appropriate genotype are said to be *incompletely penetrant*. Those genotypes that are expressed to varying degrees are said to display *variable expressivity*. Although not fully understood, incomplete penetrance and variable expressivity are presumably due to the effects of either environment or other genes. In any case, these phenomena can cloud the effects of single genes.

However, the major difficulty in ascertaining single-gene influences in behavior may be a measurement problem. The phenotypes that behavioral scientists study are often heterogeneous collections of diverse and complicated behaviors. For example, it is unlikely that a single gene can be found for complex behavioral problems such as reading disability or hyperactivity. With cancer research, we no longer talk about the cause or the cure of cancer, but rather the causes and cures of cancers, because we know that there are many different kinds of cancer. It is safe to assume that many of our behavioral phenotypes are similarly heterogeneous.

In summary, bimodal distributions are not necessary to demonstrate a single-gene effect, although it is exciting to find such distributions. A better test of single-gene influence is *segregation analysis,* which essentially looks for Mendelian segregation expected from family data. Earlier segregation analyses assumed qualitative distributions (Morton, 1958). Current sophisticated models can detect major genes for quantitative variables (Elston and Stewart, 1971). If segregation analysis suggests single-gene influences, linkage analyses (see Chapter 6) are appropriate to pinpoint the gene on a specific chromosome.

MENDELIAN CROSSES IN MICE

With experimental animals, we can conduct behavioral studies parallel to Mendel's studies of pea plant characteristics. Crosses between truebreeding parental populations yield an F_1 generation, and the members of the F_1 generation can be crossed to produce an F_2. One of the favorite subjects of behavioral genetics research is the mouse.

The Genetics of Waltzer Mice

Many studies have been made concerning the effects of a single-gene mutation on mouse behavior. For example, one of the earliest behavioral conditions studied in mammals was that of "waltzing" in mice. In spite of the name, animals exhibiting this behavior are quite ungraceful. They shake their heads, circle rapidly, and are very irritable. Waltzer mice were prized by mouse fanciers and imported to Europe and North America from Asia around 1890 (Gruneberg, 1952). Several different waltzer conditions are known. For example, the Nijmegen waltzer (Van Abeelen and van der Kroon, 1967) runs in tight circles in both directions with both horizontal and vertical head shaking. Researchers wondered whether this was the result of a single gene. When waltzer males were crossed with waltzer females, all offspring were waltzers, suggesting that the waltzers are a truebreeding population. Waltzers were then crossed with a nonwaltzing population to produce an F_1 generation. Of the 254 offspring, all were nonwaltzing. This suggests that, if a single gene is operating, the allele for waltzing is recessive to the normal allele.

If a single recessive allele causes waltzing, the F_2 generation should yield the typical Mendelian segregation ratio of ¾ nonwaltzer and ¼ waltzer. As shown in Table 4.3, the data from the F_2 are very close to the expected 3 : 1 ratio.

TABLE 4.3

Data observed and expected on the basis of a single-locus autosomal-recessive model for F_2 Nijmegen waltzer mice

	Normal	Mutant
Observed	124	47
Expected	128.25	42.75

SOURCE: After van Abeelen and von der Kroon, 1967.

There are many other examples of inherited neurological defects in mice. Most appear to be due to single-locus, recessive genes. However, some are caused by dominant genes, and a few appear to be due to the combined effects of genes at several loci. A partial list of some of these conditions is presented in Table 4.4. The highly descriptive names convey some of the diversity of behavioral anomalies that have been described. However, the study of behavioral mutants in mice has been eclipsed in recent years by induced mutation and mass screening of behavioral mutations in single-celled organisms, such as bacteria and paramecia, and in invertebrate organisms, such as round worms and fruit flies. These will be discussed in the next chapter because they illustrate the chemical nature of genes as it affects behavior. We shall now consider a couple of additional examples of effects of single-gene mutations on mouse behavior in order to introduce other concepts of Mendelian analysis.

The Genetics of Twirler Mice

Exact Mendelian ratios are not always observed, even for single-locus characters. For example, the "twirler" mouse was first described by Mary Lyon (1958) as shaking its head frequently in a horizontal plane and circling. Evidence indicated that this condition was probably due to a single dominant allele. One test of this hypothesis is to cross twirler males and twirler females. If a single dominant allele is at work, twirlers (born to twirler, crossed with nontwirler, parents) are heterozygotes, and this cross should be like an

TABLE 4.4

Partial list of inherited neurological defects in mice

Class of Syndrome	Name
Waltzer-shaker	Shaker (1 and 2), pirouette, jerker, waltzer, varitint-waddler, fidget, twirler, zigzag
Convulsive	Trembler, spastic, tottering
Incoordination	Quaking, jimpy, reeler, agitans, staggerer

F_1 cross. Their offspring should show the typical 3 : 1 F_2 segregation ratio. However, the results of the cross yielded 84 twirler offspring and 58 normal offspring. Since a 3 : 1 segregation ratio would produce 106.5 twirlers and 35.5 normal offspring, the observed ratio departs significantly from these expectations.

However, dominant mutant genes are often lethal in the homozygous condition. If individuals having the homozygous genotype die before investigators observe their behavior, the expected ratio among offspring of a monohybrid cross would be 2 twirlers to 1 normal instead of 3 : 1. When this hypothesis is tested, a slight deficiency of twirler offspring is still noted, but the departure from expectation is not sufficient to be statistically significant. Subsequent research supports this hypothesis. About 25 percent of newborn offspring resulting from twirler crosses have a cleft palate or cleft lip and palate. These pups die within 24 hours of birth and consequently would not be observed for behavioral abnormalities. These mice are believed to be the missing homozygotes, accounting for the 2 : 1, rather than the 3 : 1, ratio.

The Genetics of Audiogenic Seizures in Mice

Two other points about single-gene analyses can be made based on the example of sound-induced seizures in mice. Some mice respond to high-frequency sound with wild running, convulsions, and even death; others are apparently unaffected by it. The first point is that behavioral genetics analysis is only as reliable as the behavioral measurements on which it is based. The second point is that Mendelian crosses need not be limited to an analysis of the F_1 and F_2 generations. Predictions can be made to test the hypothesis of a single gene using other crosses, such as a cross between the F_1 generation and the parental population.

Earlier work suggested various modes of inheritance for seizure susceptibility. Researchers suggested dominance (Witt and Hall, 1949), a two-gene model (Ginsburg and Miller, 1963), and a polygenic model (Fuller, Easler, and Smith, 1950). However, a measurement problem, due to the fact that subjects were tested repeatedly, may have obscured some of the earlier results. Animals from a seizure-resistant mouse strain become susceptible to seizures if exposed to a loud sound at an early age (Henry, 1967). Thus, repeated testing of animals may have confounded the response to the first noise and the response to later noises.

When the effects of initial exposure to a loud noise are measured, the situation is greatly clarified (Collins, 1970; Collins and Fuller, 1968). Mice of the C57BL/6J strain, a strain whose members rarely convulse on initial exposure to a loud noise, were crossed with mice of the DBA/2J strain, whose members almost always convulse on initial exposure. As illustrated in Figure 4.4, several additional crosses were conducted in order to provide more extensive analyses. The F_1 animals were crossed back to the parental

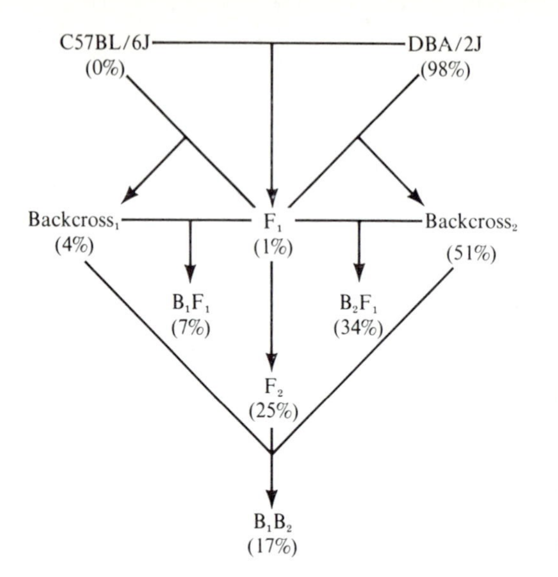

FIGURE 4.4

Various crosses used to test a single-gene model for audiogenic seizures in mice. Percentages in parentheses refer to the data in Table 4.5.

lines. These crosses are appropriately called backcrosses. B_1 is the backcross to one parental line (P_1, for example, the C57BL strain) and B_2 is the backcross to P_2 (DBA). An F_2 generation was also obtained, as well as crosses between the two backcross generations and the F_1 generation, yielding generations symbolized as B_1F_1 and B_2F_1. The two backcross generations were also crossed (B_1B_2).

Mice from these 9 groups were individually tested for initial seizure susceptibility at 3 weeks of age. Each subject was placed in a box and exposed to an electric bell that was rung until the onset of a convulsion or for a maximum of 1 minute. The proportions of seizures observed and expected, assuming a single-locus model, are presented in Table 4.5. As indicated in this table, C57BL mice had no seizures, almost no F_1 mice had seizures, but almost all DBA mice had seizures. These results suggest that susceptibility to audiogenic seizure on initial exposure to the loud noise may be determined at a single locus by a recessive allele.

In order to test the adequacy of this single-locus, recessive model, it may be hypothesized that: (1) DBA mice are homozygous recessive, and thus prone to audiogenic seizures; (2) C57BL mice are homozygous dominant; and (3) F_1 mice are heterozygous. The observed proportion of seizures by these genotypes can be used to predict the proportion having seizures in each of the 6 segregating generations. For example, in the F_2 generation, the genotypic segregation ratio should be ¼ homozygous dominant (with a seizure frequency of 0.000) to ½ heterozygous (with a seizure frequency of 0.011) to ¼ homozygous recessive (with a seizure frequency of 0.983). Thus, the expected proportion having seizures in the F_2 generation can be calculated as follows: $\frac{1}{4}(0.000) + \frac{1}{2}(0.011) + \frac{1}{4}(0.983) = 0.251$. It may be seen in Table 4.5 that the results conform closely to those expected on the basis of the model. These results are consistent with the hypothesis that the difference between these two strains in their susceptibility to audiogenic seizures

TABLE 4.5

Summary of data and genetic analysis for the incidence of initial audiogenic seizures in mice

Generation	Number of Subjects	Proportion Observed to Have Seizures	Proportion Expected to Have Seizures (Single-Locus Autosomal Model)
P_1 (C57BL/6J)	45	0.000	
P_2 (DBA/2J)	58	0.983	
F_1	89	0.011	
B_1	115	0.035	0.006
B_2	119	0.513	0.497
F_2	105	0.247	0.251
B_1F_1	128	0.070	0.128
B_2F_1	96	0.344	0.374
B_1B_2	185	0.168	0.191

SOURCE: From Collins and Fuller, "Audiogenic seizure prone (asp): a gene affecting behavior in linkage group VIII of the mouse," *Science*, 162, 1137–1139, 1968. Copyright © 1968 by the American Association for the Advancement of Science.

is due to a single gene, and that seizure susceptibility is a recessive trait.

Although these results are consistent with a single-gene hypothesis, they are also consistent with other models such as a two-gene model. A two-gene model would also predict these results if the effects of the alleles at the two loci simply added to produce the effect. A two-gene model would hypothesize that the C57BL mice are of genotype $A_1A_1B_1B_1$ and the DBA are $A_2A_2B_2B_2$. The predictions for this two-gene model would be identical to those for the single-gene model. Thus, although the data are compatible with a single-gene model, they also fit other models. One way to provide more definitive support for the single-gene model is to study the prevalence of the behavior in various inbred strains or in recombinant inbred strains. These methods are discussed in the following sections. As discussed in the section on recombinant inbred strains, recent evidence from a study using this technique suggests that the difference between the C57BL and DBA strains in susceptibility to audiogenic seizures is due to more than one gene.

The Genetics of Squeaking in Mice

Inheritance of the tendency of some mice to squeak when lifted by their tails was investigated in another test of a single-gene hypothesis. Glayde Whitney (1969) found that C57BL mice rarely vocalize when lifted by their tails, but about two-thirds of the JK strain vocalize. Table 4.6 summarizes

TABLE 4.6

Genetic analysis of handler-induced vocalization in mice

Generation	Number of Subjects	Proportion Observed to Vocalize	Proportion Expected to Vocalize (Single-Locus Model)
P_1 (C57BL)	70	0.03	
P_2 (JK)	71	0.68	
F_1	99	0.56	
B_1	47	0.26	0.29
B_2	45	0.62	0.62
F_2	80	0.46	0.46

SOURCE: After Whitney, 1969.

the results of the various crosses. F_1 mice from the cross between C57BL and JK strains vocalize somewhat less than JK mice, but considerably more than C57BL mice. Thus, if the single-gene hypothesis is correct, partial dominance is suggested. In fact, the observed proportions in Table 4.6 agree closely with the expected results.

In discussing susceptibility to audiogenic seizures, we mentioned that one way to provide more definitive support for the single-gene model is to study the prevalence of squeaking in various inbred strains. If two or more loci influence squeaking, some strains of mice should show intermediate frequencies of squeaking. When Whitney tested seven different inbred strains of mice, he obtained the results shown in Table 4.7. Inbred strains of mice are like truebreeding plants in that they have identical (homozygous) alleles at all loci. (See Chapter 8 for details.) Whitney's results support the single-gene hypoth-

TABLE 4.7

Observed vocalization on handling in seven inbred mouse strains

Strain	Number of Subjects	Proportion Vocalizing When Handled
A	134	0.00
BALB	179	0.02
DBA	109	0.04
C57BL	126	0.02
C3H	32	0.00
CBA	10	0.00
I^s/Bi	27	0.48

SOURCE: After Whitney, 1969.

esis, since the distribution of squeaking among the lines is bimodal. Six of the seven strains vocalized to about the same extent as the C57BL. One strain (I^s/Bi) vocalized almost as frequently as the JK mice in the previous experiment. Thus, there appears to be a qualitative, rather than quantitative, difference in the incidence of vocalization on handling. When the origin of these strains was considered, it was found that several of the low-vocalizing strains were related, and that the two high-vocalizing strains (I^s and JK) were related to each other, but not to the low-vocalizing strains. Together, these data support the hypothesis of a single dominant gene.

Behavioral Pleiotropism

A general rule is that a single gene affects more than just a single phenotype. *Pleiotropy* is the name given to the multiple effects of a gene. For example, PKU is attributed to a single gene most noted for causing mental retardation. However, PKU individuals also tend to have lighter hair and skin color. Genetic influences on behavior are also likely to be pleiotropic. In 1915, the first study of behavioral pleiotropism was provided by the geneticist A. H. Sturtevant, inventor of the chromosome map. He found that a single-gene mutation that alters eye color or body color in *Drosophila* also affects their mating behavior.

One of the most often studied examples of behavioral pleiotropism concerns the relationship between albinism in mice and their activity in an open field. The open-field test was first employed by Calvin Hall (1934) to provide an objective index of emotionality. The test involves placing an animal in a brightly lit enclosure and scoring its behavior. An "automated" open field is shown in Figure 4.5. Some animals freeze, defecate, and urinate when placed in this presumably stressful situation, whereas others actively explore it. Animals that have relatively low activity and high defecation scores are called "emotional" or "reactive," and those with relatively high activity and low defecation scores are considered "nonemotional" or "nonreactive." The evidence for the validity of this measure has been discussed in some detail (Broadhurst, 1960; Eysenck and Broadhurst, 1964).

It had been known for some time that several albino strains of mice obtain relatively lower activity and higher defecation scores than do pigmented mice. However, because these strains differ from one another at many loci and because some of the albino strains are closely related, it was generally assumed that the differences observed in open-field tests were due to gene differences other than those determining coat color. That is, it was assumed that the observed correlation between albinism and open-field behavior was not causal.

An extensive genetic analysis of open-field behavior initiated by DeFries determined that the greater emotionality of the albino mice was, in fact, due in part to the gene for albinism. In other words, the single gene that

FIGURE 4.5

Mouse in an open field employed by DeFries et al. (1966). The holes near the floor transmit light beams that electronically record an animal's activity. (Courtesy of E. A. Thomas.)

determines coat color (called the *c* locus) has a pleiotropic effect on open-field behavior. But how do we know that the *c* locus itself, rather than some other locus closely linked to the *c* locus, is responsible for greater emotionality? One answer is to break up any possible linkages. As we shall see in Chapter 6, nature does this for us through a process called recombination, in which chromosome pairs exchange parts. As will be explained in Chapter 10, an inbred strain is homozygous at all loci. That is, the two alleles at each locus on each pair of chromosomes are identical for all members of the strain. Because an inbred strain is homozygous at all loci, recombination of chromosomes will usually result in the same chromosomal arrangement, so that linked loci will not be altered. However, inbred strains differ from one another at many loci, and crosses between them will tend to break up linkage by means of recombination. In other words, when recombination occurs for crosses between inbred strains, their chromosomes exchange parts with different alleles, and linkages are severed. DeFries (1969) crossed a highly "emotional" albino inbred strain (BALB/cJ) and a "nonemotional" pigmented inbred strain (C57BL/6J) and bred them for many generations. He found that mice with the gene for albinism were just as "emotional" in later generations as albino animals in the F_2 generation, even though the loci linked to the *c* locus were likely to have been broken up by the process of recombination.

Rather than crossing only two inbred strains, a better method to ensure that linkages are broken up involves using heterogeneous animals (HS) derived from, for example, an eight-way cross of inbred strains (McClearn, Wilson, and Meredith, 1970). Even using HS animals, it is still possible that certain linkages might not be separated by recombination. The strongest evidence for behavioral pleiotropism is provided by comparisons of individuals with exactly the same genotype except for a newly arisen mutation. When a new mutation arises and is maintained within an inbred strain, mutant and nonmutant subjects within the strain are called *coisogenic* (Green, 1966). An albino mutant from the pigmented C57BL strain was bred to produce a coisogenic albino strain. The two strains have the same genotype except for the single locus determining coat color. If the greater "emotionality" of albino mice is a pleiotropic effect of the *c* locus itself, then these coisogenic albino mutants should differ from the pigmented C57BL mice. The answer is in favor of the pleiotropism hypothesis: The albino mutants were less active in the open field than the pigmented coisogenic C57BL mice (Henry and Schlesinger, 1967).

Additional analyses suggest that the pleiotropic effect of the *c* locus on open-field behavior is mediated by the visual system. McClearn (1960) observed that the difference in open-field behavior between an albino strain and a pigmented strain was less under a red light, which reduces visual stimulation. This led to the hypothesis that the albinos are actually afraid of the light in the open field; they may be photophobic. Data from albino and pigmented mice from the same litters support the hypothesis. (See Figure 4.6.) Under the red light there is no significant difference between albino and pigmented mice in open-field activity or defecation (DeFries et al., 1966).

The results of this study demonstrate that albino mice have lower activity and higher defecation scores than pigmented mice when tested under white light, and that this difference largely disappears when these subjects are tested under red illumination. It may be concluded that this single-gene pleiotropic effect is mediated through the visual system. If we accept the interpretation that a pattern of low activity and high defecation indicates heightened "emotionality," albinos may be regarded as being more photophobic than pigmented animals. Of course, many other genes influence a behavior as complex as open-field behavior, as discussed in Chapter 10.

RECOMBINANT INBRED STRAINS

A new strategy to uncover single-gene effects in behavior is called the recombinant inbred strain method (Bailey, 1971; Eleftheriou, 1975). As illustrated in Figure 4.7, it begins with a classical Mendelian cross between two inbred strains. An F_1 generation is produced, followed by F_2. In the F_2, genes

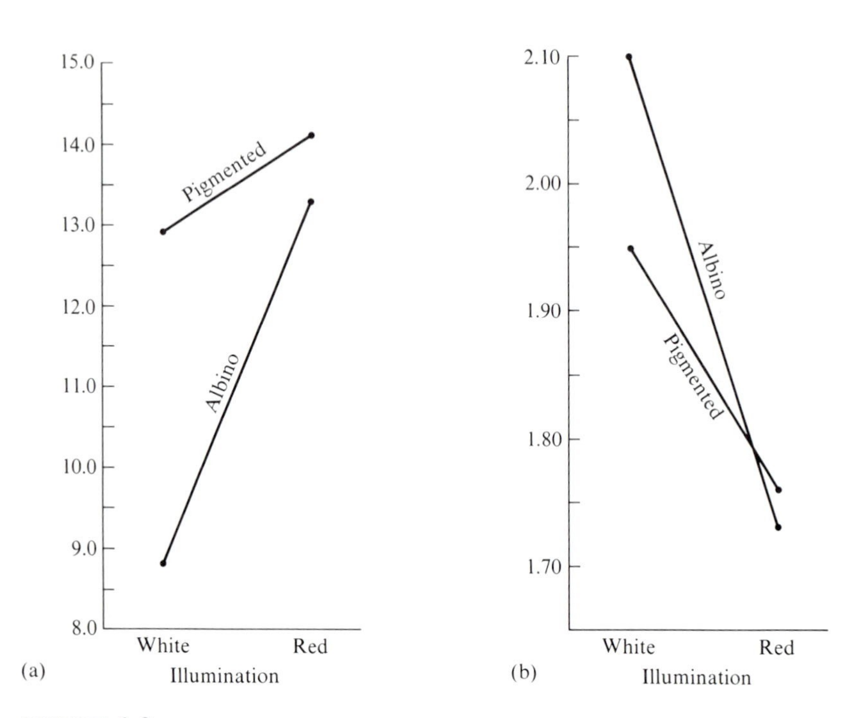

FIGURE 4.6
Open-field (a) activity and (b) defecation scores of mice tested under different illuminations. (From DeFries et al., 1966.)

assort according to Mendel's laws. Recombinant inbred (RI) strains are different inbred strains that were derived from the same F_2 generation. They are called recombinant inbred strains because they are derived from the same F_2 generation, after there has been recombination of parts of chromosomes from the parental strains. Like all inbred strains (as described in Chapter 10), RI strains are made homozygous at almost all loci by inbreeding using brother-sister matings over many generations.

How do RI strains derived from the same F_2 generation differ? For those loci with different alleles in the two parental strains, RI strains will show different combinations of alleles. For example, suppose the parental strains had different alleles at the A locus. (See Figure 4.7.) The F_1 will have one allele from each parent, but the F_2 will show segregation. The F_2 consists of homozygous and heterozygous individuals. However, the process of inbreeding increases homozygosity. Thus, each RI strain derived from the F_2 will be either A_1A_1 or A_2A_2. The same will be true for other loci. When we consider two loci at a time (the A and B loci), only ¼ of the RI strains will be $A_1A_1B_1B_1$, ¼ will be $A_1A_1B_2B_2$, ¼$A_2A_2B_1B_1$, and ¼$A_2A_2B_2B_2$. In other words, although the RI strains are homozygous at all loci, they represent new combinations of genes at different loci. While each RI strain carries half the

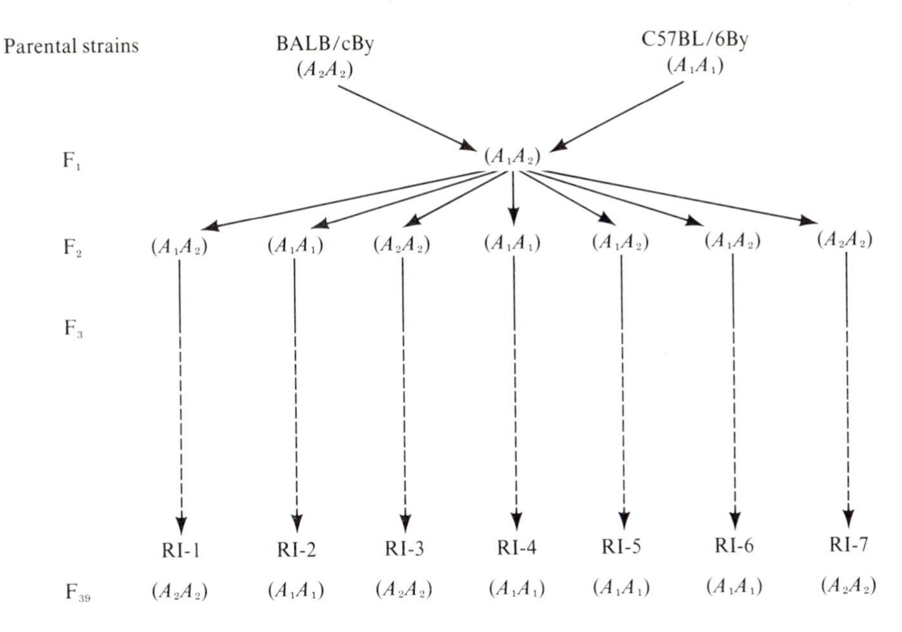

FIGURE 4.7
The derivation of recombinant inbred (RI) strains by Bailey (1971). In parentheses is an example of a locus, A, for which the parental strains differ. This example illustrates the results presented in Table 4.8, and is explained in the text.

genes of each parental strain, there is little chance that the genotype of an RI strain will be the same as that of either parental strain.

These RI strains can be used to determine whether any behavioral differences between the two parental strains are due to single major genes. If a single gene is responsible, each of the RI strains should be just like one parent or the other. In other words, there should be no intermediate phenotypes if just one locus is involved, because each RI strain will be homozygous for the allele of one of the two parental strains. However, if more than one locus is involved in a behavioral difference between two parental strains, more than two kinds of RI strains will be found. Furthermore, if there is just one gene responsible for the behavioral difference, then about half of the RI strains should be like one parent and half like the other because each RI has an even chance of receiving the allele from either parent.

D. W. Bailey (1971) crossed the BALB/cBy and C57BL/6By inbred strains, which differ for many behavioral traits. Seven highly inbred recombinant strains produced from the F_2 generation of this cross (see Figure 4.7) have since been used in many experiments. One behavioral example concerns a major gene effect for learning to avoid shock in a shuttle box (Oliverio, Eleftheriou, and Bailey, 1973). Mice can be trained to avoid a shock by changing compartments as soon as a light is turned on. C57BL mice perform this task poorly, avoiding the shock only 14 percent of the time. The seven RI strains are either low performers like the C57BL mice or high

TABLE 4.8

Percent avoidance over 250 trials (50 trials per day for 5 days) for BALB and C57BL mice and 7 recombinant inbred strains

Low Performers		High Performers	
Strain	Avoidance (%)	Strain	Avoidance (%)
C57BL	14	BALB/c	46
RI-2	13	RI-1	60
RI-4	17	RI-3	36
RI-5	7	RI-7	58
RI-6	8		

SOURCE: After Oliverio et al., 1973.

performers like the BALB mice, as indicated in Table 4.8. These results suggest that a single gene may underlie the marked difference in avoidance learning between BALB and C57BL mice. If the RI strains had been intermediate to the two progenitor strains, *polygenic* (multi-gene) influences would have been implicated. For example, this was the result found in an RI study of susceptibility to audiogenic seizures (Seyfried, Yu, and Glaser, 1979). Using 21 RI strains derived from C57BL and DBA progenitors, 13 of the 21 RI strains were intermediate to the C57BL and DBA progenitor strains. This finding suggests that susceptibility to audiogenic seizures for these two strains is due to polygenic influences, even though the classical Mendelian crosses were consistent with a single-gene model.

Although there are other ways to search for single-gene effects, the RI method is useful in screening behavioral differences between two progenitor strains. In addition to the RI strains, *congenic* strains have been created that are the same as one of the progenitor strains except for a small segment of chromosome derived from the other. The use of congenic strains and other comparisons using RI strains facilitate locating single-gene influences on specific chromosomes (Eleftheriou and Elias, 1975). For example, the major gene difference between the two progenitor strains for learning to avoid shock appears to be on chromosome number 9 (Oliverio et al., 1973). However, as we might expect, most of the behaviors submitted to RI analysis suggest polygenic effects (Broadhurst, 1978).

MENDELIAN ANALYSIS IN HUMANS

In many respects, the human species is an unfavorable population for genetic analysis: experimental crosses cannot be performed; environmental control cannot be imposed; the generation interval is relatively long; and the number of offspring per family is relatively small. However, these problems are not insurmountable. Because the human population is large, a rich store of genetic information is potentially available. Although planned Mendelian crosses are not feasible, data can be collected from families in which particular types of mating have occurred. Although environmental control cannot be imposed, it is possible to study members of different families who have been reared in more or less similar environments. Although the human generation interval is long, data from several generations can nonetheless be obtained. Finally, although the number of children in human families is relatively small, data from many families can be pooled to provide adequate samples for statistical analysis. The fact that many important advances have occurred in human genetics within the last two decades demonstrates that human populations—like that of mice, fruit flies, bread mold, and colon bacteria—can also be subjected to genetic analysis.

Single-gene analyses of human behavior are principally concerned with testing the adequacy of single-locus hypotheses. Researchers first note familial transmission of some character of interest, and then determine whether the observed pattern of transmission conforms to that expected on the basis of a simple Mendelian model. Pedigrees, or family trees, are frequently used to depict hereditary information. An example of a pedigree is given in Box 4.3.

A pedigree analysis of many families can be used to determine whether a particular behavioral character conforms to Mendelian expectations. As in the examples of mouse behavior, we can test for the adequacy of a single-gene model, as well as the mode of transmission. The following sections consider the basic expectations for certain crosses for dominant and recessive transmission. The expectations are different for genes on the sex chromosomes, as compared to those on the other 22 pairs of human chromosomes, called *autosomes*. We shall first discuss autosomal dominant and recessive expectations and then consider transmission of single genes on the sex chromosomes. Because specific predictions can be made about single-gene characters, these expectations are of practical use in genetic counseling. (See Box 4.4.)

Dominant Transmission

Testing particular hypotheses about the transmission of single genes is really a matter of logic. Table 4.9 lists the expected outcomes of certain crosses if a character is influenced by a single dominant gene. By definition, only one allele (A) will cause an individual to be affected. Thus, the only way an offspring can be affected is if at least one of the parents is affected. In other words, we should not find affected offspring with unaffected parents (Table 4.9 [1]). Also, about half of the offspring of an affected parent should be affected, because each offspring has a 50-50 chance of inheriting that allele from the affected parent. We can determine whether an affected parent is a heterozygote or a homozygote by studying the parent's parents. If an affected individual has only one affected parent, then the individual must be heterozygous. As indicated in Table 4.9 (2), half of the offspring of matings between affected heterozygotes and unaffected individuals are likely to be

Box 4.3
Pedigree Analysis

A sample *pedigree* (from *pié de grue,* "crane's foot," a three-line mark denoting succession) is illustrated below. Affected individuals—those manifesting the condition under study—are designated by solid symbols. Individuals who were probably affected are shown by a cross. Nonaffected individuals are indicated by open symbols. Females are represented by circles and males by squares. A diamond is used to represent an individual whose sex is unknown. Parents are joined by a horizontal *marriage line,* and offspring are listed below. Members of a *sibship* are connected to a horizontal line that is joined by a perpendicular line to their parents' marriage line. The *siblings* (brothers and sisters) are listed from left to right in order of birth.

In this hypothetical pedigree each generation is designated by a Roman numeral, and each individual within a generation is denoted by an Arabic numeral. For example, individuals II-5 and II-6 were siblings, but information concerning their parents was not included in the pedigree. The marriage of individuals II-2 and II-3 resulted in no children. In order to save space, a number enclosed in a large symbol can be used to indicate the number of siblings of like condition. Thus, III-1, III-2, and III-3 were unaffected males in a family of seven children. Twins are indicated by two symbols that are connected either at or just below the sibship line. Individuals III-4 and III-5 were *monozygotic* (identical) twins, indicated by the short vertical line that descends from the sibship. Individuals III-6 and III-7 were *dizygotic* (fraternal) twins. The finding of a family of interest frequently comes only after the discovery of a particular affected individual. This specific individual who first comes to the attention of the investigator is referred to as the *index case* (or the *propositus* or *proband*). The index case, individual III-6, is indicated by an arrow.

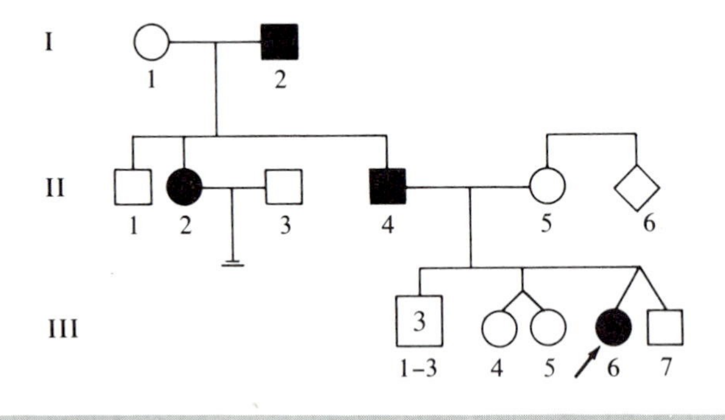

affected. Three-quarters of the offspring will be affected in matings between two affected heterozygotes. All the offspring of an affected homozygote parent will be affected. In order to determine the mode of inheritance, pedigrees such as the simple ones in Figure 4.8 must be analyzed.

Many genetic defects are single-gene, recessive characters, but dominant ones are rare. They are quickly eliminated by natural selection, as discussed in Chapter 8. However, one well-known neural disorder, *Huntington's chorea,* is caused by a single dominant gene. Huntington's disease is characterized by loss of motor control and progressive deterioration of the

Box 4.4
Genetic Counseling

The need for genetic counseling is clear. About 20 percent of infant deaths are now attributed to genetic defects (Porter, 1977), and 3 percent of all newborns have some genetic birth defect. About 1 in 200 have a chromosomal abnormality (see Chapter 7), about 1 in 250 have a single-gene defect, and at least 1 to 2 percent have problems that involve polygenic influences (Epstein and Golbus, 1977). Genetic counseling is becoming an important tool in preventive medicine (Lubs and de la Cruz, 1977).

There are three equally important steps in genetic counseling. First, a precise diagnosis must be made. Second, detailed pedigrees must be obtained. Third, the expected risk and burden must be carefully explained. The risk may be great but the burden small, as in the case of color blindness. But even when all the tools of single-gene analysis are used to make the best possible estimate of risk, predictions are often hampered by incomplete information and reduced penetrance.

An exciting new area is prenatal diagnosis involving amniocentesis (later described in Box 7.4). This technique permits an analysis of enzyme deficiencies for as many as seventy-five single-gene diseases, as well as a diagnosis of any chromosomal abnormalities of the fetus. Another promising approach to genetic counseling involves screening populations for particular single-gene problems. In the past, most people who sought genetic counseling did so because they or someone in their family already had an affected child. Now, for some single-gene, recessive diseases, it is possible to detect carriers as well as to diagnose the disease prenatally. It is important to identify carriers since 25 percent of the offspring will be affected in matings between two carriers. A number of cities have established programs to identify carriers of Tay-Sachs disease (infantile amaurotic idiocy, discussed in the next chapter). In California, 33,000 carriers were detected and 350 couples were identified who were both carriers (Epstein and Golbus, 1977). For those couples, prenatal diagnosis can determine whether a particular pregnancy will result in a Tay-Sachs child. Sickle-cell disease is another example of a single-gene, recessive condition that can be screened on a large scale, and can be subjected to prenatal diagnosis.

Another preventive approach is to screen infants, as has been discussed in the case of PKU. PKU screening laws exist in over forty states and, as a result, 90 percent of PKU babies are detected before they leave the hospital. Twenty million people have been screened for PKU and about 1,500 infants with PKU have been discovered (Scriver, 1977). In addition, fourteen other states have programs for screening newborns for diseases other than PKU (Culliton, 1976). In 1974, New York passed a bill that mandates that hospitals must screen for abnormalities if they have the knowhow. There are arguments against such screening, however. For example, screening for sickle-cell anemia at birth is questionable because there is no known treatment. On the other hand, carriers can be detected in this way.

Important new information often creates ethical dilemmas (Omenn, 1978). Genetic counseling abounds with both new information and ethical problems. Discussions of the ethical issues are complex in that they require a balance between the rights of a fetus to be born healthy, the rights of parents, and the rights of society. A negative tone often pervades such discussions because of the specter of abortion. However, in the vast majority of cases, genetic counseling provides positive information for parents. For example, 98 percent of the pregnant women who go through amniocentesis are relieved to find that their fetus has no chromosomal abnormality. Although many parents fearing a genetic disorder in their families might choose not to have children, genetic counseling and prenatal diagnosis may be able to relieve them of that fear.

TABLE 4.9

Offspring expected from various crosses for a completely dominant single autosomal gene (*A* refers to the dominant allele; thus, *Aa* and *AA* individuals are affected.)

1. Matings Between Two Unaffected Individuals

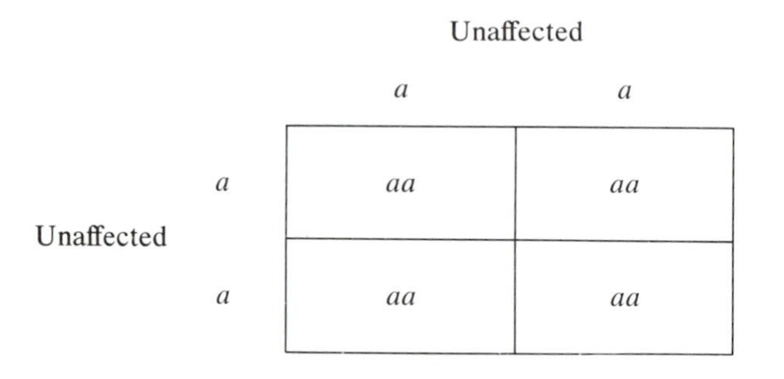

		Unaffected	
		a	*a*
Unaffected	*a*	*aa*	*aa*
	a	*aa*	*aa*

2. Matings Between a Heterozygote and an Unaffected Individual

(Shaded area indicates affected offspring.)

		Affected	
		A	*a*
Unaffected	*a*	*Aa*	*aa*
	a	*Aa*	*aa*

3. Matings Between Two Heterozygotes

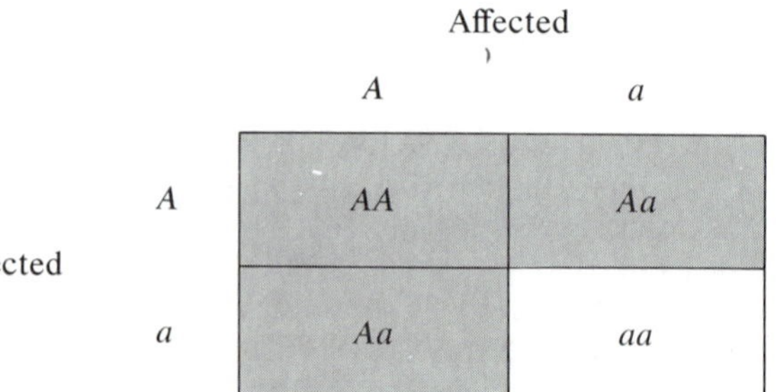

		Affected	
		A	*a*
Affected	*A*	*AA*	*Aa*
	a	*Aa*	*aa*

TABLE 4.9 (*continued*)

4. Matings Between an Affected Homozygote and an Unaffected Individual

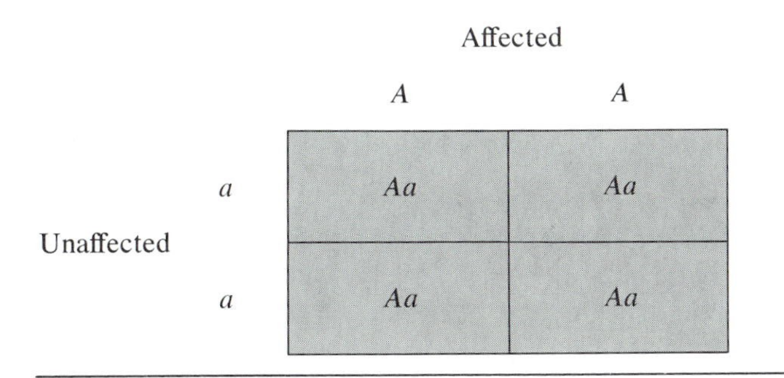

		Affected	
		A	*A*
Unaffected	*a*	*Aa*	*Aa*
	a	*Aa*	*Aa*

central nervous system. When this condition was traced through many generations of pedigrees, a consistent pattern of transmission was observed. Most afflicted individuals had a parent who was also afflicted, and approximately half of the children of an affected parent eventually develop the disease. The persistence of this insidious dominant lethal gene in the population is due to the fact that the disease is not usually expressed until after the childbearing years. This late age of onset illustrates the principle that hereditary conditions are not always manifested at birth.

Huntington's disease also illustrates the fact that the age of onset of a given condition may differ among individuals. The distribution of age of onset of Huntington's chorea in 762 patients is plotted in Figure 4.9. The usual age of onset is between thirty and fifty years. Data of this type are very useful for obtaining age-corrected incidence data. As an oversimplified example, assume that only one-half of individuals genetically predisposed to develop a disease actually express the disease by age forty. In such a case, the observed incidence of the disease among individuals at age forty should be doubled to obtain age-corrected incidence data. Another condition with variable age of onset is *schizophrenia*, discussed in Chapters 11 to 13.

Recessive Transmission

In the case of single-gene, recessive conditions, affected individuals are homozygous recessive. Carriers are heterozygotes, and thus have only one

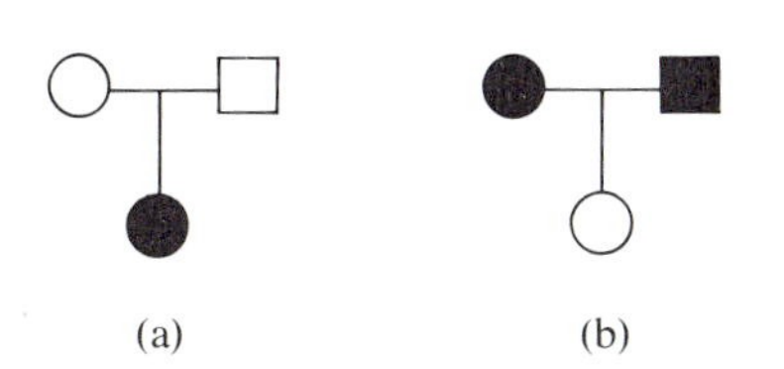

FIGURE 4.8

Hypothetical pedigrees. The condition indicated in (a) could be due to a recessive autosomal gene, but not a dominant; whereas that in (b) could be due to a dominant, but not a recessive.

FIGURE 4.9

Huntington's chorea. Distribution of age of onset in 762 patients. (From *Principles of Human Genetics*, 3rd ed., by Curt Stern. W. H. Freeman and Company. Copyright © 1973.)

allele of that type. Table 4.10 presents the expectations for offspring of different types of matings. The catch is that we often cannot tell whether an individual is an unaffected homozygote or a carrier. Natural selection cannot distinguish them either, which is the reason why recessively inherited problems persist. Sometimes pedigree information can distinguish carriers. At a biochemical level, carriers sometimes differ from unaffected dominant

TABLE 4.10

Offspring expected from various crosses for a completely recessive, single autosomal gene (*a* refers to the recessive allele; thus, *aa* individuals are affected.)

1. Matings Between Two Carriers

(Shaded area indicates affected offspring.)

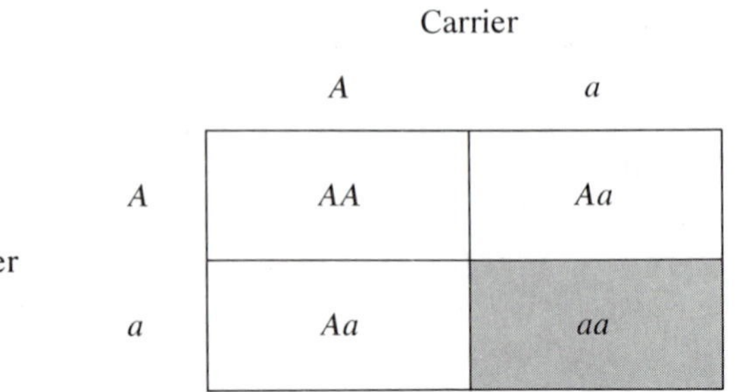

		Carrier	
		A	*a*
Carrier	*A*	*AA*	*Aa*
	a	*Aa*	*aa*

TABLE 4.10 (*continued*)

2. Matings Between an Affected Individual and an Unaffected Homozygote

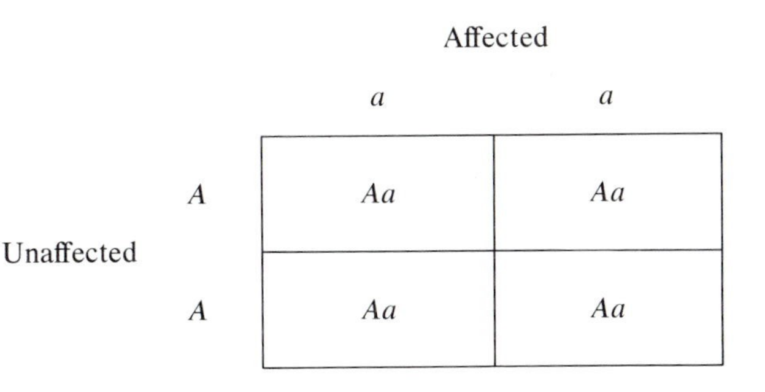

3. Matings Between an Affected Individual and a Carrier

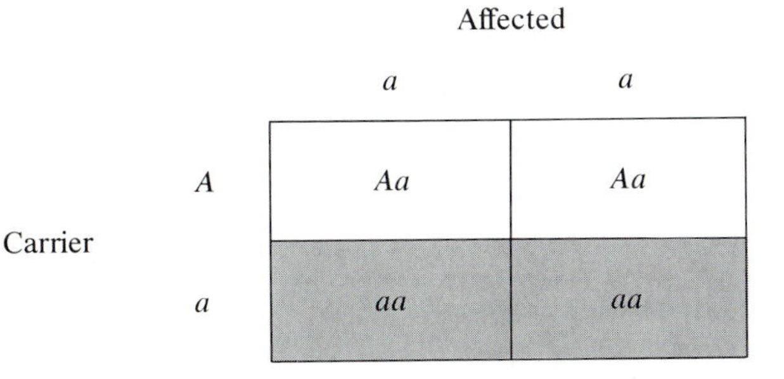

4. Matings Between Two Affected Individuals

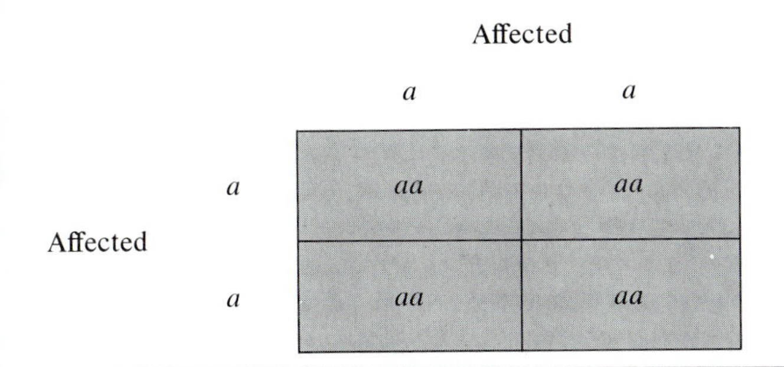

homozygotes and can be discriminated through the use of "carrier tests," as described in the next chapter.

For recessive characters, one commonly finds affected offspring with unaffected (carrier) parents (Table 4.10 [1]). One-quarter of the offspring of two carriers are affected. Also, as indicated in Table 4.10, none of the offspring of an affected parent is affected unless the spouse is affected or is at least a carrier. All the offspring of two affected individuals are affected. Another mark of a recessive character is that it occurs more frequently in offspring of matings between related individuals—that is, children of consanguineous marriages. This is called *inbreeding,* and is discussed in Chapter 8. Marriages between related individuals, such as cousins, are more likely to result in offspring with a double dose of recessive alleles that had been carried in the heterozygous state by the parents.

As mentioned previously, phenylketonuria (PKU) is a type of severe mental retardation caused by a single, recessive gene. A pedigree for PKU is shown in Figure 4.10. Individuals marked with a cross were probably affected, but their condition was not known for certain because they died young. A cousin marriage is indicated in the middle of the pedigree. As these individuals all lived in an isolated group of small islands in Norway, there may actually have been more inbreeding than indicated by this pedigree. Although PKU may manifest incomplete penetrance, the pattern of transmission indicated in Figure 4.10 nonetheless conforms to that expected of an autosomal recessive gene because affected subjects have normal parents, and an increased incidence accompanies inbreeding. Of the 18 children in generation IV, 4 were definitely affected and 2 were probably affected. If both unaffected parents of each sibship were carriers, as must be the case for a recessive condition, then only one-fourth of the children would be expected to be affected. Although this departure from expectation (6 versus 4.5) is not significant, an excess of affected individuals is frequently observed in pedigree data because of a bias called *truncate selection.* Couples who are at risk of producing affected offspring but who have not actually done so are excluded from pedigree data. This results in more affected individuals than would be

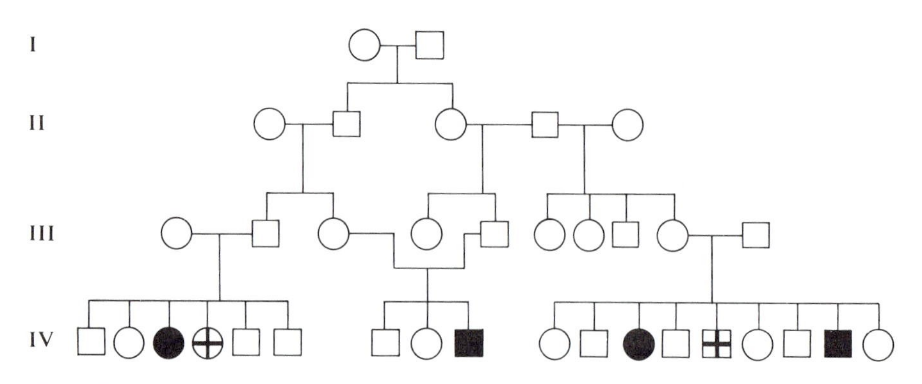

FIGURE 4.10

Pedigree of phenylketonuria and associated mental deficiency. For key to symbols, see Box 4.3. (From *Principles of Human Genetics,* 3rd ed., by Curt Stern. W. H. Freeman and Company. Copyright © 1973. After Følling, Mohr, and Ruud, 1945.)

expected from Mendelian calculations. PKU will be discussed in greater detail in the next chapter.

Sex-Linked Transmission

Pedigrees are also useful in detecting single genes located on the sex chromosomes. As discussed in Chapter 6, females have two large X chromosomes, whereas males have only one X and a small chromosome called Y. Recessive genes on the X chromosome express themselves in males when they have one such allele on their single X chromosome. Females, however, show the condition only when they have the allele on both of their X chromosomes. Thus, one of the first indications of sex-linked (meaning X-linked) recessive transmission is a greater incidence in males.

The transmission of the sex chromosomes from one generation to the next is portrayed in Figure 4.11. The figure shows that daughters inherit their father's X chromosome, but sons do not. Thus, sons cannot inherit sex-linked conditions from their father. Daughters inherit sex-linked genes from

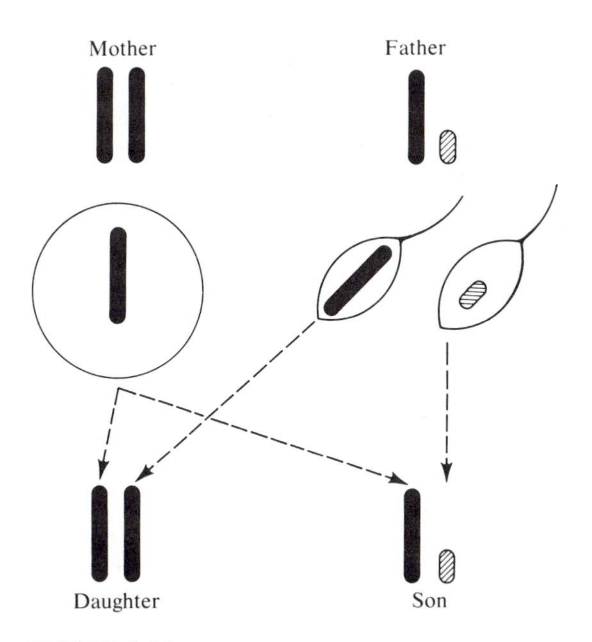

FIGURE 4.11

Transmission of the X chromosomes from one generation to the next. (From *Principles of Human Genetics*, 3rd ed., by Curt Stern. W. H. Freeman and Company. Copyright © 1973.)

their father, but they do not express a recessive condition unless they receive another such allele on the X chromosome from their mother.

Table 4.11 presents the expectations for offspring of certain matings involving a single recessive gene on the X chromosome. Affected fathers produce no affected offspring unless the mother is a carrier. However, all the daughters of an affected male are carriers (1). Half the sons of a carrier mother will be affected (2 and 3). Half of her daughters will be carriers if the spouse is a normal male (2) and half will be affected if the spouse is affected (3). Finally, all the sons of an affected woman will be affected, and all of her

TABLE 4.11

Expected frequencies of offspring for a sex-linked recessive character (X_a represents the X chromosome with the particular recessive allele; thus, X_aY males and X_aX_a females are affected.)

1. Matings Between an Affected Father and a Homozygous Normal Mother

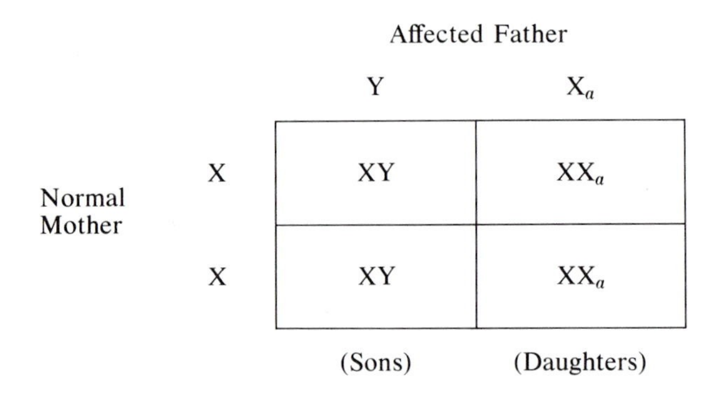

		Affected Father	
		Y	X_a
Normal Mother	X	XY	XX_a
	X	XY	XX_a
		(Sons)	(Daughters)

2. Matings Between a Normal Father and a Carrier Mother

(Shaded area indicates affected offspring.)

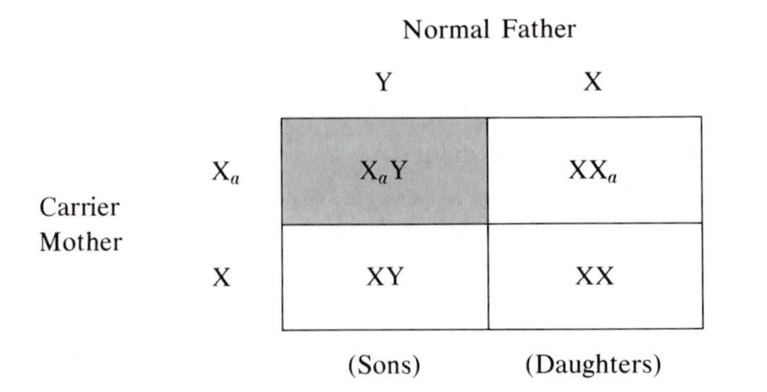

		Normal Father	
		Y	X
Carrier Mother	X_a	X_aY	XX_a
	X	XY	XX
		(Sons)	(Daughters)

TABLE 4.11 *(continued)*

3. Matings Between an Affected Father and a Carrier Mother

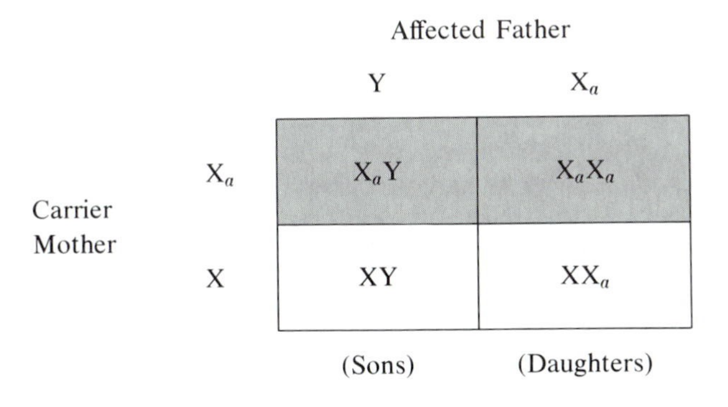

4. Matings Between a Normal Father and an Affected Mother

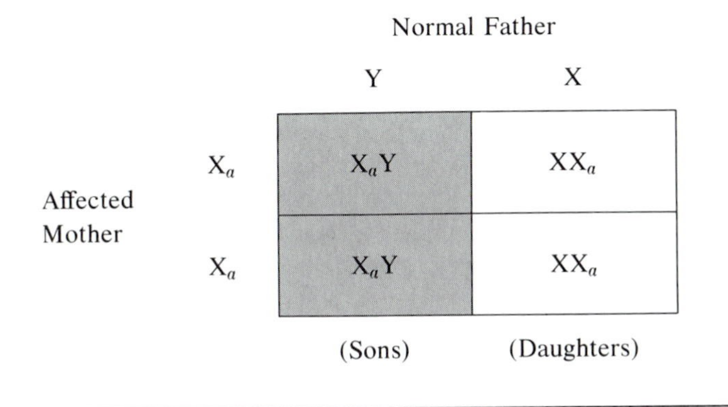

daughters will be carriers if the father is normal (4). Another peculiarity of recessive, sex-linked inheritance is the alternate-generation phenomenon, in which the maternal grandfather and his grandsons are affected, but none of those in the intermediate generation. However, half of the grandfather's daughters' sons are affected. Because the affected grandfather's daughters are all carriers, their sons have a 50-50 chance of being affected.

Five pedigrees depicting the transmission of color blindness are presented in Figure 4.12. Although a number of different forms of color blindness are known (Stern, 1973), the more common types have a similar genetic basis, and thus will not be differentiated here. The pattern of transmission evident in Figure 4.12 conforms closely to that expected of a sex-linked, recessive gene.

In the 1960s, some data pointed to the possibility that a major gene on the X chromosome influences a specific cognitive trait called spatial ability. Tests of spatial ability involve tasks such as deciding what a two-dimensional

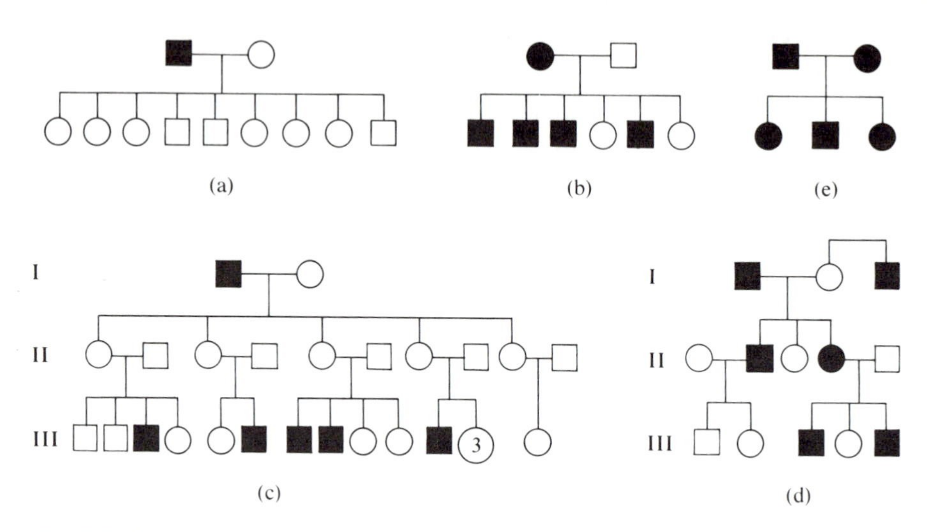

FIGURE 4.12

Pedigrees of color blindness. (a) Part of pedigree no. 406, Nettleship. (b) Pedigree no. 584. (c) Part of Horner's pedigree. (d) Pedigree, Whisson, 1778, the first known pedigree of color blindness. (e) Pedigree, Vogt. (From *Principles of Human Genetics*, 3rd ed., by Curt Stern. W. H. Freeman and Company. Copyright © 1973.)

form will look like in three dimensions, and mentally rotating two- or three-dimensional objects to determine whether they are the same or different from a standard shape. Of all the specific cognitive abilities, spatial ability shows the most consistent sex difference. Males tend to score higher on tests of spatial ability.

One possible explanation for sex differences is sex linkage. Suppose that a major recessive gene enhancing spatial ability is on the X chromosome. Males would express such an allele more frequently than females, and would thus have higher scores. Moreover, because sons never receive their father's X chromosome, fathers and sons should be less similar than mothers and daughters, mothers and sons, or fathers and daughters if X-linked genes are involved. Earlier data suggested that fathers and sons are less similar than other relatives, and revealed other expected patterns of familial similarity for spatial scores. These data seemed to support the hypothesis of a major X-linked gene for spatial ability. However, recent data from larger studies do not confirm the hypothesis of sex-linkage (Bouchard and McGee, 1977; DeFries et al., 1976; DeFries et al., 1979; Guttman, 1974; Loehlin, Sharan, and Jacoby, 1978). For example, in the study by DeFries and associates (1979), the correlation for spatial ability scores between 672 father-son pairs was 0.33 in a Caucasian group and 0.26 between 241 father-son pairs in an Oriental sample. The average correlation for other family relationships was 0.33 for the Caucasian sample and 0.25 for the Oriental sample. Thus, these results do not support the hypothesis that spatial ability is influenced by a sex-linked, recessive gene.

Scope of Single-Gene Effects in Human Behavior

Only a few of the many known single-gene behavioral effects in humans have been discussed in this chapter. An excellent reference for single-gene effects is Victor McKusick's *Mendelian Inheritance in Man* (1975). This book discusses 583 established autosomal dominant, 466 autosomal recessive, and 93 X-linked characteristics of human beings. In addition, there are over a thousand tentatively identified single-gene effects. Mental retardation is listed as a clinical symptom of 117 of the autosomal recessive characters, 19 of the dozen autosomal dominant characters, and 21 of the X-linked characters. It is interesting to note that nearly all single-gene influences on behavior seem to be deleterious in terms of mental ability. However, mental ability is only one of a large number of human behavioral characters that can be subjected to single-gene analysis. Thus, although we recognize a large number of human behavioral effects that are due to a single gene, the search for such effects has just begun.

SUMMARY

In each body cell of higher organisms, there are pairs of chromosomes that carry hereditary factors, now called genes. A gene may exist in two or more different forms (alleles). One allele can dominate the other. Genotype refers to alleles considered two at a time as they exist in individuals. Mendel's monohybrid experiments showed that alleles are not contaminated by the presence of an alternate allele in the genotype, and that the two alleles separate (segregate) completely during hereditary transmission. Mendel's dihybrid and trihybrid experiments demonstrated that the alleles for one trait assort independently of the alleles for other traits. In this way, Mendelian genetics provides a mechanism for the maintenance of vast amounts of genetic variability.

Mendelian crosses can be used to determine the mode of inheritance for behavioral characters in nonhuman animals. There are many examples of single-gene abnormalities in mice. Genes with well-known morphological effects, such as albinism, may also affect behavior (pleiotropism). The use of recombinant inbred strains provides a new strategy to uncover single-gene effects in animals.

Although humans were once thought to be poor subjects for genetic analysis, great advances in human genetics have occurred during the last twenty years. Pedigrees can provide information concerning naturally occurring Mendelian crosses in order to understand the mode of transmission of single-gene influences. Dominant, recessive, and sex-linked modes of inheritance are discussed in the chapter.

5

Mechanisms of Heredity and Behavior

Mendel was convinced that his "elements" were material units located in the gametes. The state of knowledge at the time, however, made it impossible for him to specify their physical nature in any greater detail. It was fortunate that the basic Mendelian laws could be established by treating the "elements," or genes, as hypothetical constructs, without precise knowledge of their location or structure.

In this chapter, the story will be made more complete by describing the chemical structure and function of genes. In the second part of the chapter we will consider examples of the mechanisms by which genes influence behavior and how mutations can be used to "dissect" behavior.

MECHANISMS OF GENE ACTION

In Chapter 3, we discussed the evolution of the replicators, the living material that has the critical ability to reproduce itself. These replicators are genes on chromosomes in the nucleus of each cell in our bodies. The function of genes was determined before their chemical nature was known.

In 1902, A. E. Garrod discussed a rare human condition, called alkaptonuria, in which affected individuals display the remarkable symptom of excreting urine that turns black when exposed to air. Garrod concluded that the condition was inherited, and that it obeyed the newly rediscovered Mendelian laws. As significant as it was then to have an example of Mendelian inheritance in man, the conclusion Garrod drew about the physiological basis of the disorder was of even greater importance. Analysis revealed that the urine of the affected individual contained large quantities of homogentisic acid instead of the usual compound, urea. Garrod knew that homogentisic acid is normally converted into urea. He thus suggested that the usual metabolic route that converts the acid into urea had been blocked, and called it an "inborn error of metabolism." Garrod (1908) also proposed that other defects in man, such as albinism, may be due to similar metabolic blocks. William Bateson (1909) suggested that inborn errors of metabolism might be due to the failure of the enzymes that control the normal reactions.

Other work distributed over the next twenty years on the inheritance of pigmentation in plants and animals (see Sturtevant, 1965, for a review) supported the hypothesis that genes produce some sort of biochemical substance. Another line of investigation lent support to this general proposition. The existence of human blood groups was described in 1900, and their Mendelian inheritance was shown in 1929, establishing another link between genetics and a biochemical process. While these results showed that genes could influence the physiological functioning of an organism, it was not clear whether these were typical or unusual situations.

In the 1930s, the common bread mold, *Neurospora,* was introduced into genetic research. This organism has extremely simple nutritional requirements and normally can survive on a simple medium containing salts, glucose, and a compound called biotin. From this simple diet, the organism is capable of metabolizing all the complex chemicals required for life. George Beadle and E. L. Tatum (1941) X-rayed spores of the fungus, and found that some of the organisms had undergone mutation and were no longer able to survive on this simple medium. By analyzing the nutritional requirements of these mutants, they were then able to describe the normal metabolic sequence and show that each enzymatic step in the sequence is under the control of a single gene. These results gave rise to the "one-gene, one-enzyme hypothesis." And it became increasingly reasonable to assume that the basic mechanism of gene action operates through the production of enzymes.

Later in this chapter we shall describe behaviorally relevant single genes. The main point is that the basic function of genes was understood by the 1940s. It was known that genes control the production of enzymes. However, the chemical structure of genes was not understood until the 1950s. Knowledge of the chemical structure of genes led to an understanding of how genes faithfully replicate themselves and how they control enzyme production.

THE CHEMICAL NATURE OF GENES

A great deal of research has been done in an attempt to understand the chemical nature of the gene itself. The chemical hereditary substance would have to meet several requirements:

1. It would have to be found in the nucleus of the cell, because chromosomes, which are the carriers of the genes, are found within nuclei.
2. The substance would have to be capable of self-duplication, because the genes have this ability.
3. The chemical would have to be capable of existing in various forms, because it was known that there are a large number of genes and that they occur in different allelic forms. In other words, it would have to be able to carry a variety of genetic information.

The correct answer had actually been correctly guessed about the same time that Galton wrote *Hereditary Genius*. But considerable work and many decades were required before the supporting evidence was conclusive and the detailed mechanism could be outlined.

The successful synthesis of all of the available data was accomplished by James Watson and Francis Crick (1953a, 1953b). They hypothesized that deoxyribonucleic acid (DNA) is the fundamental component of the hereditary material, and proposed a molecular structure that could account for its biological properties. This structure was confirmed by subsequent research, and its confirmation signaled the explosive growth and development of molecular biology. Briefly, and in necessarily oversimplified form, the basic

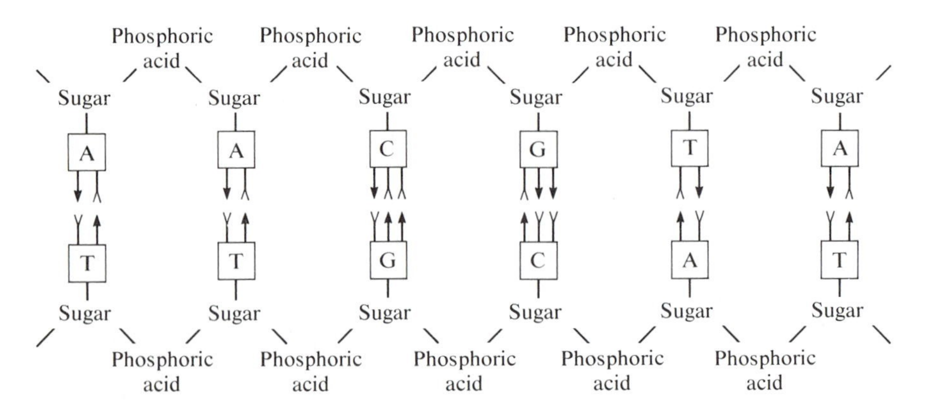

FIGURE 5.1

Flat representation of a DNA molecule. A = adenine; T = thymine; C = cytosine; G = guanine. (From *Heredity, Evolution, and Society* by I. M. Lerner. W. H. Freeman and Company. Copyright © 1968.)

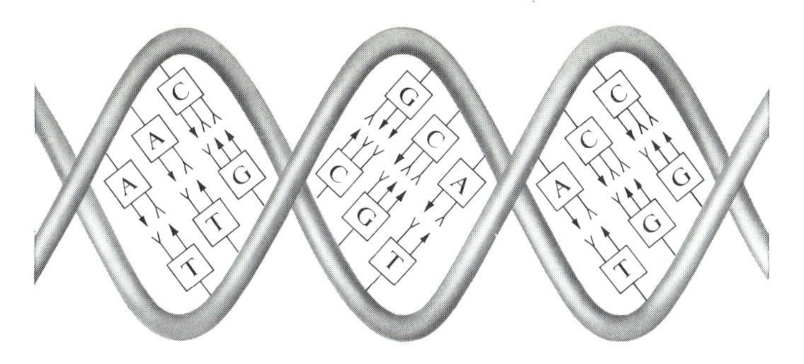

FIGURE 5.2

A three-dimensional view of a segment of DNA. (From *Heredity, Evolution, and Society* by I. M. Lerner. W. H. Freeman and Company. Copyright © 1968.)

features of the molecular structure of the gene and its action are as follows. A DNA molecule consists of two strands, each composed of phosphate and deoxyribose sugar groups. The strands are held a fixed distance apart by pairs of bases (nitrogenous compounds). There are four bases involved: adenine, thymine, guanine, and cytosine. Due to the structural properties of these bases, adenine always pairs with thymine and guanine always pairs with cytosine (Figure 5.1). The strands coil around each other to form a double helix (Figure 5.2).

The double nature of the helix and the restrictions on base pairing make possible the self-duplication of the DNA molecule. In the process of cell division, the helices of the DNA molecule unwind, the base pairs separate, and one of each pair remains attached to each strand (Figure 5.3). Within the nucleus of the cell, the raw materials necessary for the construction of new DNA are in the form of nucleotides, which consist of one of the four bases, a deoxyribose sugar, and a phosphate. Nucleotides pair with the exposed bases of the unwound strands, and ultimately form a complementary strand paired with each of the originals. By this process, two molecules of DNA are produced where there was previously but one.

GENES AND PROTEIN SYNTHESIS

One function of DNA is self-duplication. In fact, the process just described may be very similar to the way the original replicating cells duplicated themselves billions of years ago. The other major function of DNA is translation from the genetic code to enzyme production.

Although much of the biochemical functioning of the cell takes place in the cytoplasm, the chromosomes with their DNA content are located in the nucleus of the cell. Therefore, the information in the DNA molecule must be transmitted to the cytoplasm. This occurs in several steps, illustrated schematically in Figure 5.4. First, the information of the DNA molecule is

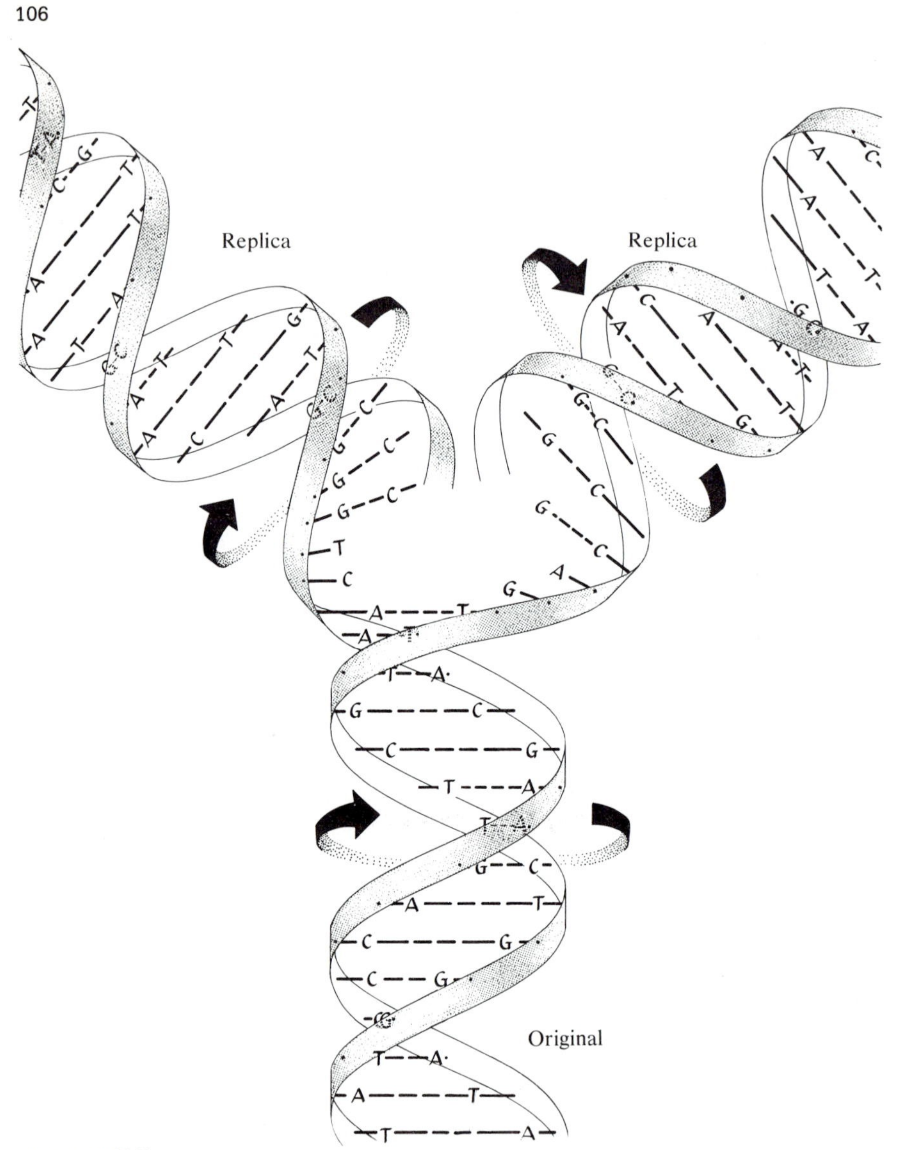

FIGURE 5.3

Replication of DNA. (After *Molecular Biology of Bacterial Viruses* by G. S. Stent. W. H. Freeman and Company. Copyright © 1963.)

transcribed onto a different sort of nucleic-acid molecule. This single-stranded molecule, ribonucleic acid (RNA), is composed of a ribose sugar, a phosphate, and the same bases as DNA, with the exception that uracil substitutes for thymine. By a process of base pairing similar to that of the duplication of DNA, a complementary RNA strand is formed using the DNA strand as a template. (In Figure 5.4, the dark DNA strand is being transcribed.) This RNA molecule, called messenger RNA, enters the cytoplasm

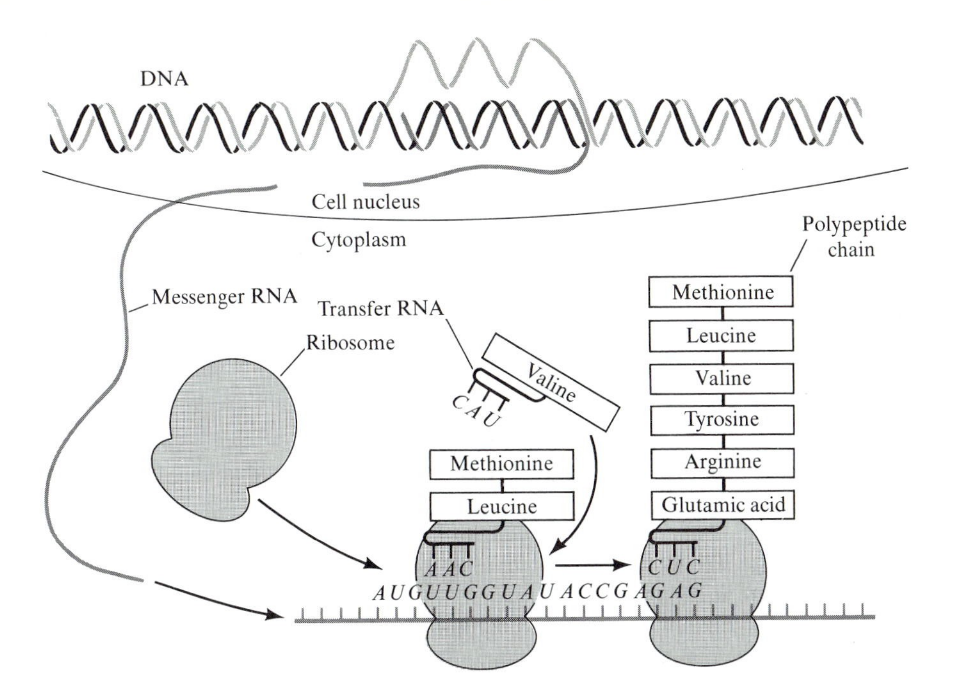

FIGURE 5.4

The "central dogma" of molecular genetics states that genetic information flows from DNA to messenger RNA to protein. Genes are relatively short segments of the long DNA molecules in cells. The DNA molecule comprises a linear code made up of four types of nucleotide base: adenine (*A*), cytosine (*C*), guanine (*G*), and thymine (*T*). The code is expressed in two steps: first the sequence of nucleotide bases in one strand of the DNA double helix is transcribed onto a single complementary strand of messenger RNA (which has the same bases as DNA except that thymine is replaced by the closely related uracil, or *U*). The messenger RNA is then translated into protein by means of complementary transfer-RNA molecules, which add amino acids one by one to the growing chain as the ribosome moves along the messenger-RNA strand. Each of the 20 amino acids found in proteins is specified by a "codon" made up of three sequential RNA bases. (From "The mechanism of evolution" by F. J. Ayala. Copyright © 1978 by Scientific American, Inc. All rights reserved.)

where it connects with ribosomes, the sites of protein synthesis. Within the cytoplasm is another form of RNA, transfer RNA. This RNA, which has a helical structure, exists in a variety of forms, each of which attaches to a specific amino acid. The transfer RNA molecules, with their attached amino acids pair up with the messenger RNAs in a sequence dictated by the rules of base pairing. The amino acids join to form polypeptide chains, which, in turn, constitute enzymes and other proteins. In this way, the genetic information of DNA becomes expressed in the production of specific enzymes.

The basic unit of the genetic code has been shown to be triplet sequences of bases, with each succeeding triplet specifying an amino acid. (See Table 5.1.) For example, three adenines in a row on the DNA molecule

TABLE 5.1

The genetic code

First Letter	Second Letter: A	G	T	C	Third Letter
A	Phe	Ser	Tyr	Cys	A
	Phe	Ser	Tyr	Cys	G
	Leu	Ser	chain end	chain end	T
	Leu	Ser	chain end	Try	C
G	Leu	Pro	His	Arg	A
	Leu	Pro	His	Arg	G
	Leu	Pro	Gln	Arg	T
	Leu	Pro	Gln	Arg	C
T	Ile	Thr	Asn	Ser	A
	Ile	Thr	Asn	Ser	G
	Ile	Thr	Lys	Arg	T
	Met	Thr	Lys	Arg	C
C	Val	Ala	Asp	Gly	A
	Val	Ala	Asp	Gly	G
	Val	Ala	Glu	Gly	T
	Val	Ala	Glu	Gly	C

NOTE: Each amino acid is coded by a triplet of three bases, as shown in the table, which is a compact way of setting out the 64 possible triplets.

The four bases are denoted by the letters A, G, T, and C. In DNA the four bases are: A = Adenine; G = Guanine; T = Thymine; C = Cytosine.

The 20 amino acids are identified as follows:

Ala = Alanine	Leu = Leucine
Arg = Arginine	Lys = Lysine
Asn = Asparagine	Met = Methionine
Asp = Aspartic acid	Phe = Phenylalanine
Cys = Cysteine	Pro = Proline
Glu = Glutamic acid	Ser = Serine
Gln = Glutamine	Thr = Threonine
Gly = Glycine	Try = Tryptophan
His = Histidine	Tyr = Tyrosine
Ile = Isoleucine	Val = Valine

SOURCE: After Cavalli-Sforza and Bodmer, *The Genetics of Human Populations*. W. H. Freeman and Company. Copyright © 1971. Data from Crick, 1966.

(AAA) will be transcribed in the messenger RNA as three uracils (UUU). When on the ribosome, this messenger RNA triplet will attract transfer RNA with the triplet sequence AAA. This particular transfer RNA is the one that "carries" the amino acid phenylalanine. Although there are 64 possible triplet codes, there are only 20 amino acids; some amino acids are coded by as many as 6 triplet codes. Since the same genetic code applies to all living

Box 5.1
Synthetic Genes and Recombinant DNA

Once the structure of DNA and the genetic code was worked out, attention quickly turned to constructing artificial genes. In 1976, after nine years of work by a team of twenty-four researchers, Nobel Laureate Har Gobind Khorana (1979) constructed a double-stranded gene that was 207 nucleotide units long. This gene codes for tyrosine transfer RNA, the RNA that transfers the amino acid tyrosine to the ribosome to become part of growing protein chains. The main body of the gene, which is 126 nucleotide units long, was synthesized in 1973. The rest of the nucleotides involve punctuation: a 56-nucleotide start signal, and a 25-nucleotide stop signal. This punctuation made the gene work in a living virus. A mutant strain of bacteriophage (a virus that invades bacteria) has a defect in this gene, so that it cannot use tyrosine and thus cannot survive. However, when the synthetic gene for tyrosine transfer RNA was added to the DNA of the mutant bacteriophage, the bacteriophage functioned normally. Khorana's work provided a technique for changing nucleotide sequences at will, and it is now technically possible to synthesize genes that will manufacture certain protein products.

Khorana's method is somewhat different from the recombinant DNA research that has come under considerable scrutiny in recent years. The recombinant DNA procedure combines naturally occurring genes from different organisms to create organisms with novel genetic combinations. Khorana synthesized a gene that already existed in the bacteriophage, as well as in all living cells. Nonetheless, his method would be much like the recombinant DNA procedure if a synthetic gene were introduced to an organism that does not normally have that gene. An example of the possibilities of such research was announced in 1977. Somatostatin is a hormone in the brain of mammals that slows down pituitary growth hormones and is used to treat growth hormone problems. Keiichi Itakura and his associates (1977) synthesized the gene for the hormone and successfully inserted it into the DNA of the bacterium *E. coli*. These bacteria then multiplied and became somatostatin factories. To realize the importance of this advance, we need to recognize that researchers who first obtained somatostatin needed brains from half a million sheep to produce a mere 0.00018 ounce of the hormone. In 1978, another example of the usefulness of recombinant DNA techniques involved synthesized genes for human insulin that were spliced into *E. coli*, thus creating insulin factories that insure abundant supplies of insulin for diabetics to metabolize sugar and other carbohydrates. Many other advances along these lines can be expected in the near future.

Another strategy for recombinant DNA research is referred to as *genetic engineering*. This procedure involves the direct transfer of a gene to an organism other than *E. coli*. For example, researchers are currently working on the problem of transplanting the beta chain of hemoglobin from one mammal to another. If such a transfer could be made from one human being to another, the hemoglobin produced by the gene transplant would allow individuals with sickle-cell anemia to live.

One of the fears associated with recombinant DNA research is that *E. coli*, the bacterium usually involved, is common in human intestines and those of all warm-blooded animals, as well as in fish and certain insects. If new bacteria were created that produced toxic substances or that were resistant to antibiotics, these bacteria could infect humans. As a result of the possible hazards, strict government guidelines have been established to regulate such research (Wade, 1976). Many articles and books have recently appeared on the topic of recombinant DNA research. One of the best is a book by Nicholas Wade (*The ultimate experiment: Man-made evolution*, Walker Press, 1977).

organisms, "breaking" this code has been one of the great triumphs of molecular biology. This information has led to the artificial synthesis of genes (Box 5.1).

A NEW VIEW OF MENDEL'S "ELEMENTS"

These developments have provided a dramatically new view of Mendel's hypothetical "elements." A gene is a functional unit that codes for a particular polypeptide. Structurally, it is composed of a stretch of nucleotide bases of DNA. The average number of nucleotide base pairs in a gene is about 900 to 1,500. Thus, we would expect that the average number of amino acids in a polypeptide is 300 to 500, because a sequence of three nucleotide base pairs codes for a particular amino acid. However, "nonsense" stretches of DNA are interspersed within many genes (Crick, 1979). RNA transcribes the DNA, nonsense and all, and then splices out the nonsense codes before the RNA reaches the cytoplasm.

Enzymes are proteins (sequences of amino acid) that serve as organic catalysts. They speed biochemical reactions that would otherwise be sluggish or would not occur at all under the conditions of temperature and pressure prevailing within an organism's body. These reactions and their timing are fundamental to the development of all the systems of an organism and to the functioning of the various organs. The influence of genes, therefore, is not exerted through some mysterious mechanism. The pathways from genes to behavior run through the skeletal system; the muscles; the endocrine glands; the digestive, respiratory, and excretory systems; and the autonomic, peripheral, and central nervous systems. Investigations of these pathways, therefore, involve studies in molecular biology and biopsychology.

Although the connection between genes and behavior can be understood as a general theoretical proposition, the specific details have been elucidated in only a limited number of cases. However, this area of research is becoming increasingly popular, and we can predict that increased knowledge of the biochemical, physiological, and anatomical mechanisms of genetic influence on behavior will be very rapid in the near future.

PHENYLKETONURIA

Phenylketonuria (PKU) illustrates the way in which a single gene can influence behavior. It is the earliest and best-understood condition of genetic involvement in mental retardation. The trail of research that led to our current understanding of PKU had its beginning in Norway in 1934. A dentist with two retarded children was distressed because they exuded a peculiar odor that so aggravated his asthmatic condition that he was unable to stay with them in a closed room. He had them examined by Asbjörn Fölling, who began the search for the cause of the peculiar odor by analyzing the urine of the children. His search quickly paid off in the isolation of phenylpyruvic acid. Fölling postulated that the disease was inherited, and that phenylpyruvic acid was present in the urine because of a disturbance in the metabolism of phenylalanine, an essential amino acid. Somehow the excess of phenylpyruvic acid was related to mental retardation.

The disease came to be known as Fölling's disease, or phenylketonuria, and it became the subject of research in a number of laboratories around the world. This research revealed that the metabolic problem is caused by the absence or inactivity of a particular enzyme, phenylalanine hydroxylase. This enzyme converts phenylalanine to tyrosine. If this conversion is blocked, phenylalanine levels increase in the blood and phenylpyruvic acid accumulates in the urine. Apparently, the high level of phenylalanine in the blood depresses the level of other amino acids, depriving the developing nervous system of needed nutrients.

This knowledge concerning the biochemical origin of PKU made possible a search for rational therapies. If a particular enzyme is deficient, it might be possible to provide the necessary amounts of that enzyme. Although that approach may soon be technically feasible (Ambrus et al., 1978), an approach that has been used successfully involves minimizing the need for the enzyme. If the mental retardation of PKU is caused by a buildup of phenylalanine, the amount of phenylalanine in the diet can be reduced. Phenylalanine is found in a wide variety of foods, particularly meats. In 1953, a special diet was prepared that was very low in phenylalanine. Although it normalized levels of phenylalanine, it did not improve intelligence of older PKU children. However, when the diet was administered to very young PKU children, retardation apparently was prevented (Hsia et al., 1958). As a result of this work, routine screening programs have been established to identify PKU infants at birth.

Through these programs and related programs to test relatives of affected persons, it became possible to assay the intelligence of individuals identified on the basis of the chemical defect. Previously, most research had been conducted on individuals biochemically identified as phenylketonurics from a population of individuals already determined to be mentally retarded. The assumption had been that all phenylketonurics were probably institutionalized for mental retardation. On this assumption, calculations had been made that the mean IQ of phenylketonurics was approximately 30. With the new screening procedure, a surprising number of individuals were discovered who were biochemically phenylketonurics, but whose intelligence was in the normal range. This discovery necessitated a reevaluation of the efficacy of dietary treatment. If some of the individuals treated with the special diet would have developed normal intelligence in any case, then the report of IQs of treated subjects in the range of 80 to 90 could hardly be taken as evidence that the diet prevented retardation.

A great deal of research has subsequently been devoted to the problem of the efficacy of the low phenylalanine diet, and it remains somewhat controversial. A straightforward experiment comparing the outcomes for treated and untreated groups, both identified at birth, would provide a critical test. However, this would involve withholding a potentially effective treatment from a group of patients, which is ethically dubious, and is unlikely ever to be done. An approach to this type of comparison can be made, however, by comparing the IQs of treated patients with those of older siblings, who were untreated because the dietary therapy had not yet been invented. The results of such a comparison by Hsia (1970) are given in Figure 5.5. With late treatment or no treatment at all, the distribution of IQs ranges from 10 to 110;

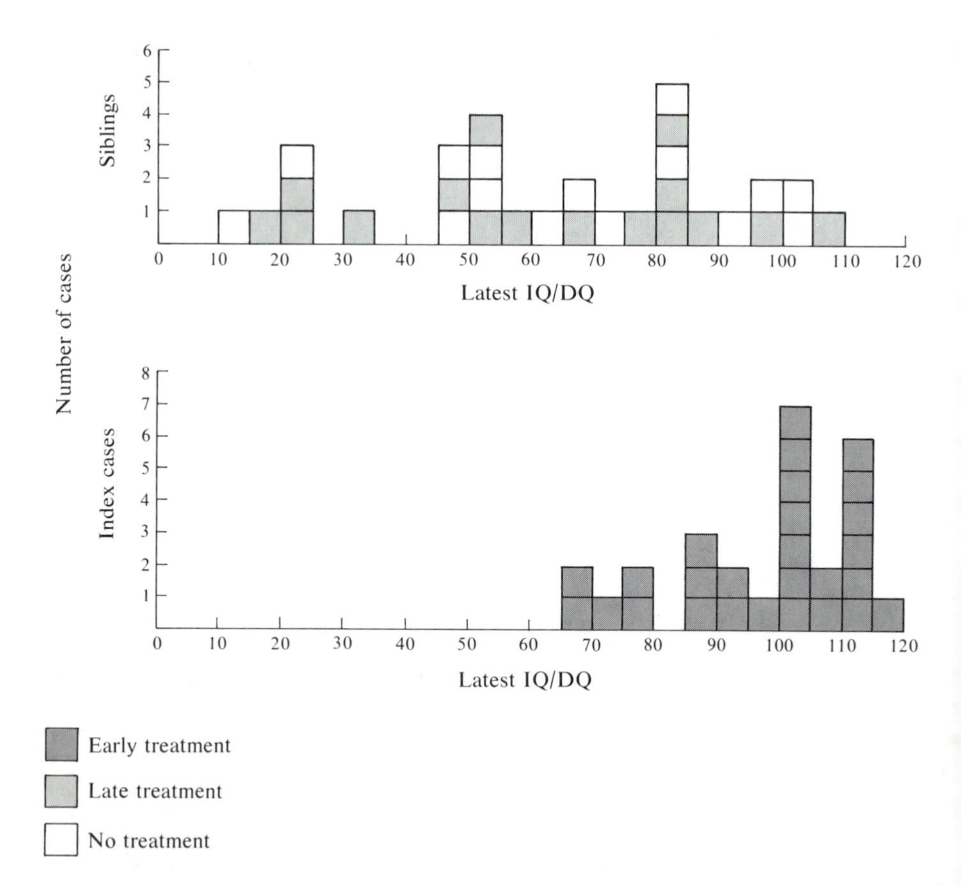

FIGURE 5.5

Frequency histograms of IQ or DQ scores of early-treated index cases versus late-treated or untreated siblings. (From "Phenylketonuria and its variants" by D. Y.-Y. Hsia. In A. G. Steinberg and A. G. Bearn, eds., *Progress in Medical Genetics*. Grune & Stratton, Inc., 1970. Used with permission.)

siblings treated from an early age have IQs ranging from 65 to 120. Furthermore, the distribution of IQs in this latter group indicates that most cases are at the higher end of the distribution. These results constitute reasonable evidence that the diet is in fact a useful therapy.

This discovery of normal IQs in untreated phenylketonurics also raises the issue of the possible heterogeneity of the condition. A number of variant forms have now been described. Hsia (1970) recommends as a working definition that patients with persistent blood phenylalanine levels in excess of 25 milligrams percent be diagnosed as "classical" phenylketonurics. Of these, approximately one-fourth may achieve normal intellectual functioning without dietary treatment. It is difficult to be sure of the diagnosis of patients with levels between 15 and 25 milligrams percent, but those with levels between 2 and 15 milligrams percent probably exhibit one of the variants of phenylketonuria that do not cause retardation. In fact, data from the newborn screening programs indicate that for every two PKU cases, there is one case of elevated phenylalanine levels not caused by PKU.

Important new information concerning PKU continues to be reported. For example, it has long been known that PKU heterozygotes are less able to convert phenylalanine to tyrosine than normal homozygotes. But only recently have researchers discovered that heterozygotes may have lower IQs than normal homozygotes (Bessman, Williamson, and Koch, 1978). This same study and others also suggest that the retardation effect of PKU may occur to some extent prenatally, rather than solely after birth.

The considerable success in elucidating the biochemical mechanisms of PKU has inspired an intensive research effort directed toward identifying and analyzing other conditions of mental retardation associated with abnormal metabolism of amino acids. At least a dozen have now been identified, but most are much more rare than PKU and none is understood nearly as well.

OTHER RECESSIVE METABOLIC ERRORS

In addition to amino acidurias such as PKU, there are single-gene metabolic errors involving carbohydrates and lipids (fats). Galactosemia is an example of a carbohydrate metabolic defect. Individuals homozygous for the autosomal recessive allele lack an enzyme to convert galactose to glucose. In many cases, early death results, and affected individuals who do survive are severely mentally retarded. However, as in the case of PKU, a rational therapy has been developed. Early identification of affected infants, coupled with replacement of milk with galactose-free substances, is quite successful. Heterozygotes have half the normal enzyme activity, which is apparently sufficient to metabolize galactose.

Tay-Sachs disease (infantile amaurotic idiocy) is a recessive condition caused by the absence of an enzyme that breaks down a lipid. Such individuals are apparently normal at birth, but begin to show symptoms of *nystagmus* (spasmodic movement of the eyes) and paralysis when a few months old. The condition steadily worsens to a state of profound idiocy, paralysis, and blindness. Death usually intervenes before two years of age. Autopsy has shown that nerve cells of the brains of affected individuals contain abnormal amounts of a lipoid substance and the neurons show degenerative changes. A carrier test has been developed, and it is also possible to diagnose affected fetuses prenatally by sampling the amniotic fluid. (See Box 7.1.)

The point of these examples is to indicate that genes affect behavior the same way that they affect other phenotypes. When we talk about single-gene influences on behavior, it is simply a convenient way to indicate that we are considering the effects of DNA production and regulation of proteins as described earlier in this chapter. Sometimes students react to these examples of single-gene influences by saying, "But genes didn't *really* cause these behavioral disturbances. The genes caused metabolic problems and the behavioral effect was only a by-product of the enzyme deficiency." However, that is just the point. Genes are not magical elements that somehow blossom into behavior patterns, as when the puppeteer pulls a puppet's strings. Genes are segments of DNA that code for protein production. In that sense, all aspects of ourselves—our bones as well as our behavior—are by-products of this process.

REGULATOR GENES

Until now, we have considered the physical basis for genes that code for particular proteins, known as *structural genes*. However, it is clear that genes are not completely active from the moment of conception. Since genes turn on and off throughout development, after birth as well as before, some mechanism must be responsible for regulating the timing and quantity of protein production. The hypothesized mechanism, suggested by Jacob and Monod (1961), is called the *operon model*. In addition to structural genes that serve as templates for protein production, there are *operator* genes and *regulator* genes. The operator gene is a short segment of DNA next to the structural gene that serves as an on-off switch, determining whether the structural gene will be transcribed for protein production. Together, the operator gene and structural gene are called an *operon*. Regulator genes switch the operator gene on and off by producing a repressor that binds with the segment of DNA referred to as the operator gene. This model suggests a mechanism for the regulation of genes. (See Figure 5.6.) The repressor produced by the regulator gene will bind with the operator gene, shutting down the structural gene. If this were the end of the story, all such structural genes would be permanently turned off. However, the repressor can also bind with a *regulatory metabolite*, a product of other operons or substances from out-

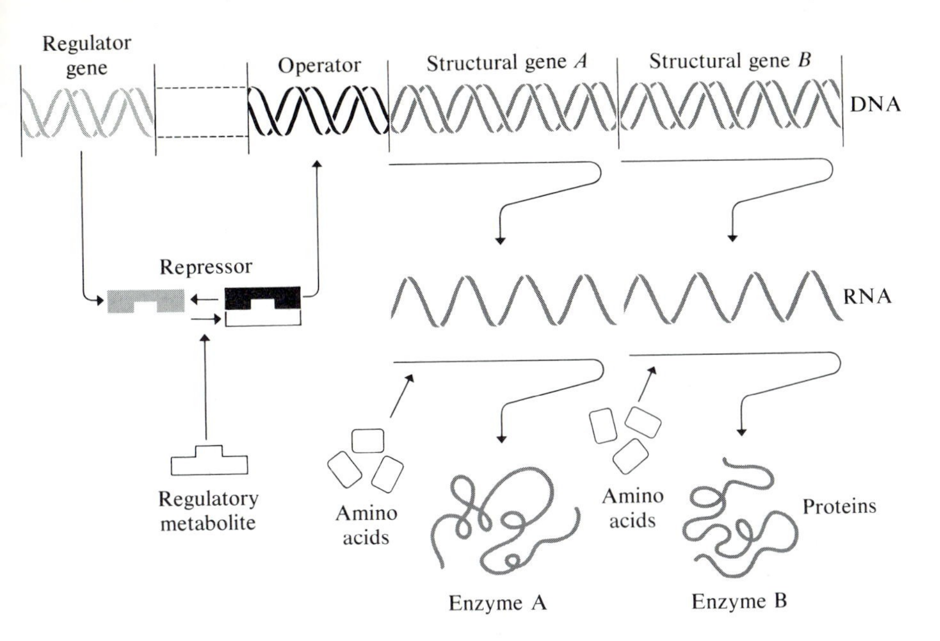

FIGURE 5.6
The operon model. When the regulatory metabolite binds with the repressor (produced by the regulator gene), the repressor does not bind with the operator. The structural gene will then be switched on and transcribe RNA, thus producing its enzyme. (From "The control of biochemical reactions" by J.-P. Changeux. Copyright © 1965 by Scientific American, Inc. All rights reserved.)

side the cell. When this happens, the repressor is, in a sense, inactivated, because it will bind with the regulatory metabolite rather than with the operator gene. The result is that the operator gene, no longer bound to the repressor, will permit transcription of messenger RNA so that the structural gene is, in effect, turned on.

Suppose that this structural gene produces an enzyme that breaks down a regulatory metabolite, and that the metabolite binds with the gene's repressor. When a structural gene is shut down by its repressor, a regulatory metabolite builds up. As the metabolite builds up, it begins to bind with the repressor until the repressor is inactivated and the structural gene turns on. The structural gene produces an enzyme that metabolizes the regulatory metabolite. As the metabolite is broken down, the repressor again begins to bind with the operator gene and eventually shuts down the operon. In effect, presence of the regulatory metabolite turns on an operon, and its absence turns an operon off.

The operon model has been demonstrated with the single-celled bacterium *E. coli*. Discussion of the details of the operon in bacteria will make the concept of regulatory genes clearer. (See Figure 5.7.) *E. coli* produces about 700 different enzymes. The enzyme that we know the most about is β-galactosidase, which metabolizes lactose into glucose and galactose. In a

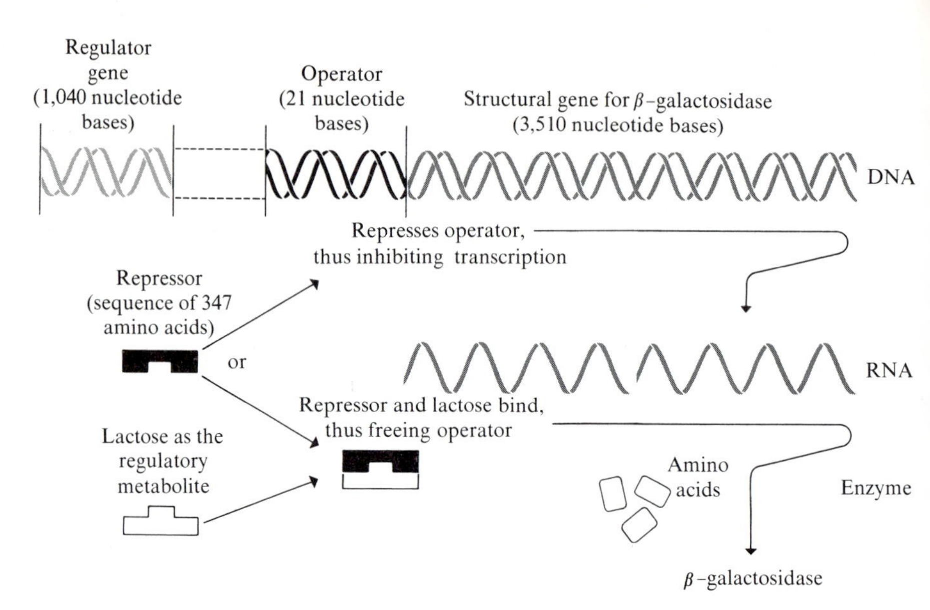

FIGURE 5.7

The operon model for β-galactosidase in *E. coli*. (Adapted from description by J. D. Watson, 1976.)

normal *E. coli* cell, there are about 3,000 β-galactosidase molecules because a normal cell lives in a lactose-rich environment. The cell needs the enzyme to metabolize lactose, its major energy source. However, when *E. coli* are placed in environments without lactose, there are as few as three β-galactosidase molecules. This is a good example of how enzyme production is rigidly controlled, not simply pumped out at a constant rate.

Synthesis of messenger RNA is critical to this control. We know this because messenger RNA exists for only 2 to 3 minutes, and variation in its synthesis is related to protein production. For example, in a lactose-rich environment, an average cell will have 35 to 50 β-galactosidase messenger RNA molecules at any given time. In the absence of lactose, the same cells will have less than one messenger RNA molecule on the average. Thus, the presence of the short-lived messenger RNA that is translated into the enzyme is dependent on the need for that enzyme. Synthesis of messenger RNA is controlled by the operon system.

The regulator gene for β-galactosidase involves 1,040 nucleotide bases, and the repressor that it codes for is a sequence of 347 amino acids. This repressor binds to another segment of DNA (the operator gene), which is only 21 nucleotide bases long. When the repressor binds to the operator, it blocks transcription of messenger RNA for the structural gene (3,510 bases) that codes for the β-galactosidase enzyme. In this way, the operon for the enzyme is switched off. However, the repressor is inactivated when lactose binds with it. When lactose builds up, the repressor no longer shuts down the operon, and the β-galactosidase enzyme is again produced. This enzyme metabolizes lactose and, when its job is done, it is again repressed. As a

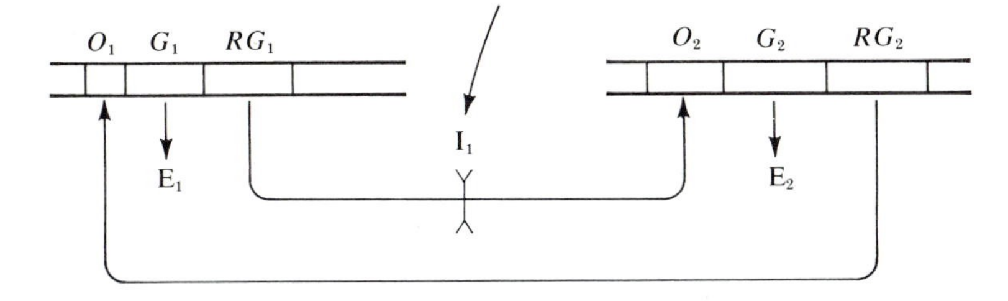

FIGURE 5.8
An operon circuit that would switch the production of one enzyme off and another on. (From *General Genetics*, 2nd ed., by A. M. Srb, R. D. Owen, and R. S. Edgar. W. H. Freeman and Company. Copyright © 1965.)

sidelight, it is known that the repressor for β-galactosidase also represses two neighboring structural genes that produce enzymes crucial to the metabolism of lactose.

Interactions among operons can provide mechanisms for relatively permanent changes in gene functioning. Figure 5.8 shows two operons that each contain one structural gene that is, in fact, a regulatory gene (*RG*) for the other operon. Thus, if operon 1 is active, the product of RG_1 will turn off operon 2. If some inducer from outside the system, I_1, a regulatory metabolite for the repressor coded by RG_1, appears even briefly, operon 2 can be switched on. G_2 will produce enzyme 2 and RG_2 will produce its repressor, which switches off operon 1. Operon 2 is now locked on and operon 1 is locked off until such time as an inducer for operon 2 appears. Far more complex types of interactions could be postulated. But the foregoing should make it clear that operon-type systems could be responsible for most of the sequential regulation of genes and the selective workings of genes in different tissues that accompany developmental processes.

Since the operon model is based on research with microorganisms, the extent to which it applies to more complex organisms remains to be established. Nevertheless, it would seem to be a reasonable working hypothesis that this or a similar mechanism functions in higher organisms. Because gene regulation is a relatively new concept, it has not been systematically incorporated into evolutionary theory or behavioral genetics. Conceptually, the impact of gene regulation will be considerable.

MUTATION

Although DNA replication is highly reliable, mistakes sometimes occur. Some of the early work in genetics concerned mutations induced by X-rays, a procedure used by Hermann J. Muller in 1927 with *Drosophila*. Since this discovery of the *mutagenic* effect of X-rays, other agents have been discovered for experimentally inducing mutations, including some chemical compounds and extreme temperatures. Thus, certain environments can produce changes in genes. It should be noted that this phenomenon differs greatly from the old notion of inheritance of acquired characters. Mutations are random and most are harmful. Such mutations are rare in nature, perhaps only one in a billion DNA replications. Given the large number of genes in an individual, however, it is clear that such mutations are also an important source of genetic change on an evolutionary time scale.

As we have seen, the average protein is a sequence of about 300 amino acids, each of which is coded by a triplet sequence of nucleotide bases. A change in any one of these bases can radically change the functioning of the gene's product through a change in the amino acid sequence. For example, sickle-cell anemia involves a change of just one of the 146 amino acids linked to form the β-chain of hemoglobin.

As we shall see in the following section, the capability of experimentally inducing mutations has proved to be of marked value in genetic research and in behavioral genetics analysis. It has contributed to our understanding of the molecular structure of genes, as well as the biochemistry of gene action. It has also been important to evolutionary theory. Recall that Darwin took great pains in considering the possible sources of heritable variation. Somewhat reluctantly, he concluded that Lamarckian mechanisms are an important source of variability. Contemporary evolutionary theory views mutation as the ultimate source of the genetic variability on which natural selection depends.

USING MUTATIONS TO DISSECT BEHAVIOR

As noted earlier, Beadle and Tatum (1941) used mutations in the common bread mold to study the operation of genes. They found mutants that were unable to convert certain substances into more complicated compounds. Their work revealed the sequence of normal metabolism, and demonstrated that each step in the sequence is controlled by a single gene—the "one-gene, one-enzyme" hypothesis.

In 1967, Seymour Benzer suggested that a similar approach could be used to dissect the physiological events underlying behavior. In the decade since Benzer's paper, hundreds of behavioral mutants have been selected in organisms as diverse as bacteria, fruit flies, crickets, and mice. Nearly all of these studies involved inducing mutations through the use of chemicals, screening individuals for behavioral abnormalities, and then searching for specific genetic causes.

In this section, we shall look at recent attempts to perform genetic dissection of behavior in single-celled organisms, such as bacteria and paramecia, as well as in more complex organisms, such as the round worm and fruit flies. The point of this section is to show how genes influence behavior. Genes affect behavior in the same way that they affect any phenotype—by controlling the production of enzymes.

Bacteria

Although the behavior of bacteria was studied in detail in the early part of this century, until recently this research area has been neglected. To be sure, the behavior of bacteria is by no means attention-grabbing—in fact, most researchers ascertain if bacteria are dead or alive only by determining whether they reproduce. However, they do behave. They move toward or away from many kinds of chemicals (called *chemotaxes*) by rotating their four to eight tiny propellerlike flagella. When the flagella rotate in a clockwise direction, the single-celled organism swims forward smoothly for about a second. When the flagella rotate counterclockwise, the organism tumbles and changes direction. The reason for the renewed interest in this simple behavior is that so much is known about the genetics of bacteria. This knowledge has led to rapid advances in isolating genes and the proteins responsible for various aspects of this behavior (Adler, 1976; Parkinson, 1977).

Great progress has been made since the first behavioral mutant in bacteria was isolated in 1966. Behavioral mutants are isolated by screening huge numbers of bacteria for deviant behavior. Specific behavioral mutants that can accomplish only certain aspects of a response are especially important. For example, some mutant bacteria can swim well, but cannot recognize certain chemical stimuli. One mutant strain can swim and recognize chemicals, but cannot metabolize galactose. Another strain's only shortcoming is that it cannot metabolize ribose. When these strains are put on a plate with ribose and galactose, each consumes the sugar it can metabolize, and each strain moves independently of the other, as shown in Figure 5.9.

The normal swimming of bacteria involves steps much like those in more complex behaviors of higher organisms. The bacterium cell must detect a stimulus and then transmit this information to produce an appropriate response. Genetic dissection using the mutant strains indicates that the first step, recognizing the chemical stimulus, probably involves about 20 genes and at least 9 genes are needed just to detect different types of sugars. The genes produce proteins that bind with particular substances, such as sugars and amino acids. The second step, signaling, involves using the products of at least 3 genes that recognize the extent of binding that has occurred. Finally, there are many genes involved in rotating the flagella. Some genes influence the rotor, other genes determine whether the flagella rotate clockwise or counterclockwise, and still others control the duration of the behavior. These genes have been mapped on the single circular chromosome of *E. coli,* as illustrated in Figure 5.10. This figure emphasizes the genetic complexity of an apparently simple behavior in a simple organism.

This illustrates the fact that most normal behavior is influenced by many genes. As pointed out in the previous chapter, there are many single genes that can seriously disrupt normal behavioral development. However, even if many genes influence a particular behavior, a single gene can still

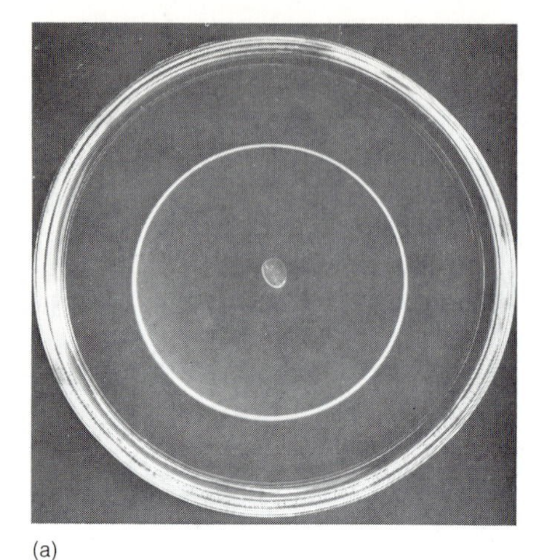

(a)

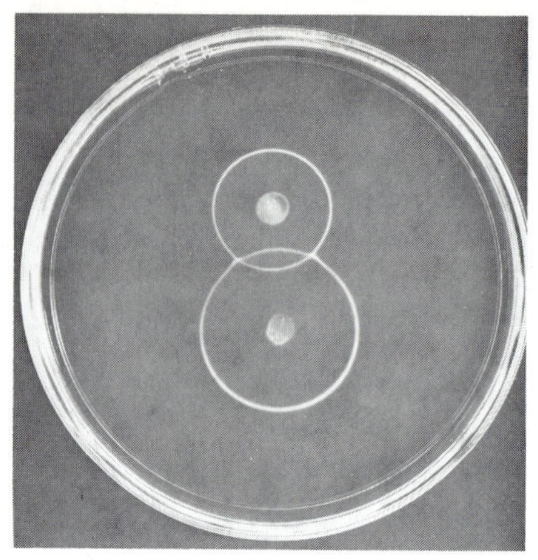

(b)

FIGURE 5.9

Bacteria on the move in normal and mutant strains. (a) Normal bacteria placed in the center of a plate with galactose radiate outward, forming a ring. (b) This plate has both galactose and ribose, and two mutant strains—one that cannot metabolize galactose and one that cannot metabolize ribose. Each strain moves independently of the other, forming two separate rings. (Courtesy of Julius Adler.)

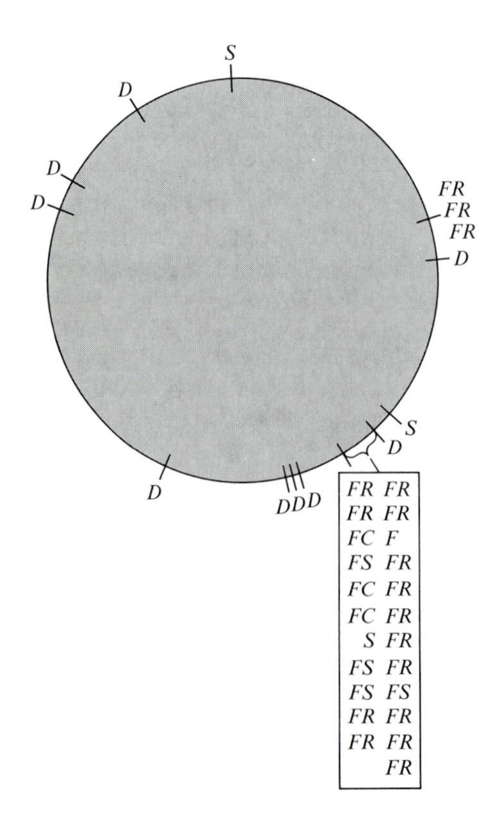

FIGURE 5.10

The single, circular chromosome of *E. coli* indicating the position of the genes that influence swimming. *D* refers to genes that produce proteins that bind with certain substances, thus detecting chemical stimuli. *S* are loci that signal a response by detecting the proportion of binding sites occupied. *F* loci are mutants that influence the rotor itself (*FR*), the switch that determines the direction that the rotor turns (*FS*), and genes that control the switch (*FC*). (From "Behavioral genetics in bacteria" by J. S. Parkinson. Reproduced, with permission, from *Annual Review of Genetics*, 11, 397–414. Copyright © 1977 by Annual Reviews, Inc. All rights reserved.)

disrupt the behavior. An automobile, which requires thousands of parts for its normal functioning, makes a good analogy. If any one part breaks down, the car may not run properly. In the same way, single genes can drastically affect behavior that is normally influenced by many genes.

Paramecia

Paramecia, like bacteria, are one-celled organisms, but they are larger and their movement is more obvious. They are covered by cilia, which propel the cell forward or backward. Paramecia avoid certain chemical and thermal stimuli by backing up and then swimming forward in a new direction. Paramecia are also interesting because of their unusual mode of reproduction. They are diploid (that is, they have a pair of chromosomes), unlike bacteria, which have only one chromosome. Diploidy would complicate molecular genetic analysis in paramecia, except for the fact that they can reproduce by a process called *autogamy*. Autogamy is self-sexual reproduction, in which identical gametes are produced and fertilize each other, producing completely homozygous organisms in just a few hours.

A mutagenic agent is fed to paramecia and they are allowed to reproduce by autogamy, thus passing on mutations to their offspring. Behavioral mutants are screened from thousands of such organisms. Over three hundred behavioral mutants have been isolated, and 20 genes have been implicated in the avoidance behavior of paramecia (Kung et al., 1975). For example, in some solutions, wild paramecia show repeated avoidance reactions. Some mutants, however, cannot swim backward; they are called *pawn* mutants (after the chess piece that can only move forward). Over 150 different pawn mutants have been identified, involving 62 different mutants at only three loci. Other mutants include: *paranoiac* (prolonged backward movement), which involves mutations at any of five different loci; *spinner* (spins in place in a certain solution), involving one locus; and *sluggish* (very slow mover) involving one locus. Figure 5.11 shows tracks of the movements of these strains during periods of about 10 seconds.

Analyses of avoidance behavior in paramecia have focused on the membrane and its electrical properties. For example, the pawn mutants have a defect in the permeability of the membrane for a particular chemical involved in the electrical response. Although researchers have done some molecular analyses of specific gene products (enzymes) underlying these behavioral mutants, such studies have not advanced as far as those of bacteria.

Round Worms

The Nematode round worm is intermediate in complexity between single-celled organisms and complex metazoans, such as fruit flies, mice, or

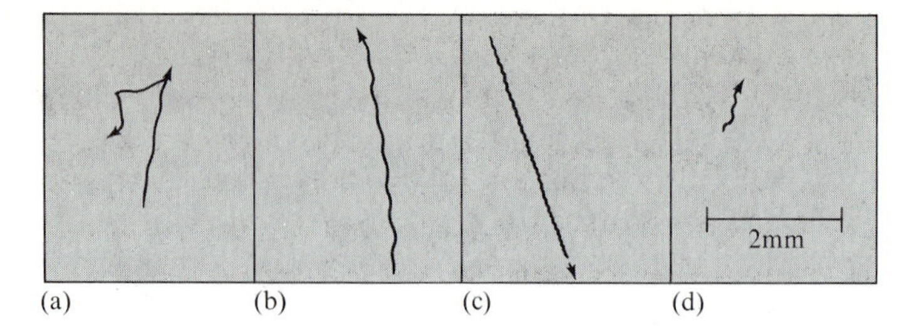

FIGURE 5.11

Movements of various paramecia strains. (a) Wild-type: spontaneous avoidance reactions, which cause turns. (b) Pawn: forward motion with no avoidance reactions. (c) Paranoiac: sustained, rapid backward motion. (d) Sluggish: very slow movement. (Adapted from "Genetic dissection of behavior in *Paramecium*" by C. Kung et al. *Science,* 188, 898–904. Copyright © 1975 by the American Association for the Advancement of Science.)

human beings. It is about 1 millimeter long, and it has about six hundred cells, half of which are nerve cells. Most are self-fertilizing, and about 100,000 descendants can be produced in one week. Genetic dissection of the Nematode's nervous system has given this lowly worm the distinction of being the only organism for whom we know the entire design of the nervous system. Behaviorally, about forty chemotaxis mutants have been isolated, involving about fifteen loci (Ward, 1977). About half of these behavioral mutants have observable differences in the anatomy of their nervous systems. As in paramecia, the gene products underlying these behavioral mutants are beginning to be identified.

Drosophila

Benzer studied behavioral mutants in *Drosophila* because of the great store of available genetic information. Long lists of *Drosophila* behavioral mutants have been compiled (e.g., Grossfield, 1975). They include: *sluggish* and *hyperkinetic* mutants; *wings-up* cannot fly because of a defect in the muscle; *easily shocked,* in which jarring produces behavior similar to an epileptic seizure; *paralyzed* collapses whenever the temperature goes above 28° C; *drop dead* walks and flies normally for a couple of days and then suddenly falls on its back and dies in just a few hours. More complex behaviors are also being studied. One such group involves courtship behavior in male *Drosophila.* Males follow other flies and then vibrate their wings prior to mounting for copulation. Behavioral mutants for aspects of courtship have been found. One male mutant, called *fruitless,* courts males as well as females and does not copulate with the females. Another male mutant cannot disengage from the female after copulation and is given the dubious title *stuck* (Hall, 1977a).

Benzer used an ingenious trick to narrow the search for mutant genes to the X chromosome. His trick also made possible the use of the important tool referred to as genetic mosaics, which we will discuss in the next section. Although the method is somewhat complicated, the goal is simply to create *Drosophila* with mutant genes on the X chromosome. Like human beings, male *Drosophila* have one X and one Y chromosome, and females have two X chromosomes. However, unlike the human situation, flies with two X chromosomes *and* one Y chromosome are fertile females. We will consider the special case of two *attached* X chromosomes that are inherited as a pair.

Suppose that we induce mutations among male *Drosophila* and cross them with nonmutated females with a double-X chromosome and a Y chromosome. Table 5.2 shows the types of offspring from such a cross. Two kinds of offspring do not survive—ones that receive the double-X from the mother and an X from the father, and offspring that receive the Y from the mother and the Y from the father. This leaves only two kinds of offspring. One type includes XY males with the X chromosome from the father, which thus show any recessive mutations induced on the X chromosome of their father. Note that this particular male receives the Y from his nonmutated mother. The other type of offspring is a female like her mother (double-X and Y). She received the double-X from her mother, and cannot express mutations on the X chromosome because she has not received an X chromosome from her father. However, such a female can express mutations on the other chromosomes from her father. But these characters will be expressed only if they are dominant because she receives only one autosome of each pair from her mutated father. In summary, the males from such a cross express all of the recessive mutations on the X chromosome and the females express only the dominant mutations on the autosomal chromosomes.

In his early work, Benzer (1967) screened these offspring for their response to light. Normal, wild-type *Drosophila* are positively *phototactic*—that is, they move toward light. Two mutant male offspring from the above cross were found to be nonphototactic (did not move toward light). Analyses indicated that a single gene on the X chromosome was responsible for this strange behavior. For example, when these males were mated to double-X and Y females, all of their sons exhibited the same behavior. The mechanism here is the same as that illustrated in Table 5.2. The only viable male off-

TABLE 5.2
Offspring from mating of normal male and attached-X female *Drosophila*

		Male Gametes	
		X	Y
Female Gametes	$\widehat{XX}$	$\widehat{XX}$X (nonviable)	$\widehat{XX}$Y (attached-X female)
	Y	XY (normal male)	YY (nonviable)

spring from such a cross received the X chromosome from their father, and express the recessive mutated gene on the X chromosome. Females do not receive the X chromosome from their father. These results indicated that nonphototaxis was due to a gene mutation on the X chromosome. Benzer and Y. Hotta (1970) extended this work by describing a series of nonphototactic mutants with problems of the retina. All of the mutant genes were located at five loci on the X chromosome. Mutant analysis has now revealed more than one hundred different genes involved in the structure of the *Drosophila* eye (Ready, Hansen, and Benzer, 1976). Other behavioral mutants have also been studied. For example, 48 mutants defective in flight behavior have been found (Homyk and Sheppard, 1977). The mutations were mapped to 34 different sites on the X chromosome.

Genetic Mosaics

Another method developed by Benzer to dissect behavior more precisely is the analysis of *genetic mosaics*. Genetic mosaics are individuals with different genes in various cells of the body. Normally, of course, life begins as a single cell, and the genetic material in that cell is replicated in every other cell of the body. In addition to the female with double-X and Y chromosomes, *Drosophila* have another genetic curiosity that permits mosaic analysis. There is an unstable, ring-shaped X chromosome that is frequently lost soon after fertilization. (See Figure 5.12.) Females with one normal X and one unstable X become mosaics because some cells lose the unstable X, and thus have only one X chromosome, compared to other cells with two X chromosomes. The cells with a single X chromosome will express all the genes on that chromosome. Moreover, in *Drosophila*, individuals with a single X chromosome are male and those with two X chromosomes are female. Thus, these mosaics are composites of male and female cells, and are called *gynandromorphs* (from the Greek, meaning "characteristics of both sexes"), or XX-XO mosaics.

The XX and XO cells of gynandromorphs are not randomly intermingled; large continuous areas with many cells are of the same genotype. By comparing many such XX-XO mosaics, it is possible to isolate the parts of the body responsible for certain behaviors. For example, work by Hotta and Benzer (1976) with 477 mosaics indicates that if the head is male, courtship behavior is male. A mosaic with a male head will follow other flies and vibrate its wings regardless of the sex of other parts of its body. Subsequent work indicates that only half of the brain must be male to produce male courtship behavior (Hall, 1977b). Hotta and Benzer also examined 130 mosaics that were successful in following and vibrating their wings and in copulation. The mosaic analysis suggested that successful copulation requires a male thorax (the part of an insect's body between the head and abdomen), as well as a male head. Of course, sex isn't all in the head; male genitals are also required for successful copulation.

Step one: Establish males with mutagenized X chromosome. (See Table 5.2.)

Step two: Establish female mosaics from females with unstable $X(X_u)$ chromosome.

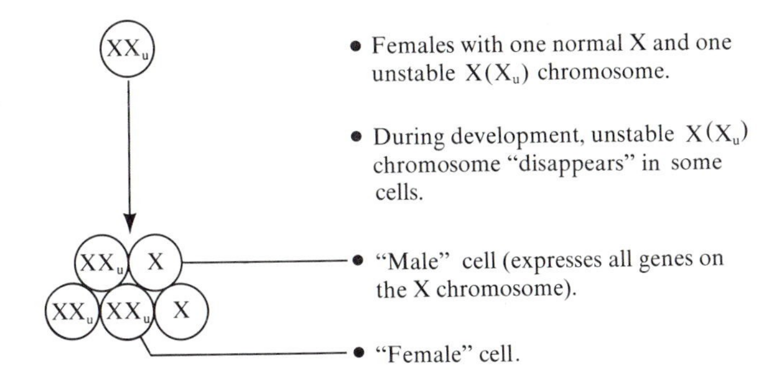

Step three: Mate males with mutagenized X and females with unstable $X(X_u)$.

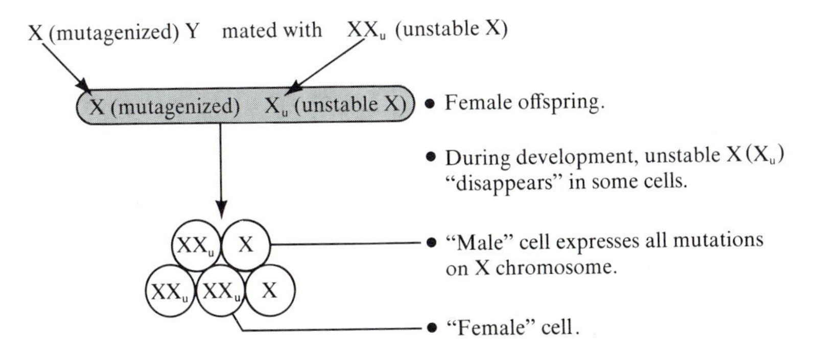

FIGURE 5.12
Mosaic analysis using *Drosophila*.

The final step is to mate the females with the ring-shaped X chromosome to the mutant males described above. (See Figure 5.12.) These males pass on their X chromosome, with its mutant gene, to the female offspring. Half of the time, these female offspring also receive one of the ''disappearing'' X chromosomes from their mother. In this way, the mosaics are not just male-female mosaics. They are also mosaic for the recessive mutated genes on the X chromosome. ''Marker'' genes for body color, for example, can also be inserted on the X chromosome to permit easy identification of the particular ''male'' and ''female'' cells.

This method has been applied to many behavioral mutants to analyze which body parts must be mutant for the mutant behavior to occur. In an early study, mosaics were created using a gene that causes a defective visual response to light. Although there is considerable variability among flies, they normally move away from gravity (negative geotaxis) and toward light (positive phototaxis). Thus, in a dark tube, most normal flies will climb straight up the tube whether or not there is a light at the top. A certain mutant gene produces a diminished response to light. When this mutant gene is present in mosaics, some flies will have one mutant eye and one normal eye. When these mosaics are put in a dark tube, they also climb straight to the top because their geotaxis is not affected. (See Figure 5.13a.)

However, when a light is placed at the top of the tube, the mosaic flies attempt to equalize the light coming into the two eyes, and thus keep their defective eyes toward the light. As a result, the mosaic flies climb up the tube in a spiral, keeping their mutant eyes closer to the light (Figure 5.13b). Spiral climbing occurs regardless of the amount of normal female tissue present, as long as one eye is normal and the other is mutant. Thus, the mutation is specific and isolates the defective visual response to the eye itself. As Benzer has indicated, phototaxis is a complex behavioral response: "Light is absorbed by a pigment in the receptor cell, producing neural excitation, transmission at synaptic junctions, integration in the central nervous system involving comparison with other inputs, and generation of appropriate motor signals such that the fly walks in a particular direction" (1967, p. 1118). A mutation resulting in a defect in any of these structures or processes could lead to a change in phototaxis. As indicated earlier, mutants for more

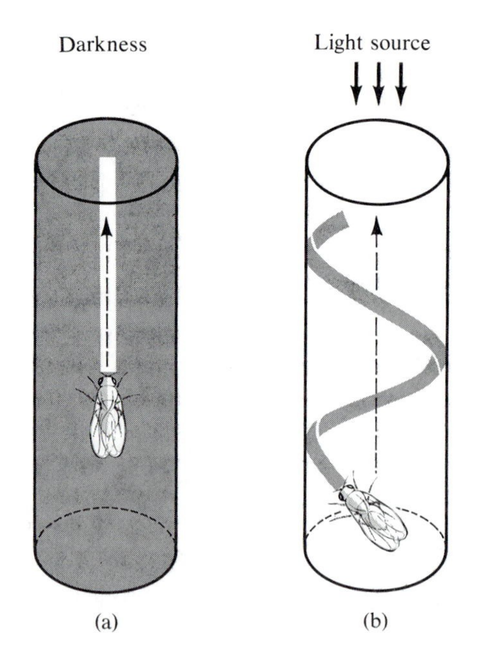

FIGURE 5.13

(a) Behavior in darkness displayed by a mosaic fly with normal vision in one eye and defective vision in the other. Since the fly is negatively geotactic, it climbs straight up. (b) When light is shining from above, the same fly turns its defective eye toward the light and climbs in a helical path. (From Hotta and Benzer, 1970.)

than one hundred different genes affecting the eye have been isolated. These are now used in mosaic analyses to dissect the visual system and to provide information about the normal sequence of events.

Mutant and mosaic analysis has also been applied to learning. A *Drosophila* mutant defective in avoidance learning has been isolated (Dudai, et al., 1976). Flies were taught to avoid a particular odor that had been coupled with shock. Although flies avoided the odor only 30 percent of the time, the learning lasted a day (Quinn, Harris, and Benzer, 1974). After screening about five hundred lines of flies with mutagenized X chromosomes, one mutant line was found that could not learn to avoid the odor associated with shock even though it had normal sensor and motor behavior. The mutant, called *dunce,* is now being subjected to mosaic analysis in order to pinpoint the source of the learning problem.

Genetic Dissection and Development (Fate Mapping)

The method of genetic dissection of behavior promises to assume even greater importance in behavioral genetics. Although mice are much slower breeders than bacteria, paramecia, round worms, and *Drosophila,* genetic dissection of the nervous system of mice has also begun (Caviness and Rakic, 1978). The more fine-grained mosaic analysis has been limited primarily to *Drosophila,* however, because it is easy to generate XX-XO mosaics (Hall, 1977a).

A recent review of genetic dissection research concludes that the next phase of research will begin to unravel the developmental interaction between genes and behavior: "Many of the mutants that have been selected for behavioral alterations are altered in the development of parts of the nervous system. This means that eventual understanding of the effects of these mutants must be sought by understanding the gene control of development. This is the most challenging problem confronting the analysis of neurological mutants" (Ward, 1977, p. 444).

One step in this direction is mosaic fate mapping. In 1929, Alfred Sturtevant (see Benzer, 1973) proposed that mosaics could be used to "map" what happens to the cells of the *blastoderm*—the surface of the blastula that is a hollow ball consisting of a single layer of cells surrounding the yolk in the young embryo. In XX-XO mosaic blastoderms, there is a dividing line between the areas of XX and XO cells. The orientation of this boundary is random, which means that the likelihood that this boundary will pass between two cells is proportional to the distance between them. We also know from histological (tissue) analysis that the location of a cell in the blastoderm determines its fate throughout development (Baker, 1978). Thus, in the adult mosaic, body parts that are usually of different genotypes (XX versus XO) are likely to be farther apart in the blastoderm. Sturtevant scored pairs of body parts in 379 mosaics for the frequency with which one part was XX and the other XO. The more often this happened, the farther apart were the ancestor cells in the blastoderm, as determined by histological analysis (Gehring, 1976). Although much remains to be learned about the genetic changes involved in such differentiation of tissue, one recent model suggests that the genetic mechanism involves irreversible repression of previously active genes, rather than selective gene activation (Caplan and Ordahl, 1978).

Benzer and Hotta extended this approach to behavior (Benzer, 1973). First, using 703 mosaics, they constructed an anatomical fate map, as shown in Figure 5.14. The distances between the origins of various parts are scored in *sturts,* in honor of Sturtevant. One sturt represents a 1-percent probability that the two structures will be of different genotypes. For example, legs 1 and 2 were of different genotypes 10 percent of the time; legs 2 and 3 differed 10 percent of the time; and legs 1 and 3 differed 20 percent of the time.

Benzer and Hotta then began to relate behavioral characteristics to this fate map. For example, the hyperkinetic mutant shakes all six of its legs when anesthetized. They showed that each leg's shaking was independent of that of the other legs. When they mapped the shaking of the legs in the

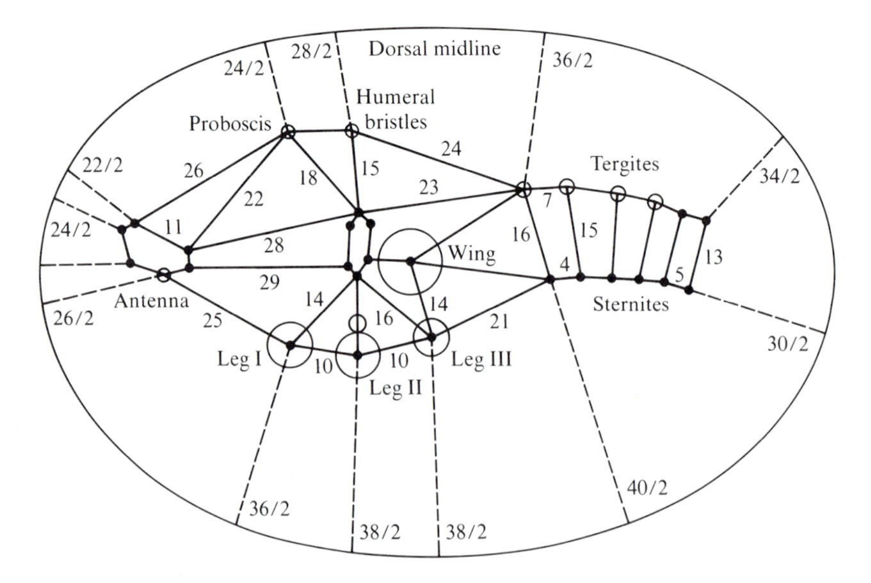

FIGURE 5.14

A blastoderm fate map of the external body parts of *Drosophila*, based on the probability that two parts of the body are of different genotypes in mosaic individuals. (From "Genetic dissection of behavior" by S. Benzer. Copyright © 1973 by Scientific American, Inc. All rights reserved.)

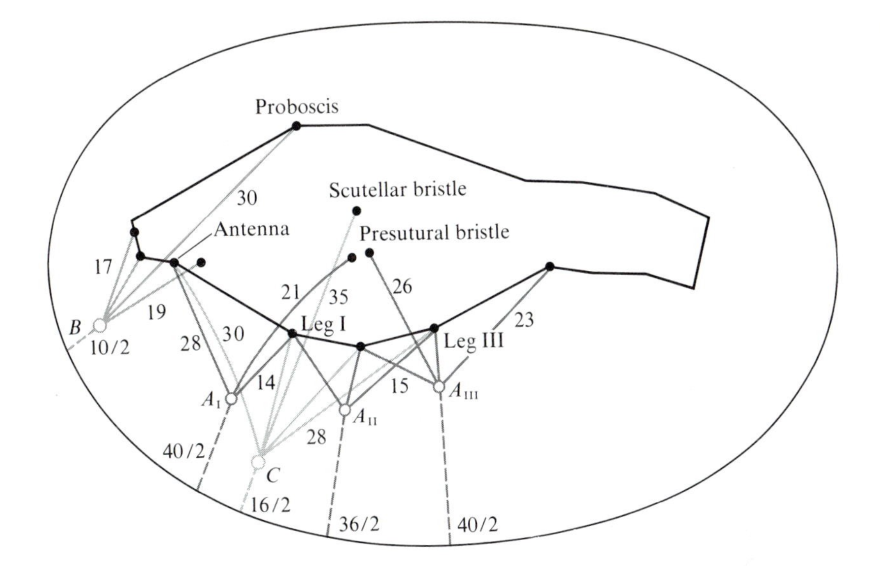

FIGURE 5.15

A fate map showing the sites on the blastoderm at which mutant genes (A) affect leg shaking in the hyperactive mutant. A similar map exists for the other side of the blastoderm, which affects shaking of the three legs on the other side of the body. The blastodermal site of the (B) drop-dead and (C) wings-up mutant genes are also shown. (From "Genetic dissection of behavior" by S. Benzer. Copyright © 1973 by Scientific American, Inc. All rights reserved.)

hyperkinetic flies, they found that the location of the original cells in the blastoderm is near the location for the appropriate leg, but always below it, in a region that is the origin of part of the nervous system. (See Figure 5.15.) The embryonic locations of the *drop-dead* and the *wings-up* mutant genes mentioned earlier are also shown in Figure 5.15. Mutant and mosaic analyses have been conducted for other behaviors, such as flying (Homyk, 1977), and we can expect many more applications of these important techniques in the future (Flanagan, 1977).

Benzer has summarized this approach as follows:

> In tackling the complex problems of behavior the gene provides, in effect, a microsurgical tool with which to produce very specific blocks in a behavioral pathway. With temperature-dependent mutations the blocks can be turned on and off at will. Individual cells of the nervous system can be labeled genetically and their lineage can be followed during development. Genetic mosaics offer the equivalent of exquisitely fine grafting of normal and mutant parts, with the entire structure remaining intact. What we are doing in mosaic mapping is in effect "unrolling" the fantastically complex adult fly, in which sense organs, nerve cells and muscles are completely interwoven, backward in development, back in time to the blastoderm, a stage at which the different structures have

> not yet come together. Filling the gaps between the one-dimensional gene, the two-dimensional blastoderm, the three-dimensional organism and its multidimensional behavior is a challenge for the future. (Benzer, 1973, p. 15)

GENES AND INDIVIDUAL DIFFERENCES

In the previous chapters on evolution and Mendelian genetics, a basic message has been genetic variability. The knowledge that genes are sequences of DNA nucleotide bases also adds to our understanding of variability. The ultimate source of genetic variability is mutation, random changes in nucleotide base sequences. Although most mutations are harmful, they are the price paid for genetic variability and the possibility of evolutionary change. We also know that all genes are not simply templates of protein production. Regulator genes control the production of proteins, and mutations of these genes can result in additional genetic variability.

As mentioned in the previous chapter, over one-third of human structural genes have two or more alleles. We know this because these genes produce detectably different enzymes. Moreover, we are learning that even more genetic variation exists for structural genes that do not detectably alter the function of enzymes, at least given existing environmental demands. Harry Harris and his coworkers frequently discover variation in amino acid sequences for functionally equivalent enzymes. For example, in one early series of observations, 3 of 12 enzymes that were examined were polymorphic—that is, several different alleles code for variants of these enzymes. In summarizing results from his work and the work of others, Harris concluded:

> Thus quite a high degree of individual differentiation in enzymic makeup can already be demonstrated from this limited series of examples, and it is of interest that most of this is attributable to variation in molecular structure of the enzymes. This must surely be only the tip of the iceberg, and one may plausibly imagine that in the last analysis every individual will be found to have a unique enzymic constitution. (Harris, 1967, p. 211)

SUMMARY

Perhaps the most exciting advance in biology in this century has been in understanding Mendel's elements. Early in the century, inborn errors of metabolism suggested the "one-gene, one-enzyme hypothesis." Knowledge of the double-stranded DNA molecule followed. The structure of DNA relates to its dual functions of self-replication and protein synthesis. Genes are sequences of nucleotide pairs long enough to code for a specific sequence of amino acids in a particular polypeptide. In addition to structural genes that serve as templates for protein production, there are also regulator genes that control the transcription of structural genes.

Mutations are changes in the nucleotide bases of DNA. Much of our early knowledge of genes was the result of experimentally induced mutations. Mutations are now being used to dissect behavioral sequences in many organisms, such as bacteria, paramecia, round worms, and *Drosophila*. The combination of mutants and mosaics in *Drosophila* permits powerful analyses of the fine structure of behavioral development.

6

Chromosomes

Advances in the field of cytology—the study of the cell and its contents—led to a major breakthrough in understanding the physical nature of heredity. In the mid-nineteenth century, it was generally accepted that cells are the basic units of living organisms. Aided by new knowledge about the chemistry of dyes, cytologists were able to stain the contents of cells to make them more visible for study. It was soon found that a portion of the cell, the *nucleus,* contains a number of small rod-shaped bodies called *chromosomes* ("colored bodies") because they can be stained by particular dyes. The number of chromosomes is the same in all cells of an organism, except for the sex cells (the sperm and eggs, which are called gametes). In the non-sex cells (called somatic cells), chromosomes come in pairs. The total number of chromosomes in each somatic cell is called the diploid number. In sex cells, only half of the chromosomes—one member of each pair—are represented. The number of chromosomes in each sex cell is referred to as the haploid number. All individuals of a species have the same number of chromosomes in each cell, although the number varies widely from one species to another. For example, the pea plant has 7 pairs of chromosomes, wheat has 21, mosquitoes have 3, fruitflies have 4, carp have 52, mice have 20, dogs have 39, and humans have 23. It gradually became clear that chromosomes were involved in heredity.

MENDEL AND CHROMOSOMES

Mendel did not know that his elements were parts of chromosomes. His law of independent assortment states that the elements, or alleles, for two different traits are inherited independently. However, in Chapter 4 we noted that Mendel would have been in for a surprise if he had happened to study two traits affected by genes located close together on the same chromosome. In this case, the traits would have been inherited together, and he would not have found independent assortment. The dihybrid segregation ratio of 9 : 3 : 3 : 1 would have been closer to a 3 : 1 ratio. As indicated in Figure 6.1, the ratio would be 3 : 1 if the two loci are so closely linked that no recombination occurs. If the A and B loci were less tightly linked, some recombinants (such as $A_1A_1B_2B_2$) would occur in the F_2 progeny.

It is now known that the pea plant has seven pairs of chromosomes. So given that Mendel studied seven traits, the probability is very small (0.006

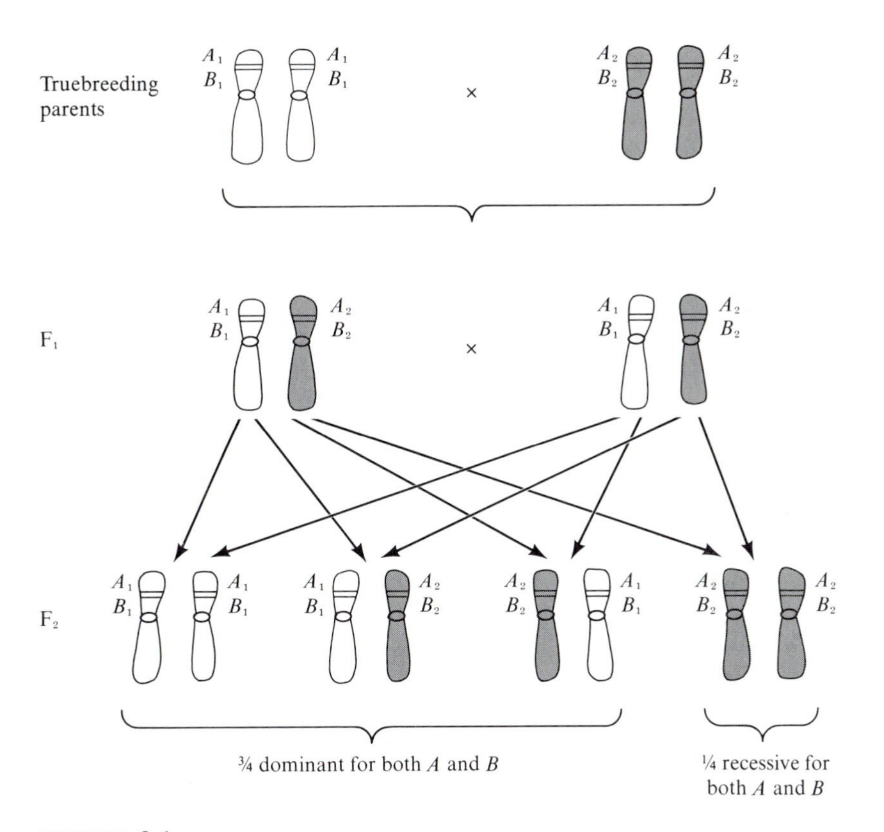

FIGURE 6.1

Dihybrid segregation ratio if genes are very closely linked on the same chromosome. (A_1 and B_1 are dominant.)

Box 6.1
Why Didn't Mendel Find Linkage?

Mendel studied seven single-gene traits of the pea plant. The pea plant has only seven chromosomes. The probability that each of the seven genes selected at random would be on a different chromosome is only 6 in 1,000. Therefore, why didn't Mendel observe any dihybrid segregation ratios other than 9:3:3:1?

The answer to the paradox is *not* that the seven genes that Mendel studied are on seven different chromosomes (Blixt, 1975). All seven genes have now been mapped to specific chromosomes, and we know that two of them are linked closely on chromosome 4. These loci are linked sufficiently close to yield a significant departure from a 9:3:3:1 ratio. However, given seven traits, there are twenty-one possible dihybrid crosses. Only one of these would have shown this departure from the expected 9:3:3:1 dihybrid segregation ratio. Mendel apparently did not perform that particular dihybrid cross.

In addition to the two tightly linked genes, Mendel reported dihybrid ratios for some other genes that were on the same chromosome. However, unless genes are close together on the same chromosome, they will recombine due to the process of *crossing over,* in which chromosomes exchange parts (as discussed in detail later in this chapter). The greater the distance between loci, the greater the frequency of recombination. The genes that Mendel happened to study are, in fact, so far apart on their chromosomes that they would not have shown a significant departure from the expected dihybrid segregation ratios for unlinked genes.

As a sidelight and prelude to Box 6.3, we should note that departures from expected dihybrid segregation ratios are not just a nuisance in Mendelian genetics. They provide an important tool for determining whether genes are, in fact, located on the same chromosome. Departures from expected segregation ratios suggest that the genes are linked. The less recombination that occurs, the closer together are the genes on the chromosome.

that all seven genes were located on different chromosome pairs. Why, then, didn't Mendel find dihybrid segregation ratios that indicated linkage? The answer to that question may be found in Box 6.1.

CELL DIVISION

Cell division involves two processes. The first involves duplication of cells. We begin life as a single cell and end up with about 10^{14} cells in our adult bodies. Second, a sample of half of our genes is transmitted from one generation to the next in our sex cells (sperm and eggs). The two types of cell division involved in these processes are *mitosis* and *meiosis,* respectively. (See Figure 6.2.)

Mitosis

In the process of growth, cells divide into two "daughter cells," each of which then later divides into two more, and so forth. Many cells in our bodies go through a typical cell cycle lasting about a day. (See Box 6.2.) On the other hand, some cells, such as nerve cells, rarely divide. Others (e.g., liver cells) divide only when injured; cancer cells divide extremely rapidly.

Study of the chromosomes revealed that a remarkable series of changes takes place during cell division. The major features of this process of mitosis are illustrated in the left half of Figure 6.2. Prior to the splitting of the cell, the chromosomal material is replicated, and spindle fibers become attached to the centromere of each chromosome (a genetically inactive chromosome region). During cell division, half of the material goes into one daughter cell, and half into the other. Since different chromosomes are somewhat distinctive in shape and size, researchers were able to determine that each daughter cell receives an equivalent chromosomal complement. This distinctiveness of chromosomes also resulted in the conclusion that chromosomes exist in pairs. In Figure 6.2, one chromosome of each pair is of paternal origin; the other is of maternal origin. Thus, two pairs of homologous chromosomes, or matched sets, are shown in the figure.

Meiosis

The process by which one set of chromosomes is contributed by each parent is called *meiosis*. Meiosis essentially consists of the splitting of a cell into two, without the prior doubling of chromosomal number that occurs in mitosis. (See Figure 6.2.) More precisely, meiosis begins with doubled chromosomes. Each cell divides twice, so that each original diploid cell yields four gametes, each with a haploid number of chromosomes. One member of each pair of homologous chromosomes is drawn into each haploid daughter cell before the division is complete.

The haploid set of chromosomes included in any one gamete, however, is not necessarily the same haploid set that the individual received from its mother or father. A reshuffling occurs, so that an individual transmits to its offspring some of the chromosomes received from its mother and father. This reshuffling of chromosomes can create considerable genetic variability. For example, your gametes (sperm or eggs) can have any of over 8 million possible haploid combinations (2^{23}) of your 23 pairs of chromosomes. Any of these haploid combinations can fertilize, or be fertilized by, the 8 million possible combinations of gametes of your mate. This means that you and your mate can potentially create 64 trillion chromosomally different zygotes. Meiosis results in even more genetic variability by means of the process of crossing over, discussed in the following section.

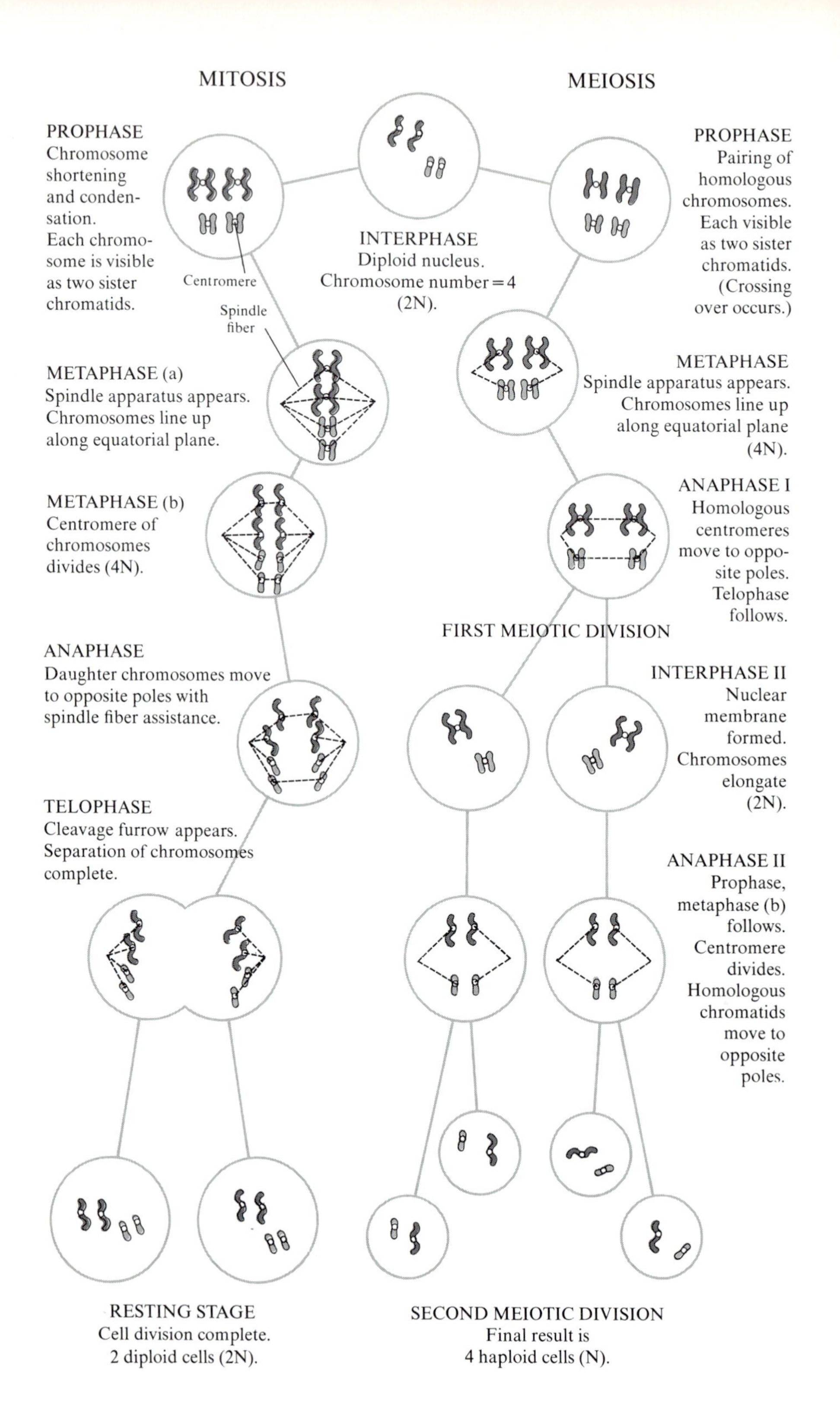

MITOSIS
MEIOSIS
PROPHASE
Chromosome shortening and condensation.
Each chromosome is visible as two sister chromatids.
Centromere
Spindle fiber
INTERPHASE
Diploid nucleus.
Chromosome number = 4 (2N).
PROPHASE
Pairing of homologous chromosomes.
Each visible as two sister chromatids.
(Crossing over occurs.)
METAPHASE (a)
Spindle apparatus appears.
Chromosomes line up along equatorial plane.
METAPHASE
Spindle apparatus appears.
Chromosomes line up along equatorial plane (4N).
METAPHASE (b)
Centromere of chromosomes divides (4N).
ANAPHASE I
Homologous centromeres move to opposite poles.
Telophase follows.
FIRST MEIOTIC DIVISION
ANAPHASE
Daughter chromosomes move to opposite poles with spindle fiber assistance.
INTERPHASE II
Nuclear membrane formed.
Chromosomes elongate (2N).
TELOPHASE
Cleavage furrow appears.
Separation of chromosomes complete.
ANAPHASE II
Prophase, metaphase (b) follows.
Centromere divides.
Homologous chromatids move to opposite poles.
RESTING STAGE
Cell division complete.
2 diploid cells (2N).
SECOND MEIOTIC DIVISION
Final result is 4 haploid cells (N).

Box 6.2
Typical Cell Cycle

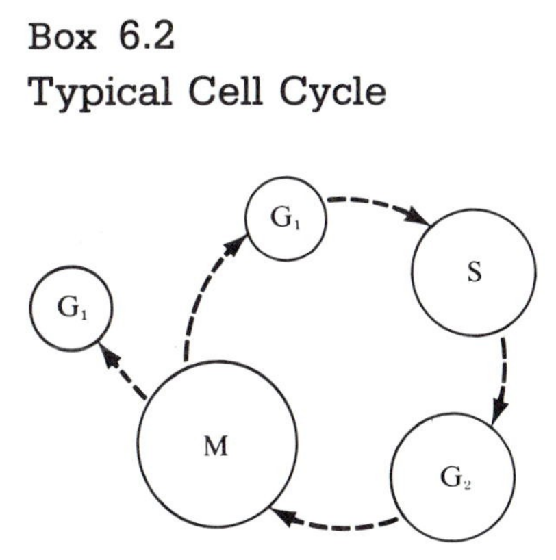

G_1 is a period of initial cell growth, lasting about 8 hours.

S is a period (about 6 hours) in which chromosome replication occurs. Since genes are carried on chromosomes, and since DNA is the genetic material, this is a period of active DNA synthesis.

G_2 is a second period of growth, which lasts about 5 hours.

M is the final phase, in which cell division occurs. This is the shortest phase (about 1 hour) and yields two identical daughter cells that begin the cycle over again. During this phase, the chromosomes become visible under the light microscope and various stages of mitosis may be observed.

In short, mitosis occurs in almost all cells of the body. Its function is to duplicate cells. Meiosis occurs only in the ovaries and testes, which produce the sex cells. Its function is to shuffle chromosomes and produce gametes with single chromosomes rather than chromosome pairs.

CROSSING OVER

Understanding of the chromosomal basis of heredity helped to delineate exceptions to the law of independent assortment, as mentioned in Box 6.1. It was evident long ago that there must be more genes than there are chromo-

---

FIGURE 6.2

A comparison of mitosis and meiosis. N is the number of chromosomes of each type. (From James D. Watson, *Molecular Biology of the Gene*, 3d ed., Benjamin/Cummings Publishing Company, Menlo Park, CA. Copyright 1976.)

somes. In other words, each chromosome must contain a number of loci. If two characteristics are determined by loci on the same chromosome, the alleles at the two loci may not assort independently.

Genes on the same chromosome are said to be *linked* if their alleles do not assort independently. However, linkage does not mean that the traits occur together in a population. Linkage refers to loci, not to alleles for segregating loci. For example, hemophilia and color-blindness are both determined by recessive alleles of genes on the same (X) chromosome. However, a hemophiliac is no more likely than anyone else to be color-blind because the alleles for hemophilia and color-blindness are not inherited together even though their loci are on the same chromosome.

Alleles for two loci on the same chromosome may be separated in a population by the process of *crossing over*. During the first stage of meiosis, homologous chromosomes line up pair by pair. (See Figure 6.2.) Each member of each pair duplicates, and the duplicates (chromatids) separate, except at the centromere. The chromosomes frequently come into contact and exchange parts or "cross over." Crossing over may occur within a locus, as well as between loci. That is, crossing over can break up DNA at any point, not just at some point that divides two loci. This is another important source of genetic variability, because the sequence of nucleotide bases can change if chromosomes trade bases within a locus (thus creating new alleles).

The "life expectancy" of a genetic unit is related to its size. A whole chromosome will survive intact only one generation if the chromosome, on the average, undergoes one crossover every time a gamete is formed. A smaller genetic unit, say 0.002 of a chromosome (close to the size of the average gene), has a 0.9998 chance of arriving intact from either parent.

One way of understanding the important contribution that crossing over makes to genetic variability is to consider your parents. Your father had the same 23 pairs of paternal and maternal chromosomes in every cell of his body, except his sperm. When he formed sperm through meiosis, some of the homologous chromosomes from his mother and father crossed over, so that chromosomes in his sperm were a unique patchwork of his mother's and father's chromosomes. The patchwork chromosomes in the sperm that fertilized the egg that produced you are now half the chromosomes in every cell of your body. Your gametes, in turn, are recombined chromosomes from your father and mother.

Figure 6.3 is an illustration of this process for one pair of chromosomes only. It should be remembered that the same events may be occurring at the same time for all other chromosome pairs. The maternal chromosome, carrying the alleles A_1, B_2, and C_1, is represented in white; the paternal chromosome, with A_2, B_1, and C_2, is gray. When each homologous chromosome duplicates to form sister chromatids, these chromatids may cross over one another. During this stage, the chromatids can break and rejoin. Each of the chromatids will be transmitted to a different gamete. Consider only the A

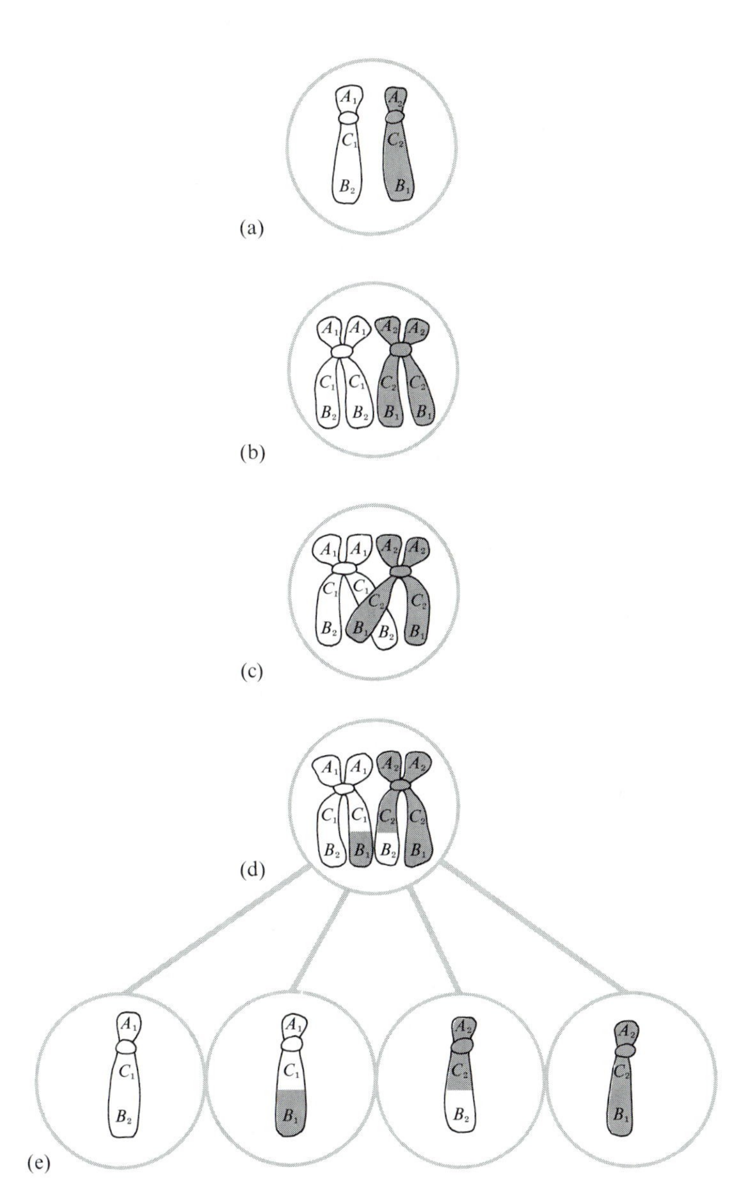

FIGURE 6.3

Diagrammatic illustration of crossing over—the mutual exchange of material by homologous chromosomes. (After "The inheritance of behavior" by G. E. McClearn. In L. G. Postman, ed. *Psychology in the Making*, Copyright © 1963. Used with permission of Alfred A. Knopf, Inc.)

Box 6.3
Mapping Genes

If two genes are closely linked on the same chromosome, they will not assort according to the 9:3:3:1 segregation ratio predicted by Mendel's law of independent assortment. Moreover, the closer the genes are on a chromosome, the less likely they are to be separated by crossing over. Thus, we can map the distance between two genes on a chromosome by determining the extent to which recombination occurs. For example, in Figure 6.3 the *A* locus must be closer to the *C* locus than it is to the *B* locus, if recombination is less frequent between *A* and *C* than between *A* and *B*. For plants and nonhuman animals, controlled test crosses can be obtained to determine the extent of recombination. For example, consider genes *A* and *B*, each with two alleles. A test cross involves matings between "doubly heterozygous" individuals and "doubly homozygous" ones. For example, consider the cross between $A_1A_2B_1B_2$ (doubly heterozygous) and $A_2A_2B_2B_2$ (doubly homozygous) individuals. When the *A* and *B* loci are on different chromosomes, there will be four types of offspring, each with equal frequency:

		Frequency of Gametes Produced by Doubly Heterozygous Parents			
		$\frac{1}{4}A_1B_1$	$\frac{1}{4}A_1B_2$	$\frac{1}{4}A_2B_1$	$\frac{1}{4}A_2B_2$
		Offspring			
Gametes for Doubly Homozygous Parents	A_2B_2	$\frac{1}{4}A_1A_2B_1B_2$ +	$\frac{1}{4}A_1A_2B_2B_2$ +	$\frac{1}{4}A_2A_2B_1B_2$ +	$\frac{1}{4}A_2A_2B_2B_2$

However, if *A* and *B* were close together on the same chromosome, the law of independent assortment would not prevail. Suppose that the doubly heterozygous parent had the A_1 and B_1 alleles on one chromosome and the A_2 and B_2 alleles on the other. The doubly homozygous parent has A_2B_2 on both chromosomes. Without crossing over, we would not find any of the "mixed" offspring, $A_1A_2B_2B_2$ or $A_2A_2B_1B_2$. If an offspring has an A_1 allele, then that offspring should also have a B_1 allele (that is, *not* B_2B_2) because the A_1 allele is on the same chromosome as the B_1 allele from the doubly heterozygous parent. Similarly, if the offspring has a B_1 allele, then that offspring should also have an A_1 allele, not A_2A_2. However, crossing over could create these recombinations. *Map distance,* the relative distance between genes on a chromosome, is simply the percentage of recombination. For instance, where *A*

and *B* loci for the moment. As shown at the bottom of Figure 6.3, one gamete will carry the genes A_1 and B_2, as in the grandmother, and one will carry A_2 and B_1, as in the grandfather. The other two will carry A_1 with B_1 and A_2 with B_2. For the latter two pairs, recombination has taken place.

Crossing over of this kind can occur at any place along the chromosome, and the probability of recombination is a function of the distance between the particular genes. In Figure 6.3, for example, crossing over has and B are on different chromosomes, the map distance (percent recombination) is 50—meaning that 50 percent of the offspring are recombinant types. This suggests that there is no linkage, and thus independent assortment. However, suppose the percentage of the two recombinant genotypes was 40 percent instead of 50 percent. The map distance between the A and B genes would be 40 map units. If there is less recombination, the map distance is smaller, and the genes are closer together on the chromosome.

In 1906, Bateson found two genes in the sweet pea that did not assort independently. By 1915, T. H. Morgan et al. summarized results for 85 genes in the fruit fly (*Drosophila*), and showed that they fall into four linkage groups. We now know that *Drosophila* has four chromosomes.

For humans, the problem is more difficult because naturally occurring test crosses must be found. Expectations for recombination must be estimated for a given set of family data. Linkage for the sex chromosome is more easily studied because males have an X and a Y chromosome and females have two X chromosomes. Recessive traits on the X chromosome show a greater incidence in males. Because the gene is recessive, a woman will show the trait only if she is homozygous for those alleles. In males, however, there is no corresponding locus on the Y chromosome, so that a single recessive gene on the X chromosome is expressed. If a gene is on the X chromosome, we would expect to find no father-to-son inheritance because fathers give their sons only the Y chromosome. Other predictions can be made concerning familial resemblances for an X-linked trait. (See Chapter 4.)

Once sex linkage is demonstrated, map distance can be determined by the "grandfather method." Females who are doubly heterozygous for two genes on the X chromosome can be detected because their fathers will show the alleles present on one of their X chromosomes. Recombination can then be studied for the sons of the doubly heterozygous females, and map distance can be determined.

In 1936, Haldane established the distance between the genes for color blindness and hemophilia on the X chromosome. Nearly one hundred genes have now been identified as being located on the human X chromosome. In 1954, J. Mohr demonstrated the first linkage between two human autosomal genes (located on a chromosome other than the sex chromosomes, X and Y), and M. C. Weiss and H. Green localized the first human gene to a specific chromosome. Linkage and mapping research continues to progress rapidly, particularly because of a new mapping technique called somatic cell hybridization. This technique is discussed in Box 6.4.

not affected the relationship between the A locus and the C locus. All gametes are either A_1C_1 or A_2C_2, as in the grandparents, since the crossover did not occur between these loci. Crossing over could occur between the A and C loci, but that would be less frequent than between A and B. Because of this, the crossover gametes occur less often than the noncrossover. Genes located on different chromosomes do, of course, assort at random. These facts have been used as a tool to map genes on chromosomes; see Box 6.3.

HUMAN CHROMOSOMES

Although the chromosomes of *Drosophila* and other organisms were subjected to detailed analyses as early as the 1930s, human cytogenetics lagged far behind. About twenty-five years ago, students were taught that the number of chromosomes in man was 48 (24 pairs). However, after using improved techniques, it was reported in 1956 that the normal diploid chromosome number in man is 46, not 48 (Tjio and Levan, 1956). Since that date, important developments in human cytogenetics have occurred with great rapidity.

Karyotypes

In order to study the chromosomal complement, or *karyotype,* of an individual, a sample of white blood cells (*leukocytes*) is usually obtained and cultured in the laboratory for two or three days. A chemical (phytohemagglutinin) is added to the culture to stimulate growth and cell division. Dividing cells are then exposed to colchicine, a chemical that inhibits the separation of doubled chromosomes. This results in the accumulation of cells at the stage of mitosis, illustrated in Figure 6.2, in which the duplicated chromosomes have shortened and thickened and are still attached at the centromeres. The cells are then washed with a saline solution, resulting in swelling of the cells and spreading out of the chromosomes. When these cells are squashed or air dried, the chromosomes tend to lie in the same optical plane. The cells are then stained and photographed under high-power magnification. The chromosomes in the photograph may then be cut out and rearranged according to their size and the location of the centromeres. The karyotype of a normal male is shown in Figure 6.4.

An international conference was held in Denver, Colorado, in 1960 for the purpose of standardizing the classification of human chromosomes. The resulting "Denver classification" is based on both chromosome length and location of the centromere. If the centromere divides a chromosome into arms of approximately equal length, the chromosome is said to be *metacentric.* If the centromere is very close to one end of the chromosome, the chromosome is referred to as being *acrocentric.* If the centromere is located somewhere between the middle and one end, the chromosome is described as being *submetacentric.*

As indicated in Figure 6.4, the 23 pairs of human chromosomes are classified into seven distinct groups. Group A includes chromosome pairs 1, 2, and 3. These are large chromosomes that can be distinguished from others on the basis of their size and the central location of the centromere. Group B includes chromosome pairs 4 and 5, which are large submetacentric chromosomes. Group C, the largest group, includes chromosome pairs 6 through 12 and the X chromosome, all of which are medium sized and

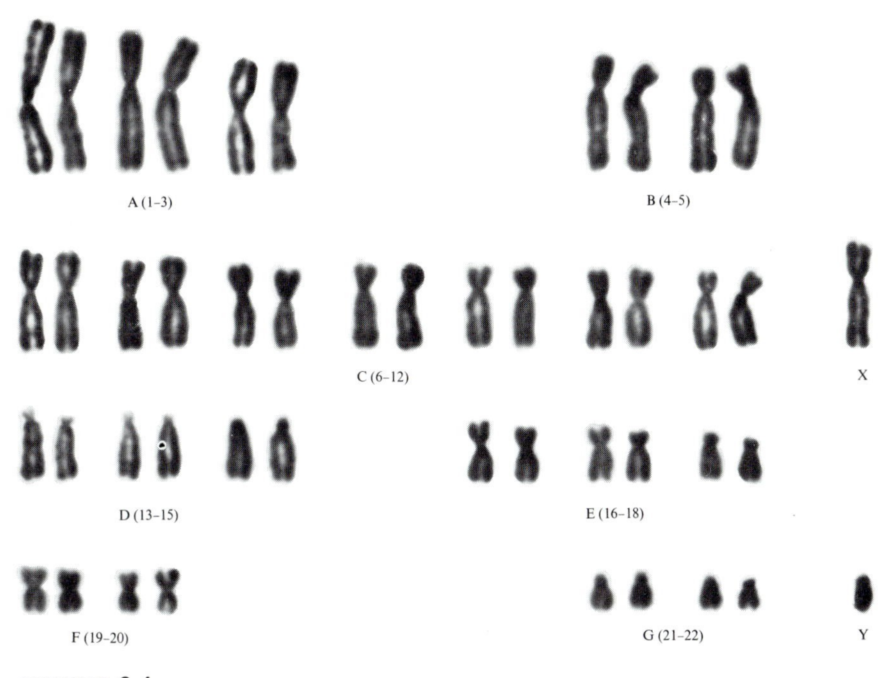

FIGURE 6.4

Male karyotype. (Courtesy of Dr. Margery Shaw.)

submetacentric. Group D includes the medium-sized acrocentric chromosome pairs 13, 14, and 15. Group E chromosome pairs (16, 17, and 18) are relatively short and metacentric or submetacentric, and those in Group F (19 and 20) are metacentric and shorter. Group G chromosomes are very short and acrocentric. This group includes chromosomes 21 and 22, as well as the Y chromosome.

A symbolic system has also been devised for describing the chromosomal complement of an individual. The total chromosome number is indicated first, followed by the sex-chromosome constitution and any autosomal abnormality. A plus or minus sign before a chromosome number or letter indicates that an entire autosome is represented an extra time or is missing, and a question mark indicates uncertainty. This system of nomenclature is illustrated in Table 6.1.

Banding

In 1968, another significant advance was made in the technology of human chromosome identification. In that year, T. Casperson and co-

TABLE 6.1

Nomenclature for human chromosome complements, including aberrant ones

Abbreviation	Description
46, XY	Normal male
46, XX	Normal female
45, X	22 pairs of autosomes, one X chromosome; one sex chromosome missing
47, XXY	22 pairs of autosomes; one extra sex chromosome
45, XY, −C	Male; one chromosome missing in group C
47, XX, +21	Female; one extra chromosome number 21
45, XX, −?C	Female; one autosome missing, probably in group C
45, X/46, XX	A mosaic, some cells like those of a normal female and some missing an X chromosome

SOURCE: After Hsia, *Human Developmental Genetics*. Year Book Medical Publishers. Copyright © 1968. Adapted from Chicago Conference, Standardization in Human Cytogenetics, Birth defects, Original Article Series II: 2. New York: The National Foundation, 1966.

workers reported that metaphase (middle stage of mitosis) chromosomes can be stained by fluorescent DNA-binding agents to yield a pattern of up to 320 light and dark bands when viewed with a fluorescent microscope. Other banding techniques have since been found that yield generally similar results. The most widely used techniques are those that do not require special equipment such as fluorescent microscopes. In 1977, it was discovered that many more bands can be detected when chromosomes are stained during earlier stages of mitotic division than in metaphase (Sanchez and Yunis, 1977). Figure 6.5 compares the patterns of human chromosomes at metaphase and prophase. In the left half of each chromosome (the left chromatid), we see the banding picture as of 1972. The chromatids at right depict the banding pattern using techniques available in 1977.

It is not clear what the bands represent. At first, they were thought to be active genes, but it is now known that the average band has 30,000 nucleotide bases, which is over twenty times more than necessary to code for most proteins. It was then suggested that the bands represent related proteins, but this view is now considered unlikely. Thus, we are left with no good explanation. Nonetheless, the bands are extremely useful in identifying chromosomes. Previous techniques often resulted in no more than the assignment of individual chromosomes to groups on the basis of their size and the location of the centromere. Now it is possible to identify each chromosome on the basis of its characteristic banding pattern. In addition, the improvement of chromosomal identification has facilitated analyses, such as a gene mapping technique known as somatic cell hybridization (see Box 6.4) and the identification of minor chromosomal abnormalities.

Banding has progressed to the point that it is possible to compare human chromosomes to those of other species. The chromosomes of the

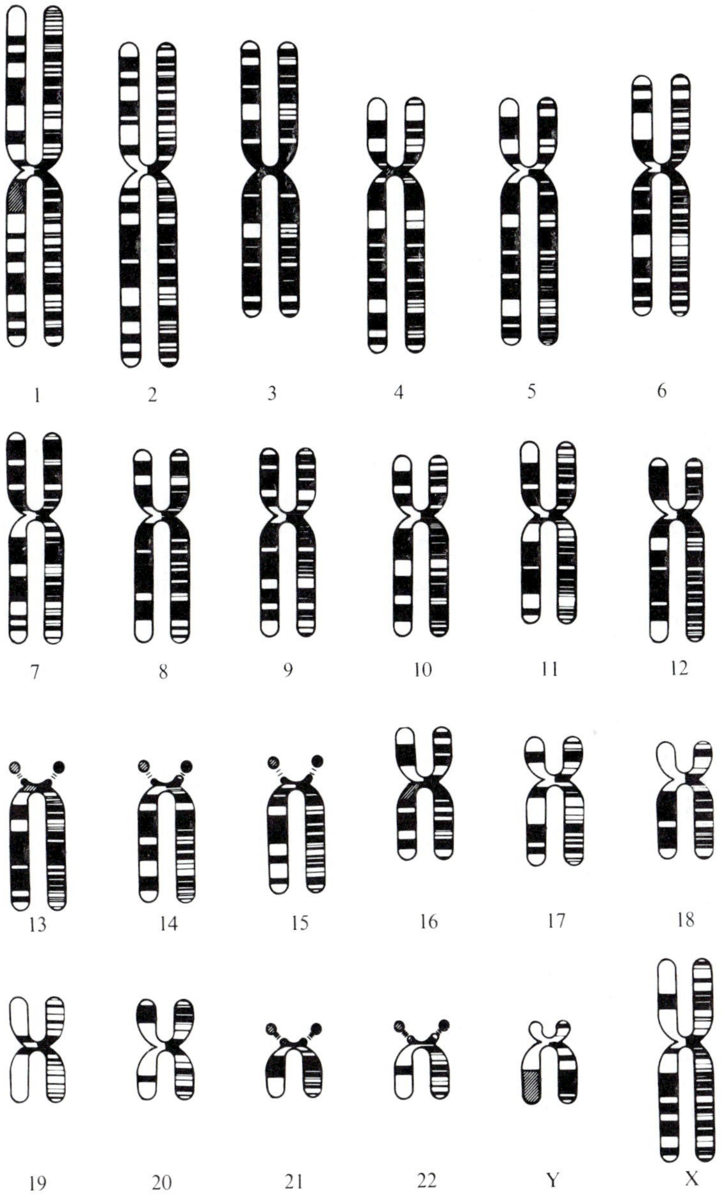

FIGURE 6.5

Representation of human chromosome bands. For each chromosome, the left chromatid is the banding pattern (320 bands) observed at metaphase. The right chromatid is the banding pattern (1,256 bands) in prophase. (From J. J. Yunis, "High resolution of human chromosomes," *Science,* 191, 1268. Copyright © 1976 by the American Association for the Advancement of Science.)

Box 6.4
Somatic Cell Hybridization

A completely different method—one that sounds like science fiction—has greatly increased our knowledge of human chromosomes. The technique is called *somatic cell hybridization* because it fuses human cells with the cells of other mammals. *Somatic* refers to cells of the body, as opposed to sex cells. The resulting hybrid cells have different assortments of human chromosomes and chromosomes from the other mammals. The figure on the facing page shows (a) the 40 chromosomes of a mouse cell, (b) the 46 chromosomes of a human cell, and (c) a man–mouse hybrid cell of 73 chromosomes, only 8 of which are human chromosomes. All of the mouse and human chromosomes are active in these hybrid cells. By comparing enzymes from replicates, or clones, of such hybrid cells, we are able to determine which chromosomes are responsible for specific enzymes. The last panel of the figure (d) shows hypothetical results for four human enzymes and three human chromosomes in five hybrid clones. Linkage is determined by noting which human enzymes and chromosomes appear simultaneously in the different hybrids. For example, in the figure, enzymes I and III appear together in all five clones and they are correlated with the presence of chromosome 2. Thus, genes for these two enzymes must be linked on chromosome 2.

Somatic cell hybridization has rapidly increased our knowledge of gene maps. The number of mapped genes increases monthly. Over one hundred autosomal loci and about one hundred sex chromosomal loci have currently been mapped for humans. Most of the genes involve proteins in the blood. These mapped genes may be useful in conjunction with mapping techniques other than somatic cell hybridization (see Box 6.3) to locate major genes responsible for behavioral characteristics.

Somatic cell hybridization. (a) The 40 chromosomes of a mouse cell. (b) The 46 chromosomes of a human cell. (c) Mouse and human chromosomes are both present in a hybrid cell formed by the fusion of two cells like those whose chromosomes are pictured here. There are 73 chromosomes in this cell, only 8 of which are human. (Courtesy of F. H. Ruddle and R. S. Kucherlapati.) (d) Hypothetical results for somatic cell hybridization. (From "Hybrid cells and human genes" by F. H. Ruddle and R. S. Kucherlapati.)

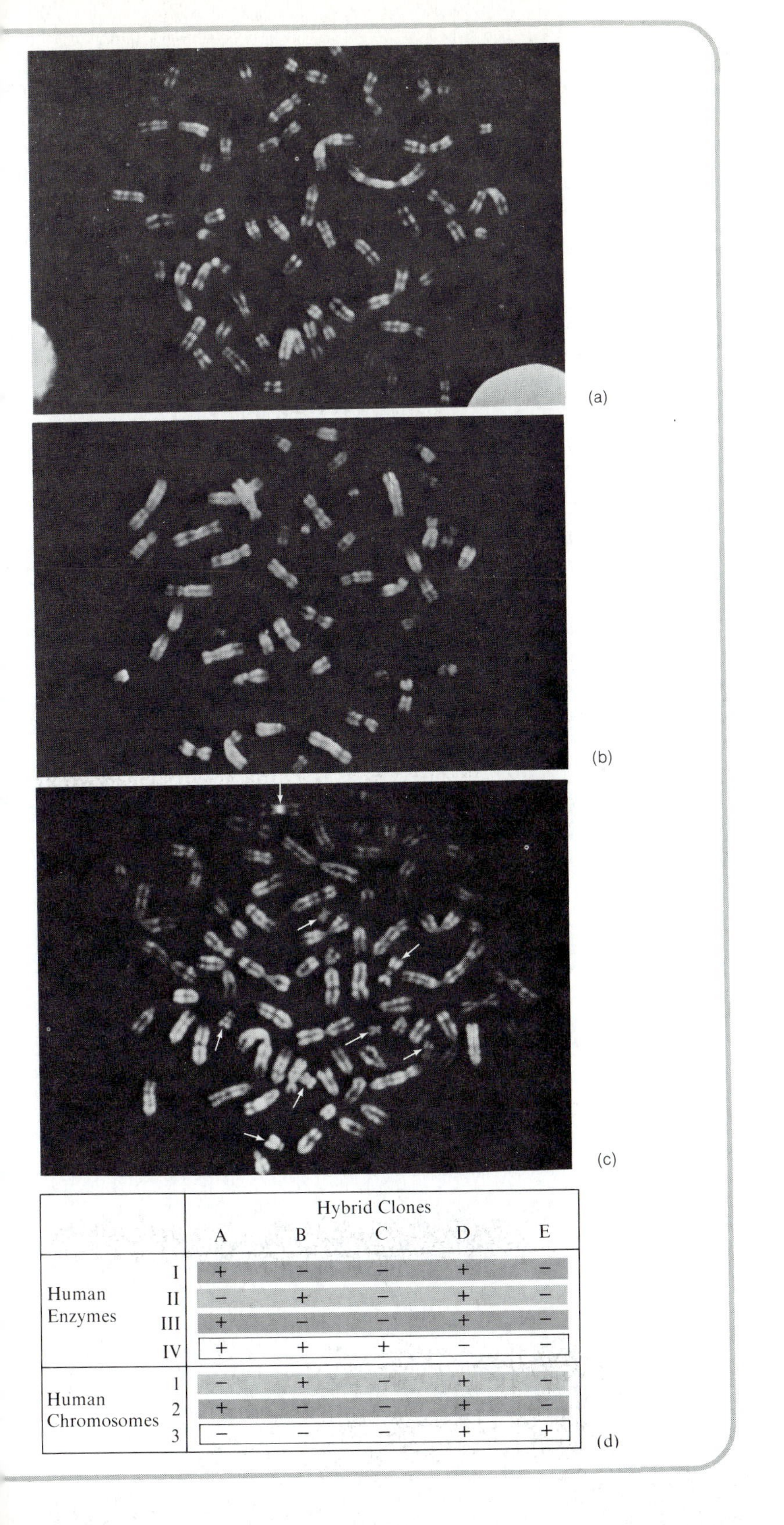

		Hybrid Clones				
		A	B	C	D	E
Human Enzymes	I	+	−	−	+	−
	II	−	+	−	+	−
	III	+	−	−	+	−
	IV	+	+	+	−	−
Human Chromosomes	1	−	+	−	+	−
	2	+	−	−	+	−
	3	−	−	−	+	+

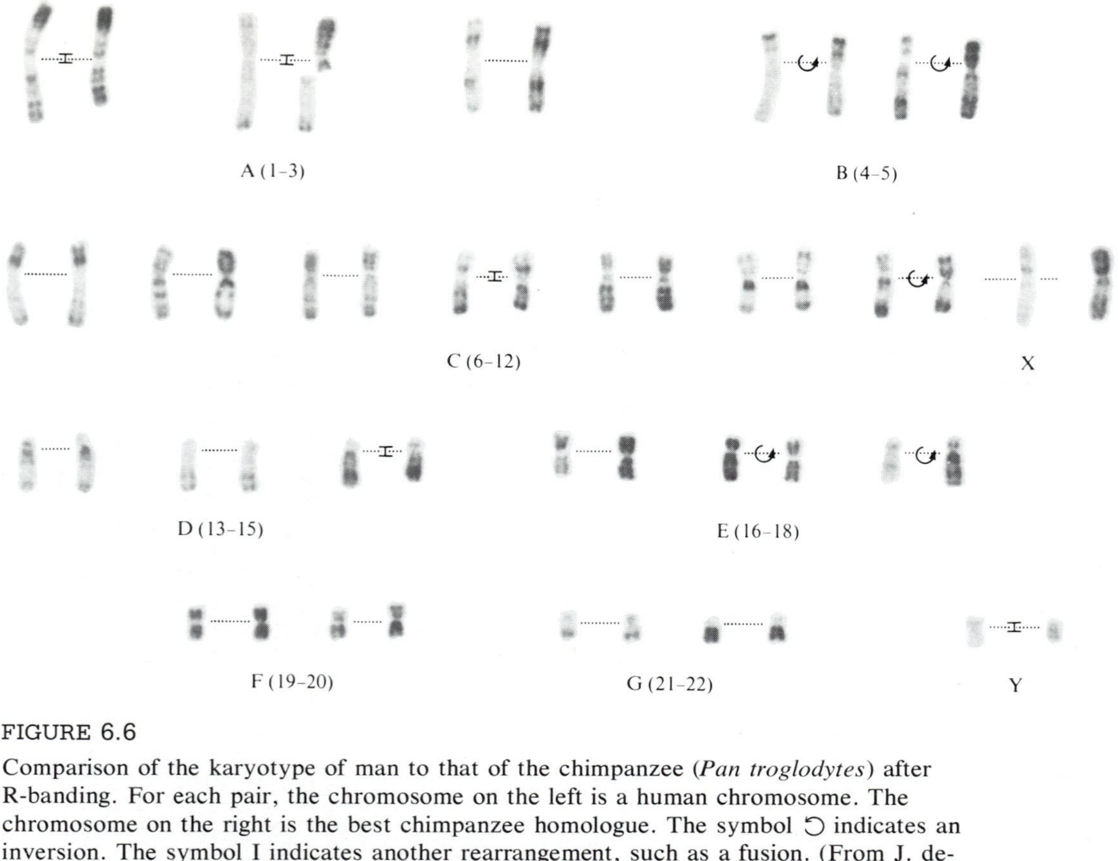

FIGURE 6.6

Comparison of the karyotype of man to that of the chimpanzee (*Pan troglodytes*) after R-banding. For each pair, the chromosome on the left is a human chromosome. The chromosome on the right is the best chimpanzee homologue. The symbol ↺ indicates an inversion. The symbol I indicates another rearrangement, such as a fusion. (From J. de-Grouchy, C. Turleau, and C. Finaz, "Chromosomal phylogeny of the primates." Reproduced, with permission, from *Annual Review of Genetics,* 12, 289–328.)

great apes (chimpanzee, gorilla, and orangutan) and humans are very similar (deGrouchy, Turleau, and Finaz, 1978). Although the great apes have 48 chromosomes and humans have 46, banding analyses have shown that the difference in chromosome number is due to the fact that two short chromosome pairs in the great apes fused to form a large chromosome pair (number 2) in the human species. Banded chromosomes for humans and chimpanzees are compared in Figure 6.6. This comparison indicates that humans are not very different chromosomally from chimpanzees. In fact, in both species the banding patterns are similar for chromosome pairs 3, 6, 7, 8, 10, 11, 19, 20, 21, and 22, and for the X chromosome. The other chromosomes usually differ by a single inversion.

CHROMOSOMAL ABNORMALITIES

Although meiosis is usually a very orderly process, irregularities occasionally occur. Chromosomes can break and result in a *deletion* of part of a chromosome. Loose, broken pieces of chromosomes tend to stick to other chromosomes. A deleted segment of chromosome can stick onto the end of the homologous chromosome, creating a *duplication* of that chromosome. Sometimes the pieces end up on the same chromosome, but in an *inverted* position. Inversions can change the location of the centromere. In other cases, the pieces may stick to a nonhomologous chromosome, which is known as a *translocation*. These types of chromosomal anomalies are illustrated in Figure 6.7.

Sometimes the chromatids of duplicated chromosomes do not separate properly during meiosis (*nondisjunction*), so that some gametes end up with extra, or missing, chromosomes. This results in two abnormal cells, one with an extra chromosome and one missing a chromosome, as indicated in Figure 6.8. These conditions are referred to as *aneuploidy*. When a gamete with an extra chromosome unites in fertilization with a normal gamete, the resulting zygote will have three of that particular chromosome. This is called a *trisomy* ("three bodies"). When a gamete missing a particular chromosome unites with a normal gamete, the result is a *monosomy* ("single body").

Chromosomal abnormalities involve gross genetic imbalance, and usually result in multiple defects, including behavioral problems such as mental retardation. J. G. Boué (1974) presented some surprising statistics concerning chromosomal abnormalities. Half of all fertilized human eggs have a chromosomal abnormality, and most of these result in early spontaneous abortions (miscarriages). At birth, about 1 in 200 babies has an obvious chromosomal abnormality. About half of these are abnormalities of the sex chromosome.

The newer banding techniques have led to the discovery that deleterious chromosomal abnormalities may occur for each of the 23 human chromosomes (Sanchez and Yunis, 1977). In addition, there are many minor

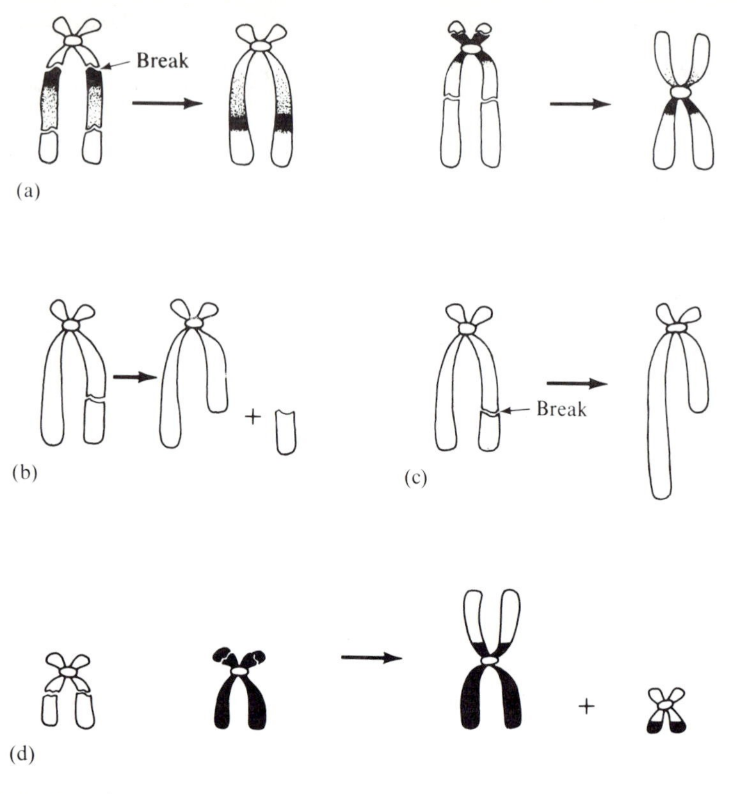

FIGURE 6.7

Common types of chromosomal rearrangements. (a) Inversions. (b) Deletion. (c) Duplication. (d) Translocation between nonhomologous chromosomes.

chromosomal anomalies with no apparent manifestations. Partial trisomies and minor deletions are most common and have been discovered in nearly every chromosome. Most complete trisomies are found only in spontaneous abortions, although cases of trisomy in live births have been discovered for chromosome pairs 8, 9, 13, 18, 20, 21, and 22, and for the sex chromosomes. Monosomies are found only in early spontaneous abortions, except for chromosome pair 21 and the X chromosome. For humans as well as *Drosophila,* the excess or missing chromosomal material in live births is almost always less than 5 percent of the total DNA. More severe abnormalities are presumably lethal.

SEX CHROMOSOMES AND THE LYON HYPOTHESIS

In 1949, it was observed that normal males and females have a striking difference in the nuclei of their cells (Barr and Bertram, 1949). In normal females, a small body, called a *chromatin* (or *Barr*) *body,* lies near the inner

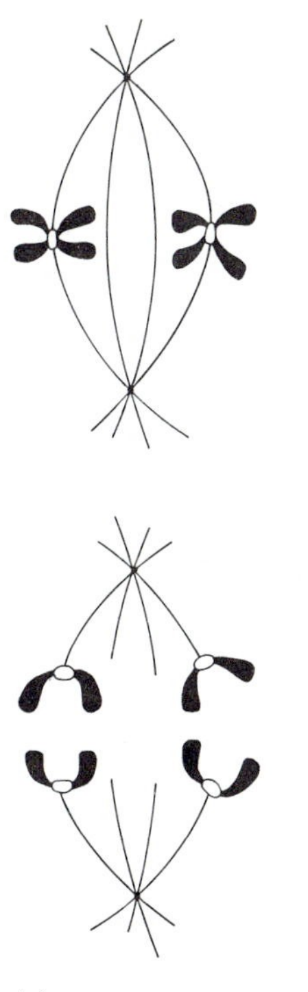

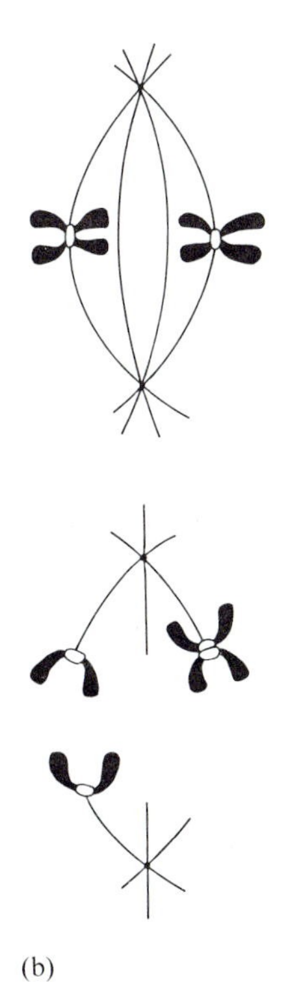

FIGURE 6.8

Diagrammatic representation of (a) normal cell division and (b) nondisjunction. (After Nadler and Borges, 1966.)

surface of the membrane of the nucleus. Staining indicates that the chromatin consists of DNA. In normal males, however, there is no Barr body. It was subsequently found that the number of these chromatin bodies is one less than the number of X chromosomes, as indicated in Figure 6.9. A male with an extra X (XXY) would thus stain positively for one chromatin body.

Later research showed that the chromatin body is a single condensed X chromosome. In 1961, Lyon hypothesized that this X chromosome is genetically inactive. Inactivation of an X chromosome in females compensates in part for the different amount of genetic material in the sex chromosomes of normal males and females. Males have only one large X and the Y chromosome, which is very small, while females have two large X chromosomes. Lyon also hypothesized that the inactive X chromosome could be either maternal or paternal in origin, even in different cells of the same individual. If the X chromosome that inactivates is of maternal origin, all daughter cells in the cell line resulting from subsequent mitotic divisions will also have an inactivated maternal X chromosome. Thus, females are mosaics for the X chromosome: in some of their cells, the X chromosome from their fathers is active; in other cells, the active X chromosome is from their mothers. This

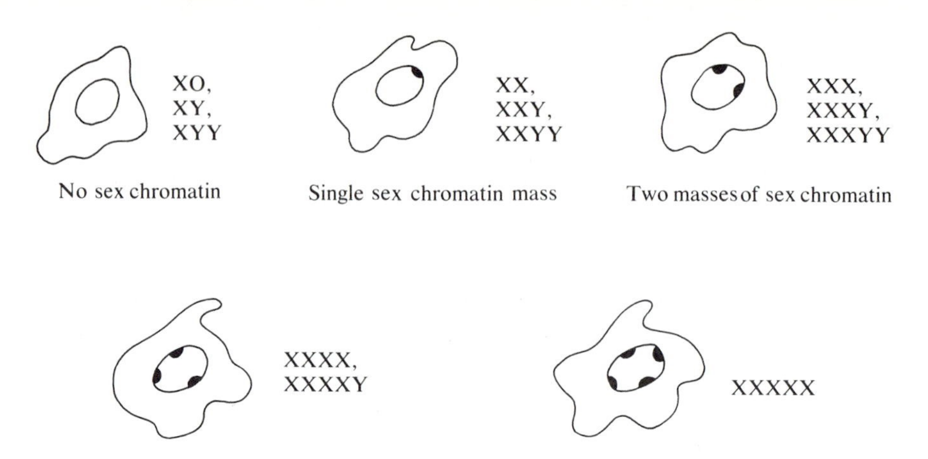

FIGURE 6.9

Correspondence between sex-chromatin patterns and sex-chromosome complements. (After Hsia, 1968.)

inactivation of the X chromosome occurs in the embryo by the third week of life. Considerable evidence has accumulated in support of Lyon's hypothesis.

The sex-chromatin test has greatly facilitated the determination of individuals with X-chromosome anomalies. This test is much simpler and less expensive than karyotype analysis. Cells may be easily obtained for examination by lightly scraping the inside of the cheek. These cells are then spread on a slide, stained, and examined microscopically for the presence of sex-chromatin masses. Because of the economy of this test, large-scale surveys have been undertaken. For example, among 8,621 mentally defective, institutionalized males and mentally handicapped schoolboys, about 0.8 percent were found to have sex-chromatin anomalies (Hsia, 1968). This is approximately twice the incidence observed in the general population.

Abnormalities of the sex chromosomes and autosomes, and their effect on human behavior are the topic of the next chapter.

CHROMOSOMES AND GENETIC VARIABILITY

We now know that Mendel's elements of inheritance are carried on chromosomes and they assort independently through the shuffling process of meiosis. Crossing over contributes even more to genetic variability by providing for recombination of genes at different loci on the same chromosome. Meiosis is responsible for the fact that we are somewhat like our parents in that we receive one chromosome of each homologous pair from each of them. However, meiosis and crossing over guarantee that we are genetically unique. Our particular combination of chromosomes and genes has never occurred before and never will occur again.

SUMMARY

Genes do not assort in a completely independent manner because they are associated on chromosomes. Crossing over breaks up linkages between alleles on chromosomes, and the rate at which this occurs has been used to locate different genes on the same chromosomes. Chromosomes replicate during the process of mitosis, creating duplicate cells. Meiosis shuffles chromosome pairs and results in haploid gametes. The 23 pairs of human chromosomes can be identified because of unique banding patterns. Abnormal karyotypes, occurring as frequently as in 1 of 200 births, are caused by deletions, translocations, inversions, nondisjunction, and other chromosomal anomalies.

7

Chromosomal Abnormalities and Human Behavior

As noted in the previous chapter, human chromosomal abnormalities are quite common. As many as half of all human fertilizations involve such abnormalities, and most of these result in early spontaneous abortion. About 1 in 200 fetuses with chromosomal anomalies survive until birth. However, some of these babies die soon after they are born. For example, only 10 percent of trisomy-18 individuals (incidence: about 1 in 5,000 births; see Figure 7.1) live for more than one year (Gorlin, 1977). Death ensues in the first month of life for 50 percent of the individuals with trisomy-13 (incidence: about 1 in 6,000 births; see Figure 7.2). Other chromosomal abnormalities are such that the individuals survive, but result in behavioral as well as physical manifestations. One of these involves the deletion of 15 to 80 percent of the short arm of chromosome 4 or 5 (Figure 7.3). The syndrome is called *cri du chat* (cry of the cat) because of a monotone cry nearly one octave higher than usual during the first month or two of life. This chromosomal abnormality usually results in severe retardation, and accounts for about 1 percent of institutionalized retardates with IQs less than 35. However, an even more important cause of mental retardation is trisomy-21, which will be discussed along with other behaviorally related chromosomal abnormalities later in this chapter.

Although behavioral effects of chromosomal abnormalities have been studied in species other than *Homo sapiens,* most research has focused on

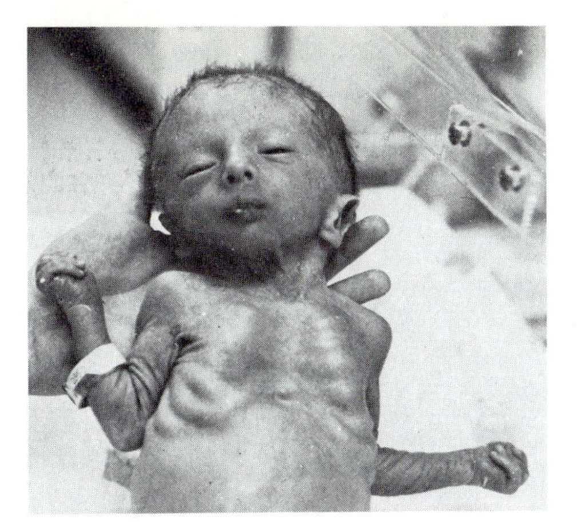

FIGURE 7.1
Patient with trisomy-18 syndrome. (Courtesy of George F. Smith, M.D.)

human beings (Borgaonkar, 1977). This research has shown that the effects of most abnormalities are broad and general. This should not be surprising considering that many genes are involved in a major chromosomal anomaly. Because the effect is general, we typically find no specific biochemical or morphological abnormalities to characterize a particular chromosomal abnormality (Smith, 1977). Behaviorally, the result of this general effect is that nearly all major chromosome abnormalities influence cognitive ability, which would be expected if cognitive ability is affected by many genes (Lewandowski and Yunis, 1977).

Recent improvements in banding techniques have led to the identification of many new chromosomal anomalies. Most of these are unique deletions and insertions of bits of chromosomal material. However, in this chapter, we shall focus on the classical chromosomal syndromes in man involving entire additional or deleted chromosomes. Down's syndrome, which is caused by the presence of an extra autosomal chromosome, was discovered

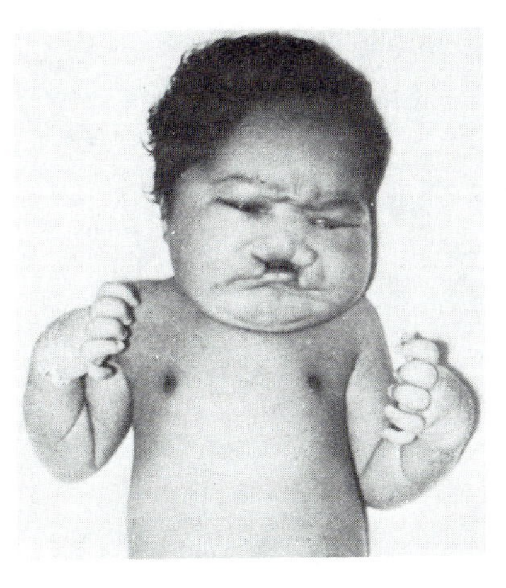

FIGURE 7.2
Patient with D-trisomy (trisomy-13) syndrome. (Courtesy of George F. Smith, M.D.)

(a)

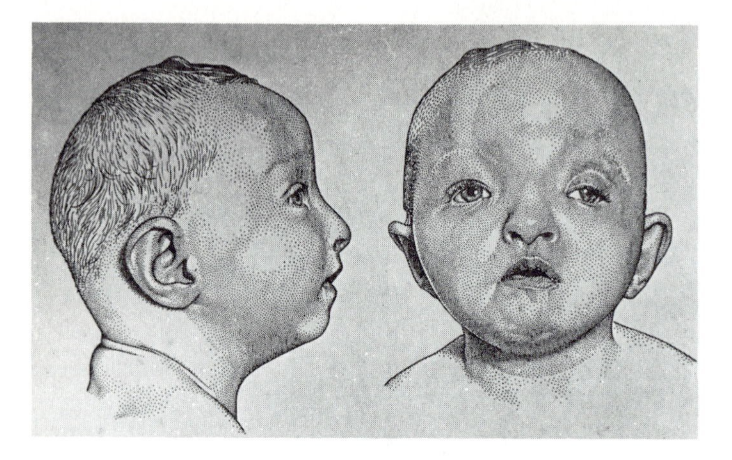

(b)

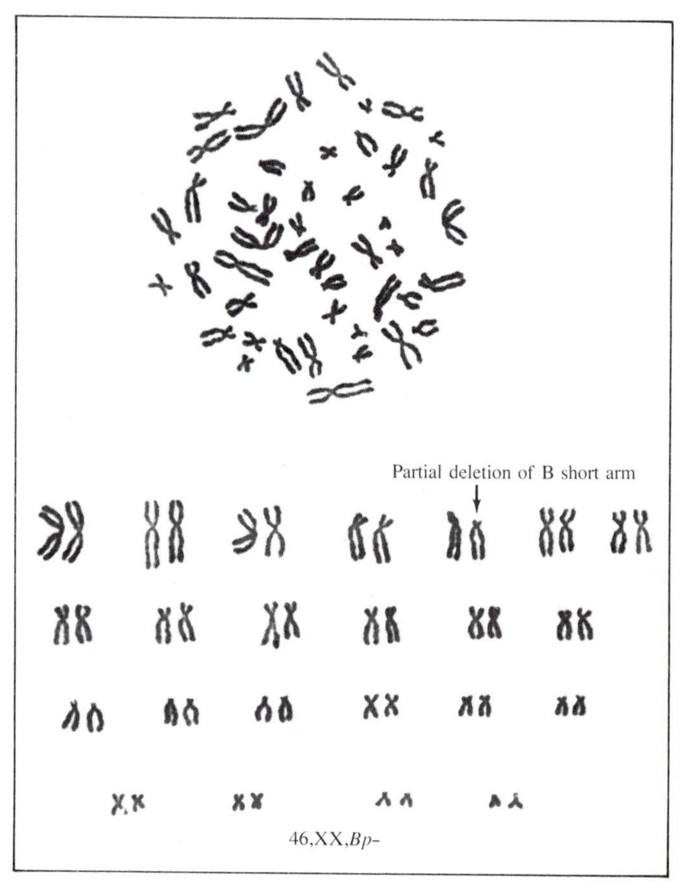

first. A disproportionate number of chromosomal abnormalities involve the sex chromosomes rather than autosomes. About half of all surviving individuals with chromosomal abnormalities have a problem with the sex chromosome. We shall discuss four of the best-known sex chromosomal anomalies: Turner's females with only one X chromosome, females with one or more extra X chromosomes, Klinefelter's males with one or more extra X chromosomes, and males with one or more extra Y chromosomes.

DOWN'S SYNDROME

The first human autosomal abnormality was discovered in 1959. Patients with *Down's syndrome,* or *mongolism,* were found to have 47 chromosomes instead of the normal 46. One of the small chromosomes in group G is present in triplicate, rather than duplicate, yielding the karyotype shown in Figure 7.4. Another name for this condition is trisomy-21, because the trisomy was thought to involve the next to the smallest autosome (number 21 by the Denver system of enumeration). We now know that the smallest autosome is the one in triplicate. Thus, trisomy-21 should really be called trisomy-22, but it has become too firmly entrenched in the literature for us to change the numbering system. We shall also continue to refer to the condition as if it involved chromosome 21.

Down's syndrome is so common (an incidence of about 1 in 700 newborns) that its general features are probably familiar to everyone. Infants affected with Down's syndrome are often quiet and uncrying during the early weeks of life. Their characteristic physical traits include upward and outward slanting eyelids and small folds of skin over the inner corners of the eyes. (See Figure 7.5.) As with other chromosomal abnormalities, however, there are many other effects. For example, the iris of the eye is speckled in about 85 percent of Down's syndrome individuals. Hearing problems are also common, as are general skeletal problems that occur in about 80 percent of Down's individuals. Instances of respiratory infections, heart problems, and leukemia are more frequent among individuals with trisomy-21. In the past, these problems have resulted in high mortality during the first few months of life and an average life span of only 20 years, although modern medical intervention has decreased the mortality rate. The pervasive effects of trisomy-21 make it unlikely that we will discover any single drug treatment to intervene in the development of Down's syndrome.

FIGURE 7.3

(a) Patient with *cri du chat.* (Courtesy of George F. Smith, M.D.) (b) Karyotype of a patient affected with *cri du chat,* showing partial deletion of the short arm of number 5. (Courtesy of Arthur Robinson, M.D.)

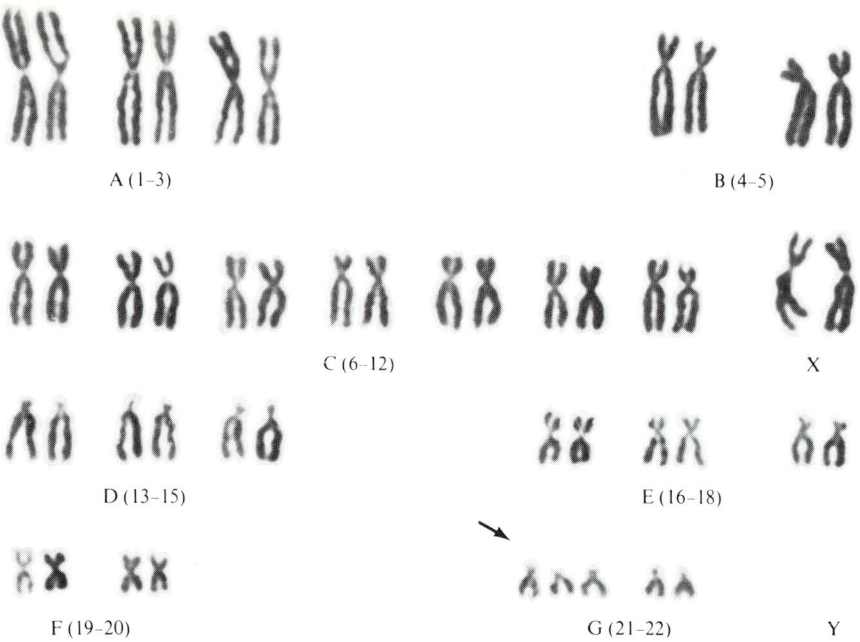

A (1–3)
B (4–5)
C (6–12)
X
D (13–15)
E (16–18)
F (19–20)
G (21–22)
Y

FIGURE 7.4

Karyotype of a Down's (trisomy-21) female. (From *Slide Guide for Human Genetics* by R. A. Boolootian. Copyright © 1971 John Wiley & Sons, Inc. Reprinted by permission of John Wiley & Sons, Inc.)

One of the most striking features of Down's syndrome is the severity of mental defects. The average IQ among institutionalized Down's syndrome patients is below 50, and 95 percent have IQs between 20 to 80 (Connolly, 1978). The traditional cutoff criterion for retardation is an IQ of 70, and the average IQ in the population as a whole is 100.

For 95 percent of individuals with Down's syndrome, the trisomy is a result of nondisjunction during meiosis, as described in the previous chapter. (See Figure 6.8.) Essentially, the chromatids fail to separate, so that one gamete ends up with both chromatids, while another is missing one. In the case of Down's nondisjunction, the gamete with an extra chromosome 21 frequently unites with a normal gamete containing one chromosome 21, thus producing the trisomy. No individuals have been found with only one chromosome 21, which would occur if the gamete with no chromosome 21 united with a normal gamete. Therefore we can assume that this monsomy is lethal. Too little genetic material is usually more damaging than extra material.

15/21 Translocation Down's Syndrome

Sometimes the extra chromosome 21, or a large part of it, attaches to one of the chromosomes in the D group. This particular translocation (see

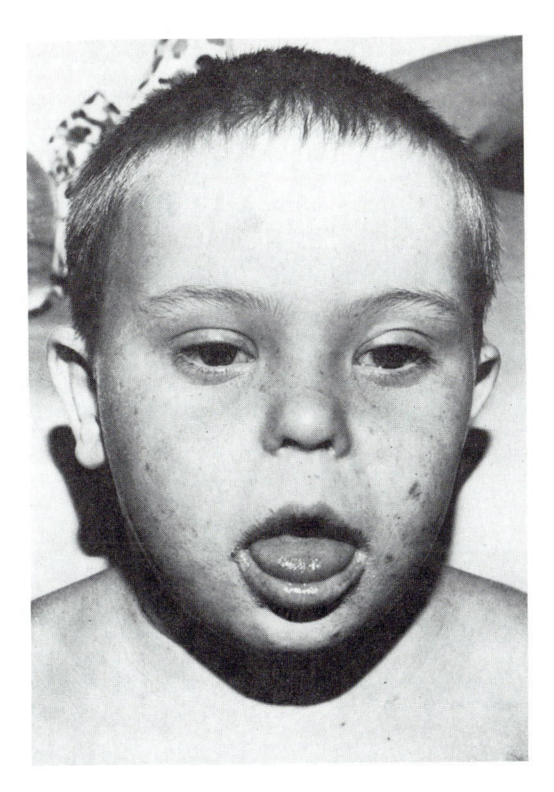

FIGURE 7.5

Patient with Down's syndrome. (Courtesy of George F. Smith, M.D.)

Figure 6.7) is called 15/21 translocation, even though we now know that chromosome 14 is more frequently involved in the translocation. It will also lead to trisomy-21 if the individual has the normal pair of chromosome 21 in addition to the translocated chromosome 21. Such Down's syndrome individuals appear to have 46 rather than 47 chromosomes. The 15/21 translocation version of Down's is different from that originally described, in that it is often inherited. Nondisjunction trisomy-21 is rarely inherited because few Down's individuals reproduce. In fact, in almost all cases, nondisjunction Down's individuals have normal parents. In contrast, Down's individuals with the 15/21 translocation usually have a parent with the same condition. As illustrated in Figure 7.6, the parent's translocation is balanced in the sense that there is a normal amount of chromosomal material, and the individual is phenotypically normal. However, the gametes produced by this parent include balanced and unbalanced translocations, as well as normal gametes. When these gametes are fertilized by normal gametes, the zygotes may be of four types: inviable, normal, balanced translocation, and 15/21 translocation. For genetic counseling purposes, it is important to determine whether the siblings of a child with 15/21 translocation Down's syndrome carry a balanced translocation. Those who do would have a substantial risk of bearing a child with this syndrome.

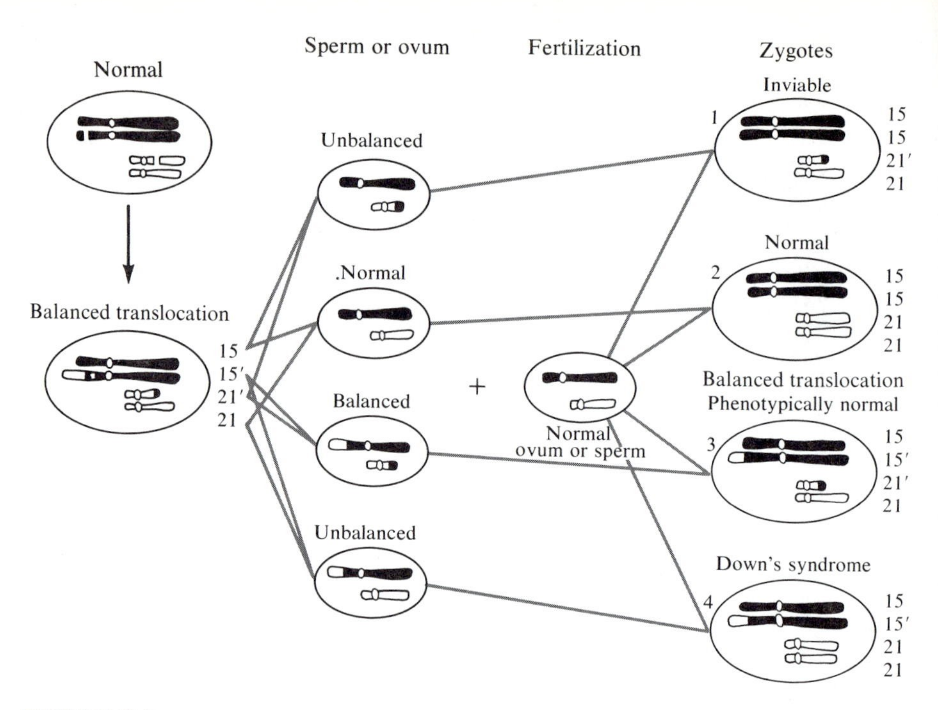

FIGURE 7.6

Diagram showing suggested origin of 15/21 translocation and its genetic consequences. (After Polani et al., 1960.)

Relation to Maternal Age

Down's syndrome, first described by Langdon Down in 1866, the same year that Mendel published his classic paper, defied explanation for many years. Although it was occasionally found to be familial, it was clearly not due to a simple dominant or recessive gene. Another factor that confounded researchers was that Down's syndrome is related to the age of the mother. Its higher incidence among children of older mothers gave rise to environmental explanations such as reproductive exhaustion. However, we now know that children of older mothers have more chromosomal anomalies in general, including trisomy-21.

All the immature eggs of a female mammal are present at birth. These have a diploid number of chromosomes, and the meiotic process that produces haploid gametes occurs periodically throughout the female's fertile years. Nondisjunction during this meiotic process is more likely to occur as the female grows older.

The prevalence of Down's syndrome increases significantly in children born to mothers 35 years of age and older. In Table 7.1, the percentage of Down's infants (9,441 cases) born to mothers of various ages is compared with the percentage of normal babies born to women in the same age-groups. The column on the far right is the ratio of Down's births to normal births for each age-group. Up to 30 years of age, the ratio does not change much. From

TABLE 7.1

Distribution of Down's syndrome by mothers' ages

Mother's Age	% Down's Infants Born	% of Normal Births	% Down's Births / % Normal Births
−19	1.9	4.9	0.39
20–24	10.5	26.1	0.40
25–29	14.5	30.9	0.47
30–34	16.6	22.1	0.75
35–39	27.0	12.0	2.25
40–44	25.2	3.7	6.81
45+	4.3	0.3	14.33
Total	100.0	100.0	1.00
Mean age	34.43	28.17	

SOURCE: After Penrose and Smith, 1966.

30 to 34 years of age, the ratio increases slightly. After 35, it rises abruptly. For example, women from 35 to 39 years of age produce only 12 percent of all normal births, but 27 percent of Down's infants. Even more striking, women from 40 to 44 produce only 4 percent of all normal births, but 25 percent of Down's infants.

The percentages in columns 2 and 3 of Table 7.1 may also be thought of as probabilities. For example, the probability that the mother of a child with Down's syndrome was between the ages of 35 and 39 when that child was born is 27.0 percent. The probability that the mother of that child was 35 or older is equal to 27.0% + 25.2% + 4.3% = 56.5%. In contrast, the probability that the mother of a normal child was 35 or older at the time of birth is only 12.0% + 3.7% + 0.3% = 16%. Thus, more than 56 percent of Down's infants are born to mothers 35 or older, although only 16 percent of normal children are born to mothers in this age range. This indicates that the number of Down's infants born would be reduced by more than half if all women completed their childbearing before the age of 35.

This is an important social problem. As many as 10 percent of all institutionalized mentally retarded individuals have Down's syndrome, making it the single most important cause of retardation. In terms of expense, the societal cost is at least $10,000 per year for institutional care for the average lifespan of 20 years, or about $200,000 for each Down's individual. In 1971, there were 44,000 Down's individuals in the United States. The lifetime cost of the care of these Down's syndrome patients now alive will thus be about $8.8 billion. However, the financial cost is minor compared with the emotional cost to parents.

The risk for Down's is about 1 in 300 for women 35 to 39 years old, 1 in 100 for women 40 to 44, and 1 in 50 for women over 44. In the 1960s and 1970s, many women delayed childbearing, and it is likely that the 1980s will

Box 7.1
Amniocentesis

As mentioned in Box 4.4, chromosomal anomalies and certain single-gene problems can be detected before birth. Amniocentesis is the procedure by which fluid with cast-off fetal cells is obtained from the amniotic cavity. (See the figure opposite.) After these fetal cells are grown in culture, they can be karyotyped to detect chromosomal abnormalities and analyzed for enzyme deficiencies characteristic of about 75 single-gene disorders (Epstein and Golbus, 1977). Unfortunately, sufficient amniotic fluid for the test does not exist until after thirteen weeks of pregnancy, and it then takes two weeks to grow the cells in culture. Thus, pregnancy will progress to four months before a diagnosis can be made. However, this is still early enough for an abortion if the parents so choose.

There were fears that the fetus might be injured by the hypodermic needle that extracts the amniotic fluid. But these early fears have been quelled by research indicating that fetuses are not injured, nor is there increased spontaneous abortion following amniocentesis. A four-year study of more than one thousand amniocentesis cases and a control group of pregnant women who did not undergo amniocentesis indicates that the technique is safe (Culliton, 1975). Only 3,000 amniocenteses were performed in 1974, but the number is likely to increase as the medical profession becomes aware of this important tool for diagnosing chromosomal anomalies prenatally.

The positive effects of amniocentesis cannot be denied. In one survey of 10,754 amniocenteses on women of all ages, a total of 209 chromosomal abnormalities were found. Many of these could have produced mental retardation. Also diagnosed were 199 X-linked diseases, 106 biochemical defects, and 86 neural-tube defects (Epstein and Golbus, 1977).

see an increase in reproduction by women in the higher risk age categories. Many women worry about reproducing later in life because of such chromosomal abnormalities as Down's syndrome. Much of the worry of pregnancies later in life can be reduced by amniocentesis, a procedure that permits karyotyping of the fetal chromosomes, as described in Box 7.1. In one study of amniocenteses performed on 3,012 pregnant women 35 years of age and older, 79 (2.6 percent) had fetuses with chromosomal abnormalities (Epstein and Golbus, 1977). The other 2,933 women (97.4 percent) no longer had to worry about the possibility that the pregnancy might result in a child with a chromosomal anomaly.

TURNER'S SYNDROME (XO)

The degree of mental defect in individuals with an extra small autosome in group G (trisomy-21) is more severe than that of individuals with X chromosome anomalies, assuming they have at least one X chromosome. This lesser

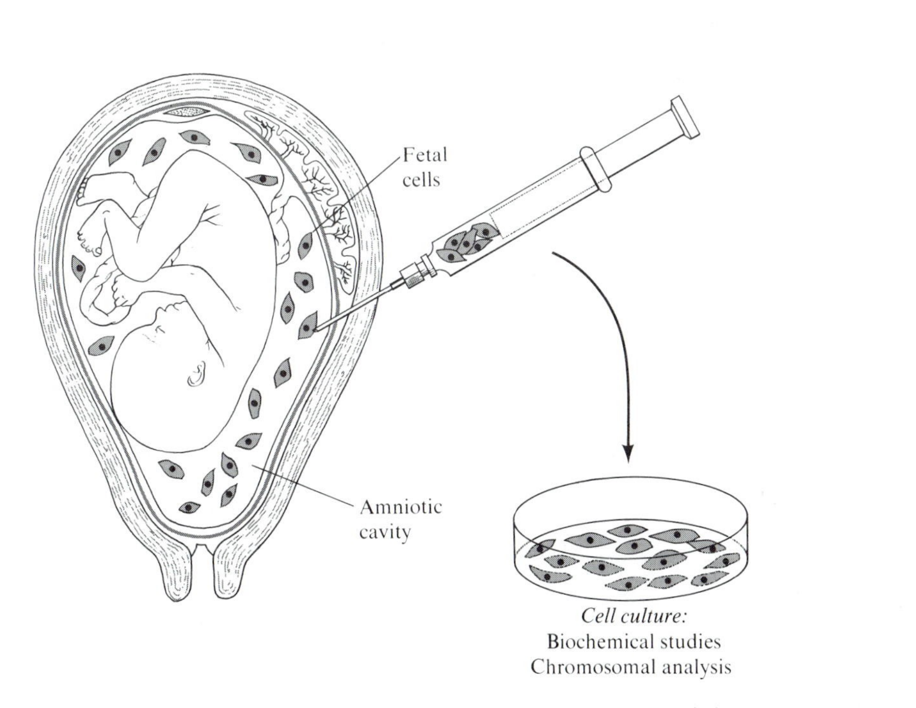

Amniocentesis. (From "Prenatal diagnosis of genetic disease" by T. Friedmann. Copyright © 1971 by Scientific American, Inc. All rights reserved.)

deficit in individuals with X chromosome anomalies is apparently due to the inactivation of all but one of the X chromosomes in each cell, which was described as the Lyon hypothesis in the previous chapter. Thus, although individuals with trisomy-21 have a less deviant total amount of DNA than individuals with sex chromosome abnormalities, their genetic imbalance is actually greater. This is also the reason why so few autosomal trisomies involving the larger chromosomes have been found; they are presumably lethal.

Turner's syndrome is a particularly interesting exception to the rule that chromosomal abnormalities cause general retardation. Turner's syndrome occurs only in females and nearly always involves sterility and limited secondary sexual development. (For example, fewer than 10 percent menstruate.) There are also other physical stigmata, such as short stature and a webbed neck. (See Figure 7.7.)

The incidence of Turner's syndrome is about 1 in 2,500 births, and it is also frequently found among spontaneous abortions. About 60 percent of women with Turner's syndrome have only 45 chromosomes, and their cells

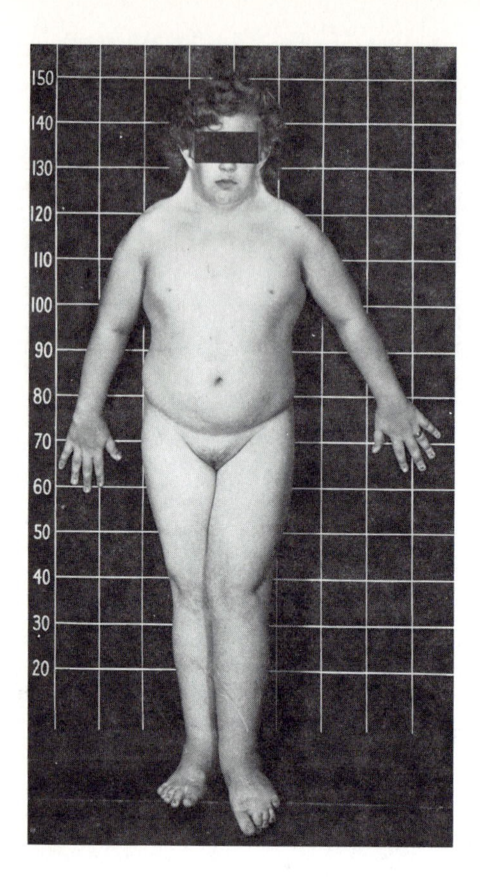

FIGURE 7.7
Patient with Turner's syndrome. (Courtesy of John Money.)

have no Barr bodies (see Chapter 6), even though they are females. Their karyotypes indicate that they have only one X and no Y chromosome, as shown in Figure 7.8. Classical Turner's syndrome is caused by nondisjunction, but shows no increase with parental age. Another type of Turner's is caused by the loss of an X chromosome early in embryonic development, so that some cells have two X chromosomes and others have only one. These are called XO/XX mosaics. Another cause of the syndrome is deletion of part of one X chromosome.

Although it was once thought that Turner's females were below average in IQ, we now know that this is not so. For example, there are no more Turner's females in institutions for the mentally retarded than we would expect on the basis of the frequency of these females in the population at large. In fact, most of the problems faced by Turner's females are cosmetic ones caused by their short stature and failure to develop sexually (for which they can now receive hormonal therapy), and problems caused by sterility. However, Turner's females are likely to have a highly specific cognitive defect. In 1962, it was first reported that Turner's females have low "perceptual organization" scores, although they have nearly normal overall IQ scores (Shaffer, 1962). Their average scores on verbal sections of IQ tests are about 20 points higher than their scores on nonverbal, performance tests.

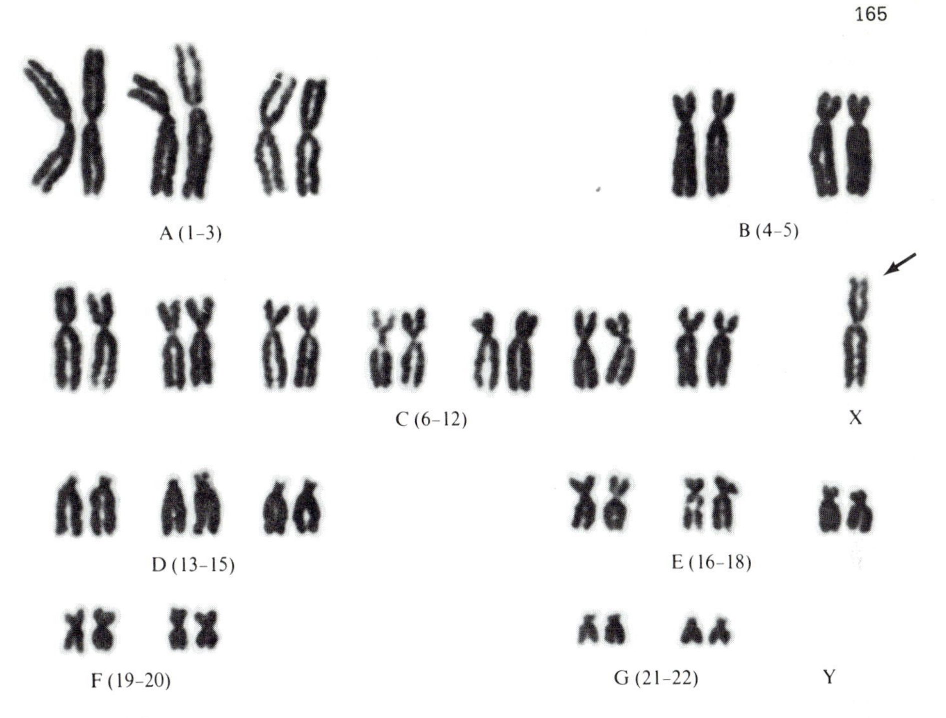

FIGURE 7.8

Karyotype of a Turner's female, 45, X. (From *Slide Guide for Human Genetics* by R. A. Boolootian. Copyright © 1971 John Wiley & Sons, Inc. Reprinted by permission of John Wiley & Sons, Inc.)

Subsequent research by John Money (1964, 1968) has shown that the most serious deficiency in Turner's females is in spatial ability and directional sense. For example, many have difficulty in copying a geometric design or following a road map.

Turner's females also display X-linked recessive characteristics just as males do. In Chapter 4, we discussed the hypothesis that spatial ability was influenced by a major recessive allele on the X chromosome. This hypothesis was developed to explain the greater spatial ability of males. We concluded that the evidence to date is against this hypothesis, and we can now add to this negative conclusion the data from the Turner's females. If the hypothesis were correct, Turner's females should have higher spatial ability scores than other females because they have only one X chromosome, just like normal males. The fact that their spatial ability scores are lower negates this hypothesis.

FEMALES WITH EXTRA X CHROMOSOMES

One of the X chromosomes is partially inactivated in each somatic cell of normal females, and additional X chromosomes in the nucleus are apparently also inactivated. Those females with extra X chromosomes often show more than one Barr body. (See Figure 6.9.) Although the partial inactivation greatly decreases the imbalance due to the extra genetic material, some problems do occur, which worsen with increasing numbers of additional X chromosomes.

Most females with an extra X chromosome have 47,XXX karyotypes, as in Figure 7.9. However, 48,XXXX karyotypes and 49,XXXXX karyotypes have been found. The incidence of trisomy-X is about 1 per 1,000 births. The characteristics of trisomy-X females are not distinctive. Earlier studies had sampled individuals from institutionalized populations, and thus tended to find retardation. More recent studies (Tennes et al., 1975; Gorlin, 1977), however, have found that about one-fourth are essentially normal, one-fourth have mild developmental lags, one-fourth have some type of congenital problem, and one-fourth have possible cognitive or emotional problems. Overall, some tendency toward retardation is suggested by the finding that the incidence of trisomy-X is about 4 per 1,000 in institutions, whereas it is about 1 per 1,000 in the general population.

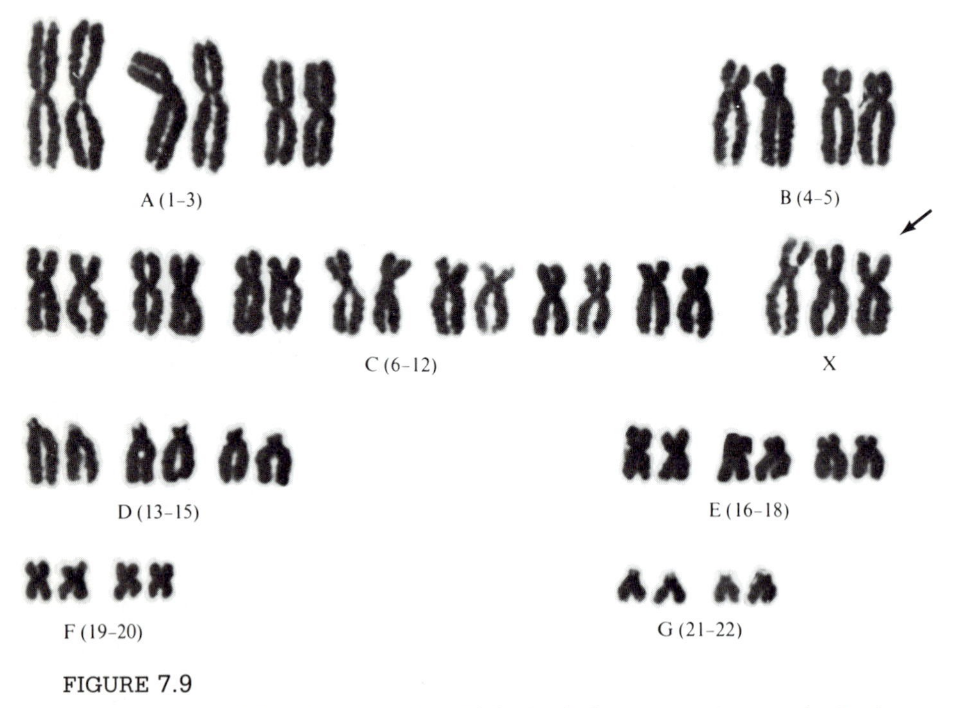

FIGURE 7.9

Karyotype of an XXX female. (From *Slide Guide for Human Genetics* by R. A. Boolootian. Copyright © 1971 John Wiley & Sons, Inc. Reprinted by permission of John Wiley & Sons, Inc.)

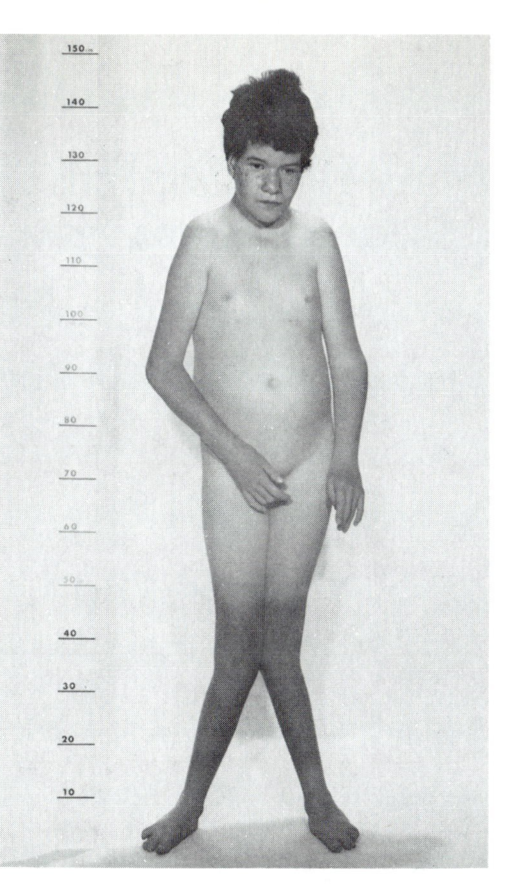

FIGURE 7.10
An XXXXX female. (From "The 49, XXXXX chromosome constitution: Similarities to the 49, XXXXY condition" by F. Sergovich et al., *Journal of Pediatrics*, 1971, 78, 285–290.)

Problems tend to multiply as additional X chromosomes are found. For example, Figure 7.10 is a photograph of a rare case of pentasomy X (that is, 49,XXXXX). This individual has a host of problems, including severe retardation, uncoordinated eye movements, undeveloped uterus and breasts, and many skeletal defects. Among about two dozen individuals who have been found to have at least four X chromosomes, mental retardation was a common feature.

KLINEFELTER'S SYNDROME

Individuals with Klinefelter's syndrome (see Figure 7.11) are phenotypic males with extra X chromosomes. They represent nearly 1 percent of males institutionalized for retardation, epilepsy, or mental illness, although their incidence in the general population is about 2 per 1,000 newborn males. Clinical features include the presence of abnormally small testes after puberty, low levels of the male hormone, testosterone, and sterility. Some of these individuals are mentally retarded (although three-quarters have IQs

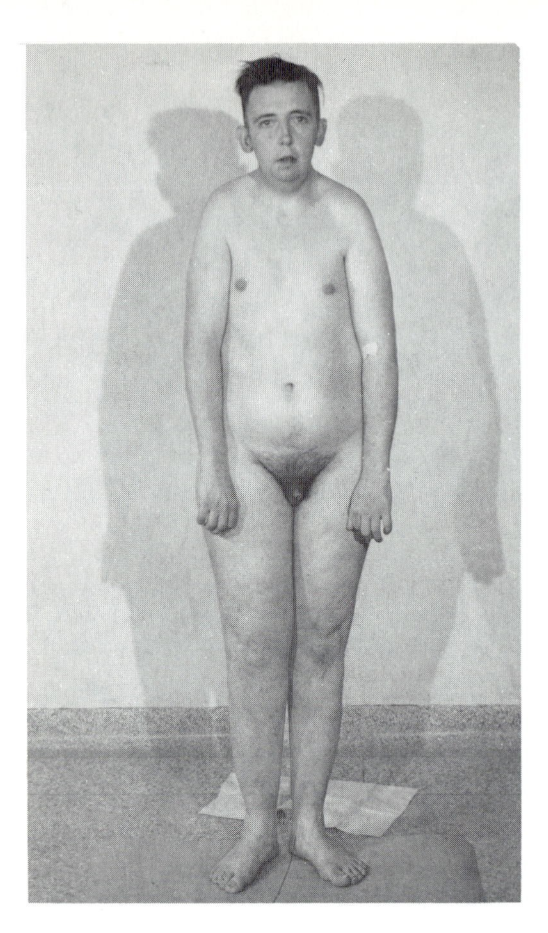

FIGURE 7.11
The XXY Klinefelter's syndrome. (Courtesy of R. J. Gorlin, M.D.)

within the normal range), and seem to have a variety of personality problems, such as passiveness and reclusiveness (Money et al., 1974). Like XYY males discussed in the next section, Klinefelter's males are somewhat taller than average.

Even though individuals with this condition are phenotypic males, they usually test positively for the presence of a Barr body. In about two-thirds of the cases, the karyotype is 47,XXY, as shown in Figure 7.12. However, 48,XXXY; 49,XXXXY; 48,XXYY; 49,XXXYY; and various other arrangements, including mosaicisms, have been described. As in other chromosomal abnormalities, the symptoms become more severe as more genetic material is added. For example, in over 90 cases of 49,XXXXY males that have been described, nearly all were severely retarded, and had severe sexual deformities, as well as other anatomical problems.

Like most chromosomal abnormalities, except Turner's syndrome, there is an increased risk of Klinefelter's syndrome among children of older mothers, although this association is less marked than that of Down's syndrome. Klinefelter's syndrome is generally thought to be caused by nondisjunction during meiosis, resulting in a gamete that has an extra X chromo-

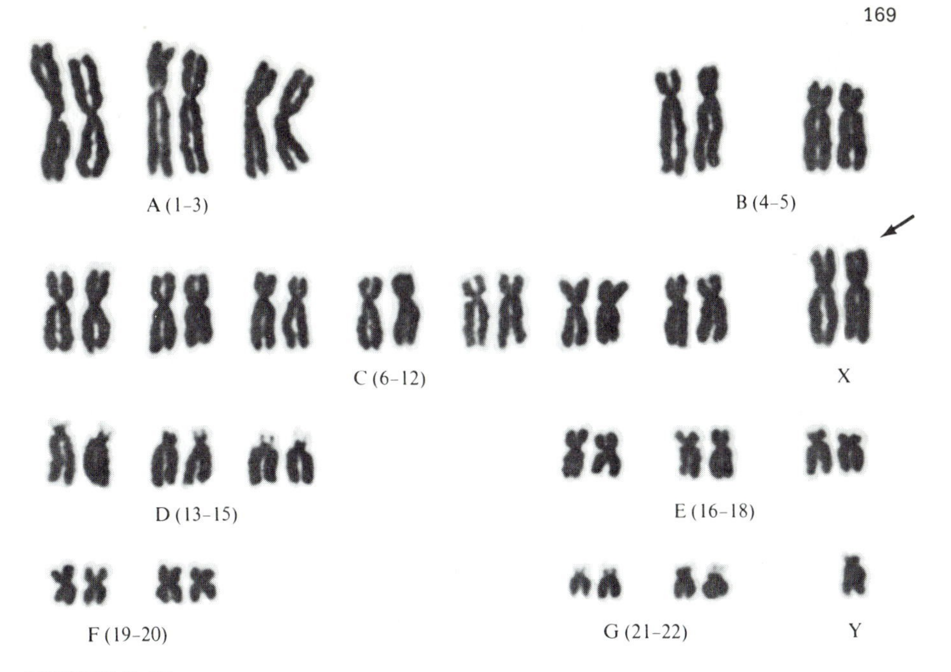

FIGURE 7.12

Karyotype of the XXY Klinefelter's syndrome. (From *Slide Guide for Human Genetics* by R. A. Boolootian. Copyright © 1971 John Wiley & Sons, Inc. Reprinted by permission of John Wiley & Sons, Inc.)

some. Fertilization of a normal X-bearing egg by an XY-bearing sperm, or fertilization of an XX-bearing egg by a normal Y-bearing sperm results in offspring with Klinefelter's syndrome. It is possible, however, that errors during early cell division in a normal zygote may also occasionally produce this syndrome.

Klinefelter's syndrome is not usually detected until after puberty, when some of the effects may be irreversible. Most of the problems are secondary to the low levels of hormones essential to proper development at puberty. Thus, there is a need for early identification so that hormonal therapy can begin soon enough to alleviate the condition.

MALES WITH EXTRA Y CHROMOSOMES

The XYY chromosomal anomaly, first described in 1961, has received considerable publicity since 1965, when it was suggested that XYY males may be predisposed to commit violent acts of crime. It seems that fantasies were triggered by the notion of a "supermale" with exaggerated masculine characteristics.

An XYY karyotype is shown in Figure 7.13. Extra Y chromosomes are the consequence of nondisjunction during meiosis in the father. Nearly 1

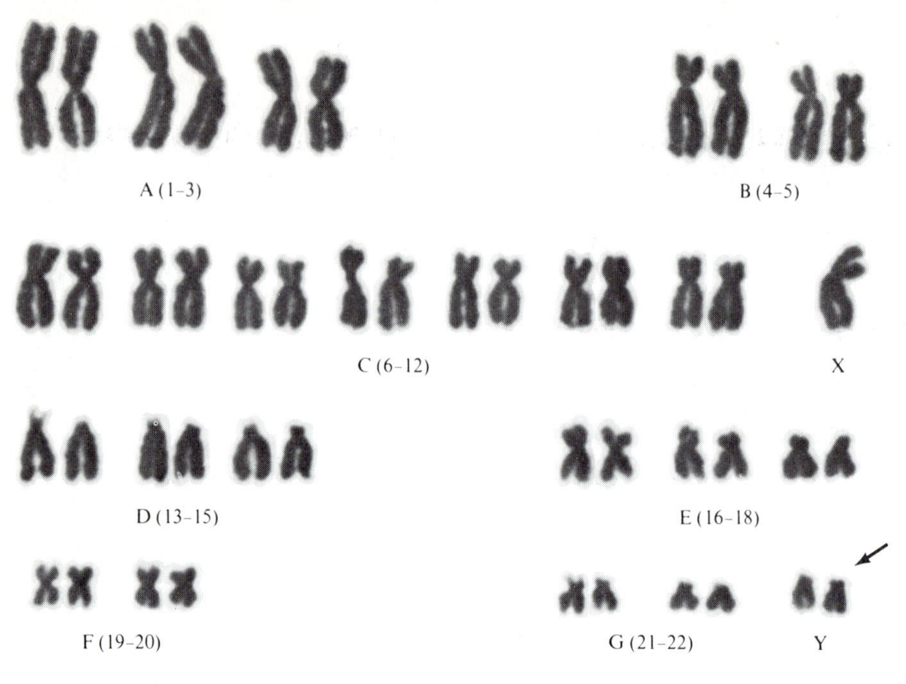

FIGURE 7.13

Karyotype of an XYY male. (From *Slide Guide for Human Genetics* by R. A. Boolootian. Copyright © 1971 John Wiley & Sons, Inc. Reprinted by permission of John Wiley & Sons, Inc.)

percent of the sperm of normal males have two Y chromosomes (Sumner et al., 1971), but the incidence at birth of XYY males is closer to 1 in 1,000. This suggests considerable selection against sperm or zygotes with two Y chromosomes. A few cases of XYYY, and even XYYYY have been reported. Research on the Y chromosome abnormality is more difficult than investigation of X anomalies, because the simple Barr body test that reveals an inactivated X chromosome will not reveal Y chromosome anomalies. Until recently, the much more costly karyotype analysis was necessary to detect Y chromosomes. For this reason, prevalence of the XYY condition in the general population has been difficult to estimate, and much of the early research focused on institutionalized males.

In 1965, Jacobs and co-workers reported that the incidence of chromosomal anomalies among individuals institutionalized because of "dangerous, violent, or criminal propensities" was higher than that in the population at large. Of 197 institutionalized volunteers who were karyotyped, 12 were found to have a chromosomal anomaly of some kind. One was a 46,XY/47,XXY mosaic, one was 48,XXYY, and seven were 47,XYY. Three had no sex-chromosome anomalies, but minor autosomal defects. The average height of the 47,XYY males was 73 inches, in contrast to an average height of 67 inches for the males of normal karyotypes in the institution. We now know that XYY boys are taller than 90 percent of their peers as early as six years of age.

Because of the importance of this initial discovery of a possible association between the presence of an extra Y chromosome and violent aggressive behavior, a number of related studies have been undertaken. These surveys have usually been of tall prisoners confined to special security sections because of their violent behavior. Thus, the possibility of sampling bias is clear. Nevertheless, a fairly consistent pattern of results has been obtained. Data from eighteen studies of 5,342 institutionalized males have been summarized (Shah, 1970). Of these, 103 (1.9 percent) possessed an extra Y chromosome. This outcome must be interpreted in the context of the prevalence of the condition in the general population. In a sample of 9,327 "normal" adult males, nine XYY karyotypes were identified (Price and Jacobs, 1970). This suggests a prevalence of about 1 in 1,000. Most reviews (e.g., Hook, 1973) concur in suggesting that the prevalence at birth in Caucasians is about 1 per 1,000, although it may be lower in other populations (e.g., Makino, 1975). Thus, even though it cannot be stated that the base rate in the general population has been firmly established, it is clear that the rate among institutionalized males is higher.

The possible association between antisocial behavior and XYY constitution has been widely debated. Many doubt that an important association exists (e.g., Noel et al., 1974). Most XYY males lead normal lives: less than 1 percent of XYY males are institutionalized (Kessler, 1975). Also, behaviors other than aggressiveness may be responsible for the increased rate of institutionalization of XYY males. For example, their greater-than-average height might make adjustment more difficult (Kessler and Moos, 1970). Also, poor coordination has been frequently noted. Sexual development, however, appears to be normal.

Our discussion of other chromosomal anomalies suggests another behavior as a likely source of greater institutionalization of XYY males. Because nearly every chromosomal anomaly results in some cognitive deficiency, we might expect to find similar deficiencies associated with the XYY abnormality. In fact, the results of a study by Witkin and Mednick and a large group of co-investigators (Witkin et al., 1976) in Denmark led to that conclusion. Their study had several unique controls. First, they used a normal sample selected for height, but not for institutionalization. Second, they evaluated the characteristics of males with normal and abnormal karyotypes. Third, they included a comparison group of XXY males.

All males born between 1944 and 1947 and still living in Denmark were scored for height from draft board records. The tallest 16 percent of these 28,884 men were selected for chromosomal analysis. Karyotypes were obtained for 4,139 men. Intellectual functioning was estimated by scores on an army selection test and by educational attainment tests routinely administered in Denmark. Criminal records of all subjects were detailed in terms of types of offenses.

Twelve XYYs and 16 XXYs were identified. The frequency of XYYs is higher (about 3 per 1,000) than the usual estimate of 1 per 1,000. However, these males were selected for height. Five of the XYYs (42 percent) had criminal records, as compared to three of the XXYs (19 percent) and 9 percent of all of the XY men. This indicates that the XYY males were, in fact, more likely to be incarcerated. But were they more violent? The answer is no. The nature of their crimes was no more violent than that of the control samples. Only 1 of the 12 XYY men had been convicted of a crime of violence against another person. In fact, the XYYs had been convicted of rather minor crimes and had received mild penalties.

Of the five XYYs with criminal records, four had army selection scores well below average and that of the fifth was somewhat below. It should be noted that criminality was also inversely related to intellectual functioning in the control sample of XY men. For both the army selection test and the educational index, men with criminal records had substantially lower scores. Also, the study found no support for the hypothesis that the greater incidence of criminality in XYY males is caused by adjustment problems due to their height. In the control sample, the noncriminal males were actually slightly taller than the criminal ones.

The authors conclude: "The elevated crime rate of XYY males is not related to aggression. It may be related to low intelligence" (Witkin et al., 1976, p. 547). They also caution that "The elevated crime rate found in our XYY group may therefore reflect a higher detection rate rather than simply a higher rate of commission of crimes" (p. 553). However, these conclusions must be tempered by the fact that, although the 47,XXY Klinefelter males had average test scores almost identical to those of the XYY males, their rate of criminality was considerably lower (19 percent versus 42 percent). In addition, other studies—although less well controlled—have found greater antisocial behavior in XYYs than in XXYs (e.g., Money et al., 1974) and no intellectual deficit among XYYs.

The discovery of these anomalies, particularly the finding that Down's syndrome is due to the presence of an extra chromosome, must be regarded as an extremely important breakthrough in the genetic analysis of behavior. Although several other chromosomal anomalies have since been described, it would seem that human chromosome analysis is still in its infancy. The new banding-pattern techniques are particularly well suited for the application of automatic computer recognition and classification of chromosomes (Caspersson, Lomakka, and Moller, 1971). The banding patterns can be measured photoelectrically and subjected to computer analysis, which may someday make karyotype analysis as routine as blood typing. Finally, it may be possible to identify the paternal, maternal, and even grandparental origin of each chromosome of an individual. In addition to greatly facilitating linkage studies, such a capability would have enormous potential for genetic counseling.

TABLE 7.2

Summary of chromosomal anomalies

	Type of Anomaly	Incidence per Live Births	Symptoms
Autosomal anomalies:			
Edward's syndrome	Trisomy-18	1 in 5,000	Early death; many congenital problems
D-trisomy syndrome	Trisomy-13	1 in 6,000	Early death; many congenital problems
Cri du chat	Deletion of part of short arm of chromosome 4 or 5	1 in 10,000	High-pitched monotonous cry; severe retardation
Down's syndrome	Trisomy-21; 5 percent involve 15/21 translocation	1 in 700	Congenital problems; retardation
Sex chromosomal anomalies:			
Turner's syndrome	XO or XX-XO mosaics	1 in 2,500	Some physical stigmata and hormonal problems; specific spatial deficit
Females with extra X chromosomes	XXX XXXX XXXXX	1 in 1,000	For trisomy X, no distinctive physical stigmata; perhaps some retardation
Klinefelter's males	XXY XXXY XXXXY XXYY XXXYY	2 in 1,000	For XXY, sexual development problems; tall; perhaps some retardation
Males with extra Y chromosomes	XYY XYYY XYYYY	1 in 1,000	Tall; perhaps some retardation

SUMMARY

Almost all chromosomal abnormalities influence general cognitive ability and growth. Trisomy-21 is the most common autosomal problem, and it leads to severe mental retardation. Its strong relationship to maternal age suggests an important role for genetic counseling. Sex chromosomal abnormalities include Turner's syndrome, females with extra X chromosomes, males with extra X chromosomes, and males with extra Y chromosomes. These chromosomal anomalies are summarized in Table 7.2.

8

Population Genetics

Many of the early approaches to the study of evolution, as described in Chapters 2 and 3, were more art than science. However, the theories and methods of population genetics have now provided evolutionary biology with a quantitative basis. In Chapter 3, we considered evolution as resulting from changes produced by genetic variability across the vast expanses of evolutionary time. We also mentioned that evolution can be analyzed from the viewpoint of genetic differences between species, as well as from that of genetic variability among individuals of a given species. Population genetics encompasses both perspectives. Its unique contribution is to assess allelic and genotypic frequencies in groups of breeding organisms and to study the forces that change these frequencies. In this chapter, we shall consider characters influenced by single genes. In the next chapter, our focus will shift to polygenic characters—that is, those influenced by many genes.

ALLELIC AND GENOTYPIC FREQUENCIES

When mice from a purebreeding albino strain are crossed to a black strain, all F_1 offspring are black. Thus, albinism is completely recessive to black coat color. In the F_2 generation, as expected, a classical Mendelian ratio of 3 black to 1 albino is observed. (See Chapter 4.) Now, if several mice from the F_2 generation are mated without regard to coat color, the ratio observed in the F_3 generation is again 3 : 1. In fact, as long as the sample size is large and mating occurs at random, this ratio will recur generation after generation, i.e., in the F_2, F_3, F_4, . . . F_n. How is this possible?

First, recall that 3 : 1 is a *phenotypic* ratio. The ratio of genotypes is actually $1A_1A_1 : 2A_1A_2 : 1A_2A_2$. Therefore, the *genotypic frequencies* are $0.25A_1A_1$, $0.50A_1A_2$, and $0.25A_2A_2$. What are the allelic frequencies? All of the alleles carried by A_1A_1 animals and half of those carried by A_1A_2 animals are A_1. Therefore, the frequency of the A_1 allele in an F_2 generation is 0.50 [i.e., $0.25 + \frac{1}{2}(0.50) = 0.50$]. Because there are only two alleles at this locus, the frequency of the other allele, A_2, is also 0.50; and the proportional frequencies of the two alleles add up to 1.00. In summary, in the F_2 population, the frequency of each allele is 0.50.

Now, what would be the frequencies of these two alleles and the three genotypes in the next (F_3) generation created by randomly mating F_2 individuals with one another? Common sense tells us that, if Mendel was right in suggesting that genes are discrete units, the alleles ought to show up in the F_3 population just as they did in the F_2. We can check this by studying the results of random crosses between F_2 individuals. Because the frequency of each allele is 0.50 in the F_2, eggs and sperm have an equal chance of carrying an A_1 allele or an A_2 allele. Table 8.1 describes the genotypic results of such a cross. For example, the frequency of A_1A_1 genotypes will be 0.25, which is the probability that an A_1 sperm (frequency = 0.50) will fertilize an A_1 egg (frequency = 0.50). We multiply the frequencies because A_1 sperm are just as likely to fertilize A_2 eggs as A_1 eggs (which is another way of saying that the F_2 individuals are mated at random). Thus, as in the F_2 population, the genotypic segregation ratio for the F_3 population is 1 : 2 : 1. Of course, the allelic frequencies ($0.50A_1$ and $0.50A_2$) do not change.

TABLE 8.1

Genotypic frequencies in the F_3 population

		Allelic Frequencies of Sperm Produced by F_2 Males	
		$0.50A_1$	$0.50A_2$
Allelic Frequencies in Eggs Produced by F_2 Females	$0.50A_1$	$0.25A_1A_1$	$0.25A_1A_2$
	$0.50A_2$	$0.25A_1A_2$	$0.25A_2A_2$

HARDY–WEINBERG–CASTLE EQUILIBRIUM

We have shown that genotypic and allelic frequencies remain stable generation after generation. This equilibrium is the cornerstone of population genetics. Population geneticists study departures from the equilibrium resulting from such factors as selection, migration, mutation, random genetic drift, and nonrandom mating.

Figure 8.1 summarizes the results of random mating in more general terms. The symbol p is often used to refer to the frequency of one allele, while q symbolizes the frequency of the other. If there are only two alleles for a particular locus, $p + q = 1$. When gametes with $p\ A_1$ and $q\ A_2$ frequencies randomly unite, the probability of producing offspring homozygous for the A_1 allele is p^2. The probability of homozygotes for the A_2 allele is q^2. The probability of heterozygotes is $pq + qp$, or $2pq$. All p^2 of the A_1A_1 genotypes have the A_1 allele, and half of the alleles for the A_1A_2 genotypes $[\frac{1}{2}(2pq)]$ are A_1. Using a bit of algebra, we can see that the frequency of A_1 is still p after one generation of random mating:

$$p^2 + \tfrac{1}{2}(2pq) = p^2 + pq = p(p + q) = p$$

Because $p + q = 1$, this expression reduces to p, meaning that the value of p has not changed after one generation of random mating. Therefore, with continued random mating, the genotypes will also remain stable with frequencies of $p^2 + 2pq + q^2$.

In an F_2 generation, the frequencies of the two alleles are equal. However, an equilibrium may also occur when allelic frequencies are different. Suppose that the initial frequency of one allele in a population was 0.80 and the frequency of the other was 0.20. Table 8.2 shows the resulting genotypic frequencies. After one generation of random mating, $p = 0.64 + \frac{1}{2}(0.32) = 0.80$ and $q = 1 - 0.80$, or $0.04 + \frac{1}{2}(0.32)$; either way, $q = 0.20$. The point is that the frequencies of the alleles have not changed after one generation of random mating.

This law of equilibrium for genetic variability was apparently so obvious to early geneticists that no one really wanted to take credit for discovering it. In 1908, an English mathematician, G. H. Hardy, and a German obstetrician, W. Weinberg, independently published papers describing the

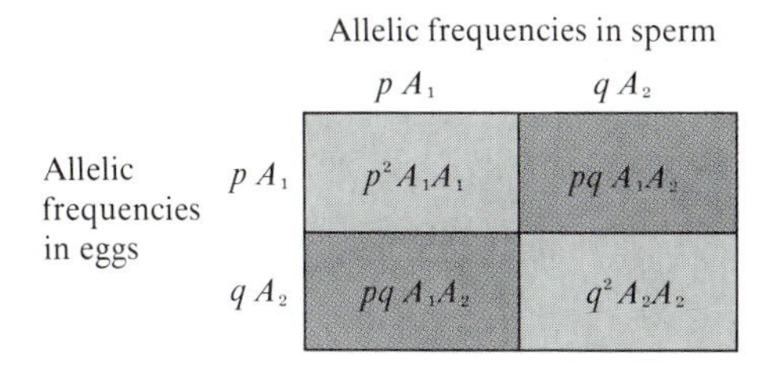

FIGURE 8.1

Genotypic frequencies after one generation of random mating.

TABLE 8.2

Genotypic frequencies after one generation of random mating when allelic frequencies are not equal

		Allelic Frequencies in Sperm	
		$0.80A_1$	$0.20A_2$
Allelic Frequencies in Eggs	$0.80A_1$	$0.64A_1A_1$	$0.16A_1A_2$
	$0.20A_2$	$0.16A_1A_2$	$0.04A_2A_2$

equilibrium. Hardy began his note in *Science* almost apologetically: "I am reluctant to intrude in a discussion concerning matters of which I have no expert knowledge, and I should have expected the very simple point which I wish to make to have been familiar to biologists" (1908, p. 49). More recently, it has been pointed out that W. E. Castle, an American geneticist, utilized and even extended this relationship in a paper published in 1903. For this reason, it has been suggested that we place in our textbooks a belated recognition of "Castle's law" (Keeler, 1968). Although, with hindsight, the law of equilibrium may seem obvious, it is the key concept of population genetics, and has many uses.

Uses of the Hardy–Weinberg–Castle Equilibrium

The Hardy–Weinberg–Castle relationship can be used to determine whether a population is in equilibrium, as well as to estimate allelic and genotypic frequencies. If a population is in equilibrium, the genotypic frequencies should correspond to $p^2 + 2pq + q^2$ for a single-locus, two-allele character. The easiest example is a codominant system in which both alleles are expressed. The MN antigens carried on the surface of red blood cells provide an example of a codominant system. Individuals with only the M antigen are MM homozygotes; individuals with only the N antigen are NN homozygotes; and individuals with both antigens are MN heterozygotes. In the United States, the frequencies of the MM, MN, and NN genotypes are 0.30, 0.50 and 0.20, respectively. What is the frequency of the M allele that produces the M antigen? It includes all of the alleles for the MM genotypes and half of the alleles for the MN genotypes. Therefore, $p = 0.30 + ½(0.50) = 0.55$. If $p = 0.55$, $q = 1 - 0.55$, or 0.45.

Now we can ask whether the MN blood system is in genetic equilibrium—that is, whether systems of mating, mutation, migration, or selection cause significant departures from the expected genotypic frequencies of $p^2 + 2pq + q^2$. If p is 0.55 and q is 0.45, then $p^2 = 0.30$, $2pq = 0.50$, and $q^2 = 0.20$. Thus, this character is in equilibrium in this population because the expected genotypic frequencies are the same as those that are observed (MM = 0.30, MN = 0.50, and NN = 0.20). Satisfy yourself that the system would not have been in equilibrium if the frequencies for the MM, MN, and NN genotypes were, for example, 0.35, 0.40 and 0.25, respectively.

If heterozygotes can be distinguished from homozygotes, as in the MN example, then allelic frequencies can be determined exactly for representative samples. When there is complete dominance, allelic frequencies can be estimated if the population is in equilibrium. Let us assume that a population is in equilibrium for a particular single-locus, two-allele trait. Then, the frequency of the homozygous dominant genotype (A_1A_1) is p^2, the frequency of the homozygous recessive genotype (A_2A_2) is q^2, and the frequency of the heterozygous genotype (A_1A_2) is $2pq$. The heterozygotes are called *carriers* because they carry only one recessive allele and thus do not display the trait. We can estimate the frequency of the recessive allele and the number of carriers in a population if we know the number of individuals displaying the trait. For example, suppose that 16 percent of the population show a recessively determined trait (as is true for the recessive blood factor called rh-negative). Because the allele is recessive, these individuals are of the A_2A_2 genotype, and have a frequency in the population of q^2. Because the frequency of the recessive allele is q, all we have to do is to take the square root of q^2 to find the frequency of q. The square root of 0.16 is 0.40. Thus, although only 16 percent of the population have the homozygous recessive genotype, the frequency of the recessive allele is 40 percent.

Given this information, one can determine the number of carriers for this recessive allele. If q is 40 percent, the frequency of the other allele is 60 percent. Given an equilibrium, the frequency of carriers will be $2pq$, which is $2(0.60)(0.40) = 0.48$. Thus, 48 percent of the population are carriers for the recessive allele. Remember, however, that these estimates are correct only if the population is in equilibrium. Using the earlier example of genotypic frequencies that are not in equilibrium ($0.35A_1A_1$, $0.40A_1A_2$, and $0.25A_2A_2$), we would mistakenly estimate q as 0.50 instead of 0.45.

Tasting PTC

As discussed in Chapter 4, about 70 percent of the Caucasians in the United States experience a very bitter taste when a solution of phenylthiocarbamide (PTC) is applied to the tongue, whereas about 30 percent find it virtually tasteless at the same concentration. If there is random mating for tasting PTC, which seems likely, then the Hardy–Weinberg–Castle equilibrium can be used to test hypotheses concerning the mode of transmission. Typically, we assume the simplest genetic hypothesis, test it, and then discard it only if there is a significant departure from the expected result. Thus, we shall retain the simplest model until we are compelled by the data to consider more complex hypotheses. The simplest genetic model to account for the family data in Table 8.3 involves two alleles at one autosomal locus, with one allele (the one for tasting PTC) completely dominant over the other.

If the population described above is in equilibrium, the frequency of the homozygous recessive genotype is q^2 (i.e., 29.8 percent). The frequency of q is thus the square root of 0.298, which is 0.546. The frequency (p) of the dominant allele for tasting PTC is $1 - 0.546 = 0.454$. Tasters can be homozygous or heterozygous for the dominant allele. The frequency of the homozygous tasters is p^2, which is 0.206, and the frequency of the heterozygous tasters is $2pq$, or 0.496. Thus, 70.2 percent of the individuals are tasters (that is, $0.206 + 0.496 = 0.702 = 70.2$ percent). Of these tasters, 29.3 percent are homozygous for the PTC tasting allele ($0.206 \div 0.702 = 0.293 = 29.3$ percent). The rest of the tasters, 70.7 percent, are heterozygous ($0.496 \div 0.702 = 0.707 = 70.7$ percent).

We can check this hypothesis by comparing the numbers of expected offspring from certain mating combinations with the actual data for PTC tasting presented in Table 8.3. One aspect of the table conforms to the hypothesis that nontasters of PTC are homozygous for a recessive allele at one autosomal locus: matings between two nontasters almost always produce nontaster offspring. The 5 taster children out of 233 could be due to variable gene expression, misclassification, or illegitimacy.

However, the rest of the data are not as straightforward, since the tasters could either be homozygous or heterozygous, which affects the expectations for the offspring. For example, if all of the tasters in the taster/nontaster matings were homozygous, then all of their offspring would be tasters. If all the tasters were heterozygous, then half of their offspring should be nontasters: they have a 50 percent chance of receiving the dominant allele from the heterozygous taster parents. However, the taster parents can actually be either homozygous or heterozygous. Thus, as expected, Table 8.3 indicates that the fraction of nontasters for this type of mating is neither 0 nor 50 percent, but rather, 36.6 percent.

TABLE 8.3

Data on the inheritance of ability to taste phenylthiocarbamide

Mating	Number of Families	Offspring		Fraction of Nontasters Among Offspring
		Tasters	Nontasters	
Taster × taster	425	929	130	0.123
Taster × nontaster	289	483	278	0.366
Nontaster × nontaster	86	5	218	0.978

SOURCE: From Stern, *Principles of Human Genetics,* 3rd Ed. W. H. Freeman and Company. Copyright © 1973; after Snyder.

In order to determine how well our model fits the data for taster/nontaster matings, we need to ask what percent of the tasters are homozygous. As indicated above, 29.3 percent of the tasters should be homozygous according to our model. Thus, of the 761 offspring from taster/nontaster matings, we would expect that 29.3 percent, or 223, result from matings between homozygous tasters and nontasters. All of these 223 children will be tasters. The other 538 children resulted from matings between heterozygous tasters and nontasters. For these matings, half of the offspring, 269, should be tasters. The other 269 should be nontasters. These calculations correspond well to the observed data for taster/nontaster matings. We have predicted that 492 (that is, 223 + 269) will be tasters and 269 will be nontasters; the observed data for these matings are 483 taster offspring and 278 nontaster offspring. If we performed similar calculations for the taster/taster matings, we would also find that our predictions correspond to the data. Thus, the data presented in Table 8.3 closely conform to that expected on the basis of a single-locus, two-allele model, where the allele for tasting PTC is completely dominant.

Although the Hardy–Weinberg–Castle equilibrium is useful for estimating allelic and genotypic frequencies, its most important use in population genetics is as a standard, like standard temperature and pressure in chemistry and physics. Forces that change allelic and genotypic frequencies are measured in relation to this standard.

FORCES THAT CHANGE ALLELIC FREQUENCY

In this section, we shall examine the changes in allelic frequency caused by migration, random drift, mutation, and selection. Because systems of mating (assortative mating and inbreeding) affect genotypic frequencies, not allelic frequencies, they will be considered in the following section.

Migration

If two populations have the same allelic frequencies at a particular locus, the frequencies will not be altered if individuals from one population randomly migrate to the other. However, if the allelic frequencies at a locus differ, then the frequencies will change due to random migration. Assume that the frequency of some allele is q in a certain population. We will designate the frequency of that allele in this original population as q_o. If individuals from another population immigrate into this population, the resulting frequency will depend on the frequency of the allele in the immigrant population (q_m) and the rate of immigration. In fact, the frequency of the allele after one generation of immigration (q_1) will be the frequency of q_m weighted by the proportion of immigrants plus q_o weighted by the proportion of natives. If

m is the proportion of immigrants and $1 - m$ is the rest of the population (the natives), we can express the effect of migration algebraically:

$$q_1 = mq_m + (1 - m)\, q_o$$

This expression can be simplified by multiplying and then factoring out m:

$$q_1 = mq_m + q_o - mq_o = m\,(q_m - q_o) + q_o$$

The change in allelic frequency (Δq) is simply the difference between the frequency after immigration (q_1) and the original frequency (q_o). If we take the above expression for q_1 and subtract q_o, we are left with:

$$\Delta q = q_1 - q_o = m(q_m - q_o)$$

Thus, the change in allelic frequency is a function of the rate of immigration (m) and the difference between the frequencies in the immigrant and native populations.

For example, assume a rate of migration of 10 percent, where the frequency of some allele is 0.20 in the natives and 0.30 in the immigrants. The change in allelic frequency will be 0.01:

$$\Delta q = 0.10(0.30 - 0.20) = 0.01$$

Although this change in frequency may appear small, sustained immigration of this magnitude over many generations would have a substantial effect.

The opposite side of the coin is emigration. Allelic frequencies will change if the frequency in the emigrant population is different from that of the remaining population. Like immigration, emigration will effect allelic frequencies in the population of individuals who remain (q_r) as a function of the rate of emigration (n) and the difference between frequencies in the emigrant (q_n) and remaining populations (q_r):

$$\Delta q = n(q_r - q_n)$$

Random Genetic Drift

Immigration and emigration are systematic migrational influences that affect allelic frequencies. Chance is also an important factor. If a population is small, the random sampling of genotypes may lead, purely by chance, to changes in allelic frequency. This type of unsystematic change in allelic frequency from one generation to the next is called *random drift* (Wallace, 1968). In a large population, random drift will not be an important source of allelic frequency change.

Mutation

In Chapter 3, we indicated that mutation is the ultimate source of genetic variability. However, our major concern here is mutation as a force affecting the frequency of a specific allele. Because the spontaneous mutation rate is low, the random event of mutation is not an important source of change for a specific allele, except when considered on an evolutionary time scale.

We can be more specific about expected changes due to mutation. Assume that there are two alleles at an autosomal locus and that A_1 mutates to A_2 with frequency u per generation. We also need to consider the possibility that A_2 can mutate back to A_1; let's call that rate v. The change in frequency of the A_2 allele is the extent to which A_1 mutates to A_2 minus the extent to which A_2 mutates back to A_1:

$$\Delta q = up - vq$$

No one knows the exact spontaneous mutation rate (u), and it can be different for various alleles. But let us assume that A_1 mutates to A_2 once in a million DNA replications. If u is 10^{-6} (i.e., 0.000001) and v is 10^{-7}, mutation will not be a major source of gene frequency change. Suppose that p is 0.90 and q is 0.10, then $\Delta q = (0.000001)(0.90) - (0.0000001)(0.10) = 0.00000089$ per generation.

Selection

Selection occurs when there are differences in reproductive rates among individuals in a population, and it can be a powerful force in changing allelic frequencies. Of course, individuals, not genes, are the targets of selection. With complete dominance of A_1 over A_2, individuals with an A_1A_2 genotype could not be distinguished from individuals with an A_1A_1 genotype. If neither of these genotypes reproduced at all, the frequency of the A_1 allele would be zero after just one generation of such severe selection. There are so few lethal dominant alleles because individuals with such alleles are quickly selected out of the population. However, recessive alleles are a different story. If recessive homozygotes (A_2A_2) did not reproduce for one generation, the A_2 allele would not be eliminated from the population since A_1A_2 heterozygotes would continue to reproduce.

Selection can also operate against both the A_1A_1 homozygote and the A_2A_2 homozygote, thus favoring the A_1A_2 heterozygote. This heterozygote advantage leads to a *balanced polymorphism*. When A_1A_2 heterozygotes mate, they continue to produce both A_1A_1 and A_2A_2 homozygotes. Balanced polymorphisms ensure genotypic variability within the population. They will be considered in greater detail after we discuss other types of selection.

TABLE 8.4

Relative fitness of genotypes for three different cases of selection

Item	Genotype					
	A_1A_1		A_1A_2		A_2A_2	
Frequency	p^2	+	$2pq$	+	q^2	= 1
Relative fitness:						
A_1 completely dominant, selection against A_2A_2	1		1		$1 - s$	
A_1 completely dominant, selection against A_1A_1 and A_1A_2	$1 - s$		$1 - s$		1	
Overdominance, selection against A_1A_1 and A_2A_2	$1 - s_1$		1		$1 - s_2$	

We can estimate the change in allelic frequencies caused by selection by starting with the Hardy–Weinberg–Castle equilibrium. The first row of Table 8.4 lists the frequencies of the three genotypes for a single-locus, two-allele character. The frequencies are $p^2 + 2pq + q^2$. Now we shall consider the effect of selection against one or more of these genotypes. However, rather than focusing on selection against a certain genotype, we will concentrate on the relative fitness of the genotype. This is the reproductive rate of that genotype compared with the genotype having the highest reproductive rate. Thus, if A_1A_1 and A_1A_2 genotypes average 20 offspring, while A_2A_2 genotypes produce only 5, the relative fitnesses of these three genotypes would be 1, 1, and ¼ (i.e., $5/20 = 1/4$), respectively. Selection against the A_2A_2 is ¾ (i.e., $1 - 1/4$). We shall label this selection coefficient s, and the relative fitness as $1 - s$. When $s = 0$, the relative fitness of all genotypes is 1, and individuals of all genotypes will contribute equally to the next generation. However, when the relative fitness of a certain genotype is less than 1, individuals of that genotype will contribute relatively less to the next generation.

The relative contribution of each genotype to the next generation is the genotypic frequency weighted by the relative fitness value. Table 8.4 describes three examples of selection. The first case involves complete dominance of the A_1 allele and selection against A_2A_2. The frequencies of the three genotypes after selection are $p^2 + 2pq + (1 - s)q^2$. The frequency of the A_2 allele after one generation of selection will be determined by the A_2A_2 individuals who reproduce $[(1 - s)q^2]$ and half of the alleles from the A_1A_2 heterozygotes ($1/2 \times 2pq = pq$). Because we express allelic frequencies as proportions, this frequency must be divided by the genotypic frequencies after selection $[p^2 + 2pq + (1 - s)q^2]$. Thus, after one generation of selection, the A_2 frequency will be:

$$q_1 = \frac{(1 - s)q^2 + pq}{p^2 + 2pq + (1 - s)q^2}$$

The term $[(1 - s)q^2]$ is equivalent to $q^2 - sq^2$. If we make this substitution in the denominator, it becomes $p^2 + 2pq + q^2 - sq^2$. Because $p^2 + 2pq + q^2 = 1$, the denominator simplifies to $1 - sq^2$:

$$q_1 = \frac{(1 - s)q^2 + pq}{1 - sq^2}$$

For example, if $q = 0.10$ (frequency of the A_2 allele) and the relative fitness of the A_2A_2 genotype is zero, the frequency of q after one generation will be 0.09:

$$q_1 = \frac{(1 - 1)0.01 + 0.9(0.10)}{1 - (1)(0.01)} = \frac{0.09}{0.99} = 0.09$$

The change in the frequency of the A_2 allele is simply $q_1 - q$. For this example, the change in frequency is -0.01.

However, the effects of selection must be considered for intervals longer than one generation. After n generations of complete selection against A_2A_2, if q_o is the frequency in the original generation, the frequency of A_2 will be:

$$q_n = \frac{q_o}{1 + nq_o}$$

As an example, consider some deleterious condition determined by an autosomal recessive allele (A_2). How much could the frequency of this recessive allele be lowered if A_2A_2 individuals did not reproduce for a number of generations? Because alleles with detrimental effects tend to have a relatively low frequency, let us assume that $q_o = 0.02$. If no A_2A_2 individuals reproduced for 50 generations, the frequency of this undesirable allele would become:

$$q_{50} = \frac{0.02}{1 + 50(0.02)} = 0.01$$

In other words, after some 1,500 years of intense selection against A_2A_2 (assuming a generation interval of about 30 years, as is the case for human populations), the frequency of this allele would only change from 0.02 to 0.01. This demonstrates the relative ineffectiveness of this form of selection if the frequency of a recessive allele is initially low.

The second example in Table 8.4 involves selection against the A_1A_1 homozygote and A_1A_2 heterozygote. Obviously, the frequency of the A_2 allele will increase, because the relative fitness of the A_2A_2 genotype is greater than that of the other two genotypes. In fact, if selection is complete ($s = 1$) against both A_1A_1 and A_1A_2, only individuals with the A_2A_2 genotype will reproduce. Thus, after one generation of such selection, no more A_1 alleles will be produced, and the frequency of the A_2 allele will become 1.00.

However, selection is not likely to be complete. The more general case can be determined similarly to the first example of selection against the A_2A_2 genotype. The frequency of the A_2 allele will come from the A_2A_2 genotypes (q^2), and half of the alleles from the A_1A_2 heterozygous individuals who reproduce [$\frac{1}{2}(1 - s)2pq = (1 - s)pq$]. In order to express this as a proportion of the total genotypic frequency after selection, we will divide this by the total genotypic frequencies weighted by their relative fitnesses [$(1 - s)p^2 + (1 - s)2pq + q^2$], as in the previous example.

Thus, after one generation of selection, the frequency of the A_2 allele will be:

$$q_1 = \frac{q^2 + (1 - s)pq}{(1 - s)p^2 + (1 - s)2pq + q^2}$$

We can simplify the denominator by multiplying through and reducing, remembering that $p^2 + 2pq + q^2 = 1$. This leaves the denominator as $1 - s(p^2 + 2pq)$. This can be further simplified because $p^2 + 2pq = 1 - q^2$. Thus,

$$q_1 = \frac{q^2 + (1 - s)pq}{1 - s(1 - q^2)}$$

This represents the more general case, in which selection may not be complete against the dominant homozygote and the heterozygote. However, to get a feeling for this equation, consider the case in which selection is complete ($s = 1.0$) against both A_1A_1 and A_1A_2. As expected, the new frequency of the A_2 allele is 1.00 because the genotypes with the A_1 allele do not reproduce.

$$q_1 = \frac{q^2 + (1 - 1)pq}{1 - 1(1 - q^2)} = \frac{q^2}{1 - 1 + q^2} = 1.00$$

In a similar manner, one generation of complete selection against the A_2A_2 genotype and the carriers (A_1A_2) would result in the elimination of the recessive A_2 allele. As in the first example, the frequency of a recessive allele is slowly changed when only the recessive homozygote is selected against. When the frequency of A_2 is low, most of the A_2 alleles will remain undetected in heterozygous carriers. However, if carriers could be detected and if they did not reproduce, it would be possible to eliminate the undesirable allele in one generation.

The third type of selection in Table 8.4 results in a *balanced polymorphism,* such that selection maintains different alleles rather than favoring just one. Heterozygotes can be distinguished from homozygotes in the case of overdominance. (See Figure 4.2.) Suppose that selection operated against both homozygous genotypes. The A_1A_2 genotypes would reproduce relatively more than the two homozygous genotypes, but they would continue to produce both homozygotes along with heterozygotes. Thus, genetic variability would be maintained.

The frequency of the A_2 allele can be determined as before. However, in this case, we need to allow for the possibility that the relative fitnesses of the two homozygotes differ. For this reason, Table 8.4 indicates s_1 for the A_1A_1 genotype and s_2 for the A_2A_2 genotype. Once again, the frequency of the A_2 allele after one generation of balanced selection will be determined by the surviving A_2A_2 genotypes $[(1 - s_2)q^2]$ and half of the alleles from the heterozygotes $[(½(2pq) = pq)]$. The denominator is the sum of the genotypic frequencies weighted by their relative fitnesses. Thus,

$$q_1 = \frac{(1 - s_2)q^2 + pq}{(1 - s_1)p^2 + 2pq + (1 - s_2)q^2}$$

Given that $p^2 + 2pq + q^2 = 1$, the denominator can again be simplified by multiplying through and reducing:

$$q_1 = \frac{(1 - s_2)q^2 + pq}{1 - s_1p^2 - s_2q^2}$$

If $p = q = ½$ and selection is equal against both homozygotes, the frequencies of p and q will remain equal. For example, substitute in the above equation: $s_1 = 1$, $s_2 = 1$, and $p = q = ½$.

If the relative fitnesses of the two homozygotes are not equal, the allelic frequencies will change. However, after many generations of such selection, an equilibrium will be reached in which the allelic frequencies no longer change. At that point, the frequency of the A_2 allele is simply a function of the two selection coefficients:

$$q_1 = \frac{s_1}{s_1 + s_2}$$

For example, if the relative fitness of the A_1A_1 genotype $(1 - s_1)$ is 0.25 and that of the A_2A_2 genotype $(1 - s_2)$ is 0.50, the frequency of q will stabilize at 0.60. (Check this by substituting $s_1 = 0.75$ and $s_2 = 0.50$ in the above equation.) Heterozygote advantage is one of several types of selection resulting in a balanced polymorphism or selectional balance. Because of the importance of balanced polymorphisms in maintaining genetic variability, let us consider this topic in more detail.

BALANCED POLYMORPHISMS

In the past, selection has often been regarded simply as a force that molded individuals to a particular environment. If this were the case, then we would expect to find very little genetic variability within a local group. To the contrary, however, it appears that at least a third of all loci are polymorphic. Although everyone now agrees that there is considerable genetic variability, there are differing opinions concerning the importance of selection in maintaining variability. *Neutralists* argue that most genetic variability has no selective value. They suggest that it is maintained simply by an equilibrium of backward and forward mutations. *Selectionists,* however, argue that the variability is maintained by selection. Although both positions are correct in some instances, in recent years several interesting examples of balanced polymorphisms have been discovered.

Heterozygote Advantage

Sickle-cell anemia in humans is a classic example of a balanced polymorphism. Although few individuals afflicted with this most serious disease (recessive homozygotes) survive to reproduce, the allele is nonetheless maintained in relatively high frequency in some African populations and among Afro-Americans. This high frequency of an essentially lethal recessive allele is apparently due to the high relative fitness of heterozygotes. Carriers seem to be more resistant than normal homozygotes to a form of malaria prevalent in certain parts of Africa. Although sickle-cell anemia is one excellent example of heterozygote advantage, other examples of this source of balanced polymorphisms are speculative.

Frequency-Dependent Selection

Predator-prey relationships may also help to maintain genetic diversity. For example, minnows will prey on the more common type of water bug, leaving the rarer forms at a reproductive advantage. Clarke (1975) has suggested that this is a general mechanism for maintaining genetic variability. Predatory birds and mammals also tend to attack more common types of prey. (See Box 8.1.)

Frequency-Dependent Sexual Selection

Another type of balanced polymorphism results from mating preference. In *Drosophila,* it has been shown that rarer males are relatively more likely to reproduce. As the rare type of male reproduces more and thus becomes more common, the reproductive edge vanishes. This rare-male advantage may be a general process for maintaining genetic variability in a population. The greater relative reproductive success of rare males was independently discovered by Claudine Petit and Lee Ehrman. Petit (1951) first discovered this phenomenon in a study of the mating success of two strains of *Drosophila*. One strain was the wild type; the other was bar-eyed, a mutant, sex-linked condition that changes eye morphology. The two strains were permitted to breed freely for a number of generations. During the course of the experiment, randomly chosen females were occasionally separated from the population and allowed to lay their eggs in individual vials. By examining the offspring of these females, it was possible to determine whether the male with which the female had mated was wild type or bar-eyed.

The reproductive success of males of a certain type is measured by the number of females mated by that type of male divided by the total number of those particular males. The coefficient of mating success (K) is useful in describing the results of experiments on frequency-dependent sexual selection. K is the ratio of the reproductive success of one type of male to the reproductive success of the other type. In Petit's experiment,

$$K = \frac{\text{no. of females mated by mutants/no. of mutant males}}{\text{no. of females mated by wild types/no. of wild-type males}}$$

Thus, K is the ratio of the number of females mated per mutant male to that of females mated per wild-type male. If the two types of males have equal reproductive success, K will equal 1.00. If K is less than 1.00, the mutant males are at a disadvantage. Conversely, if K is greater than 1.00, the mutant males mate more females than would be expected simply on the basis of their frequency.

The coefficient of mating success observed by Petit during the course of her experiment is graphed in Figure 8.2. The frequency of bar-eyed males fell from 93 percent of the total male population early in the experiment to 6 percent, due to their relatively low mating success. Although K was less than 1.00 throughout the experiment, the reproductive success of bar-eyed males was frequency-dependent, i.e., their mating success increased as they became rarer.

The results of more recent work by Ehrman have been even more striking in suggesting frequency-dependent sexual selection. Rather than relying on progeny-testing as an index of mating succes, Ehrman has employed direct behavioral observation in her studies of the rare-male advantage in *Drosophila*. (See Figure 8.3.) Females and males are placed into a mating chamber in which they are observed for several hours. Males and females from each of two strains are placed together, resulting in four possible mating combinations. Ehrman has found that the rare male is at a reproductive advantage in a number of different test situations. For example, they have the advantage when the two strains possess different chromosome arrangements, are mutant versus wild-type, and are positively versus negatively

Box 8.1
Maintaining Genetic Variability in Snails

Shell markings of a single species of land snail display great variation. (See the figure on the facing page.) The pattern of stripes and the color of the shell are controlled by different genes. Many combinations of stripes and colors are thus possible. Such genetic variability has been around for a long time; fossil snails tens of thousands of years old have similar varieties of shells.

In this case, genetic variability is maintained by selection. When snails are found in woodlands, their shells are likely to be without bands. However, snails in grasslands are likely to have banded shells. The fact that shell banding is correlated with habitat suggests that selection is at work.

Direct evidence for selection comes from an examination of the shells of snails captured by thrushes, who then smash the snails on stones to break them open (Clarke, 1975). The most conspicuous snails in a particular habitat (banded snails in woodlands, unbanded snails in grasslands) are most often preyed on by the thrushes. Given that this species of land snail occupies both woodlands and grasslands, genetic variability for shell banding will continue as the result of such selection.

Variation in the shell markings of a single species of land snail, *Cepaea nemoralis*. (From "The causes of biological diversity" by B. Clarke. Copyright © 1975 by Scientific American, Inc. All rights reserved.)

geotactic (that is, fly down rather than up). The rare-male advantage has now been demonstrated in seven species of *Drosophila,* as well as in a species of beetle and a species of wasp (Ehrman and Parsons, 1976).

As an example of Ehrman's research, we shall describe a simulated selection study that demonstrated that the rare-male advantage can result in a balanced polymorphism (Ehrman, 1970b). One hundred pairs of *Drosophila pseudoobscura* (80 pairs of orange-eyed flies and 20 pairs of purple-eyed flies in experiment 1; 20 orange and 80 purple in experiment 2) were placed in bottles for 24 hours and allowed to mate. Males were then discarded, and each female was placed in an individual vial where she laid her eggs. Eye color of resulting offspring permitted identification of the type of male with

which each female had mated. To eliminate all other selective factors, each generation utilized 100 mating pairs. The proportion of orange- to purple-eyed flies was dictated by the matings of the previous generation. Thus, if 80 females mated with 60 orange-eyed males and 20 purple-eyed males in generation n, generation n + 1 would begin with (60/80)(100) = 75 orange-eyed pairs and 25 purple-eyed pairs. The results of ten generations of such simulated selection are summarized in Table 8.5. These data indicate that the rare-male advantage occurred in both experimental populations, and continued until a stable equilibrium was achieved at intermediate frequencies.

Frequency-dependent sexual selection—at least in *Drosophila*—is limited to males. Rare females have no mating advantage. Although female

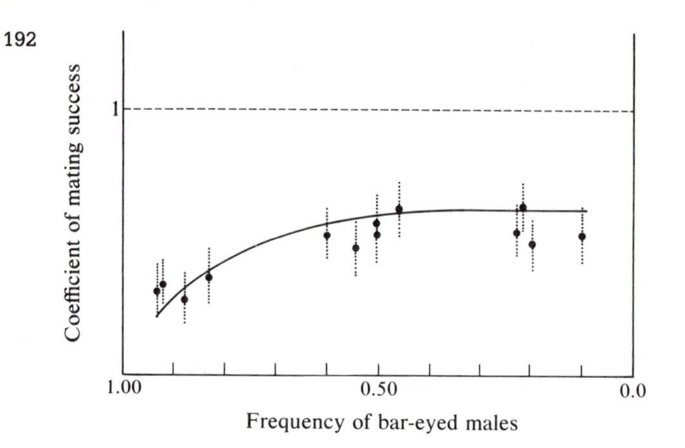

FIGURE 8.2

Sexual selection in *Drosophila melanogaster*. Reproductive success of bar-eyed males. See text for explanation. (From Petit and Ehrman, 1969.)

FIGURE 8.3

(a) The female *Drosophila pseudoobscura*, her abdomen distended with eggs, preens with both forelegs, apparently unaffected by (b) the mating couple sharing the observation chamber with her. Both females are orange-eyed mutants. The male has the dark red, nonmutant eye color of the wild-type *Drosophila*. The wings of the copulating female provide a base of support for her mate; she may even fly while carrying him. (Photograph by A. Heder. Reprinted by permission from *American Scientist*, journal of Sigma Xi, The Scientific Research Society of North America.)

TABLE 8.5

Approach to equilibrium in a simulated selection experiment due to the rare-male advantage in *Drosophila pseudoobscura*

Generation	Eye Color of Parental Pairs			
	Experiment 1		Experiment 2	
	Orange	Purple	Orange	Purple
1	80	20	20	80
2	60	40	29	71
3	68	32	38	62
4	56	44	35	65
5	31	69	41	59
6	63	37	50	50
7	62	38	52	48
8	60	40	50	50
9	50	50	44	56
10	46	54	47	53

SOURCE: After Ehrman, 1970b.

Drosophila appear to be passive during courtship, they clearly exercise discrimination. Males, on the other hand, are indiscriminately active and attempt to mate with anything resembling another *Drosophila*—including other males, females of other *Drosophila* species, dead or etherized flies, and even inanimate objects. An examination of the cues that may be involved when a female chooses one male over another points to olfactory cues as the critical factor (Ehrman and Probber, 1978).

FORCES THAT CHANGE GENOTYPIC FREQUENCIES

Migration, mutation, and selection change both allelic and genotypic frequencies. Certain systems of mating, however, change only genotypic frequencies. We shall consider inbreeding and assortative mating, both of which differ from random mating.

Inbreeding

Inbreeding is a nonrandom system of mating between genetically related individuals. If inbreeding occurs, offspring are more likely than average to have the same alleles at any locus. In this sense, inbreeding is not specific to any particular character. We shall see in the next section that the other major system of nonrandom mating, assortative mating, is character-specific.

Sewall Wright (1921) defined the *coefficient of inbreeding* as the correlation between uniting gametes. Another way of looking at the coefficient of inbreeding involves the probability that both alleles at a locus carried by an individual are identical by descent, i.e., are replicates of those carried by a common ancestor. An easier way of thinking about inbreeding is to consider the coefficient of inbreeding as the percentage decrease in heterozygosity. Inbreeding will change genotypic frequencies by reducing the frequency of heterozygotes, but in the absence of selection, it will not change allelic frequencies. Consider a self-fertilizing type of plant such as Mendel's garden peas, and a single locus with A_1A_1, A_1A_2, and A_2A_2 genotypes. If the plants fertilize themselves, the homozygotes will produce only homozygotes. However, half of the offspring of the heterozygotes will be homozygotes, as illustrated in Figure 8.4. Each generation of self-fertilization reduces heterozygosity by half. Coefficients of inbreeding for self-fertilization are indicated in Figure 8.4. Heterozygosity is reduced by ¼ in matings between full siblings, by ⅛ in matings between half-siblings, and by $^1/_{16}$ in matings between cousins.

Because inbreeding operates equally across all loci to decrease heterozygosity, it can significantly increase homozygosity in a population. The increase in homozygosity has important implications. Recessive alleles are more likely to be expressed. Because most harmful genetic traits are attributable to recessive alleles, offspring of matings between genetically related individuals are more likely to exhibit recessive genetic problems. For this reason, tradition and law have prohibited matings between closely related individuals in most populations. However, low levels of inbreeding

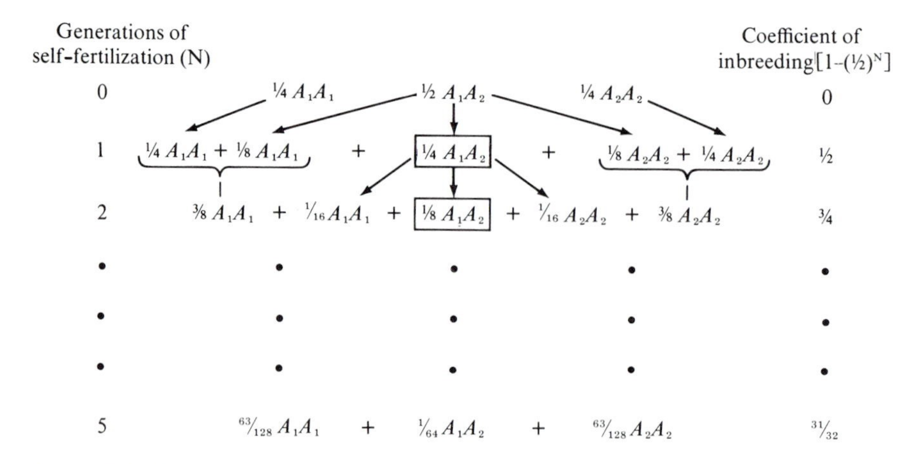

FIGURE 8.4

Reduction of heterozygosity in self-fertilizing plants. Dots represent a continuing series. See text for explanation.

occur in a few populations either by choice (for example, in Japan) or because a small population size limits options (for example, among the American Amish).

The fact that inbreeding reduces heterozygosity can be used to create strains of animals that are essentially genetically identical. For example, each generation of brother-sister matings reduces heterozygosity in the succeeding generation by one-fourth. After twenty generations of such inbreeding, at least 98 percent of all loci are fixed—i.e., all individuals have the same alleles at 98 percent of their loci. Such homozygosity means that regardless of which chromosomes are passed on from parent to offspring, the same alleles will be on each chromosome. This method has been used to create inbred strains of mice or rats. In mice, the term "inbred strain" is reserved for strains that are products of at least twenty generations of full-sibling matings. In fact, most commercially available pedigreed inbred mice have been mated brother-to-sister for fifty to one hundred or more generations. Inbred strains of mice are used in much behavioral genetic research and will be described in more detail in Chapter 10.

Many attempts to create inbred strains fail because harmful recessive alleles become incorporated into the strain by chance, and the reproductive ability of the strain drops. Even those inbred strains that survive usually have some problems. The term *inbreeding depression* has been used to describe the general malaise of inbred individuals caused by the increase in homozygosity. The other side of the coin is *hybrid vigor,* or *heterosis,* which is the increase in viability and performance when inbred strains are crossed to produce an F_1 generation. Crossing inbred strains reintroduces heterozygosity at all loci for which the two strains differ, thus masking the effects of deleterious recessive alleles (Wright, 1977).

In addition to changing genotypic frequencies, inbreeding can alter the average phenotype of a population. This occurs when there is dominance. Complete dominance means that the heterozygote (A_1A_2) will have the same phenotypic value as the dominant homozygote (A_1A_1). If inbreeding occurs, the alleles of these heterozygotes will gradually be distributed into homozygotes, as shown in Figure 8.4. Thus, the frequency of a phenotype influenced by a dominant allele will diminish over the generations because some of the heterozygotes will produce recessive homozygotes (A_2A_2). As a result, the average phenotypic value in the population will be lowered. However, if there were no dominance, the heterozygote would have a value intermediate to the two homozygotes. Because inbreeding causes the alleles of heterozygous individuals to be distributed evenly among the two homozygous types, the resulting average value in a population will not change when the alleles operate in an additive (nondominant) mode.

We can use this fact to determine whether a particular character is influenced by a dominant allele. We have already noted that inbreeding depression frequently occurs with inbred strains of mice. Inbreeding depression is a change in the average value of some trait in a population. Although we are focusing on single-gene characteristics in this chapter, there is some evidence that even a complexly determined trait (such as IQ) is somewhat affected by inbreeding depression, due to the expression of dominant alleles. In other words, inbreeding tends to result in lower IQ scores (Vandenberg, 1971). For example, the risk of mental retardation is more than 3.5 times as high among children of marriages between first cousins as among unrelated controls (Böök, 1957). In addition, children of such cousin marriages generally perform worse on subtests of the Wechsler intelligence test than children of unrelated spouses (Cohen et al., 1963; Schull and Neel, 1965). Some of these results are summarized in Figure 8.5.

A recent study (Bashi, 1977) included a representative sample of Arabs living in Israel, a group in which the frequency of marriages among relatives is about 34 percent. Because such consanguineous marriages are encouraged, even marriages between "double-first cousins" are fairly common (about 4 percent). Double-first cousins are children of siblings who are married to another pair of siblings. Raven's Progressive Matrices, a test of general reasoning, was administered to large samples of children of both first cousins and double-first cousins, as well as to fourth- and sixth-grade children of unrelated marriages. The results in Table 8.6 indicate a slight depression for children of cousin marriages, and a greater depression for children of double-first-cousin marriages. This demonstration of inbreeding depression again suggests that dominant alleles at some loci affect IQ scores.

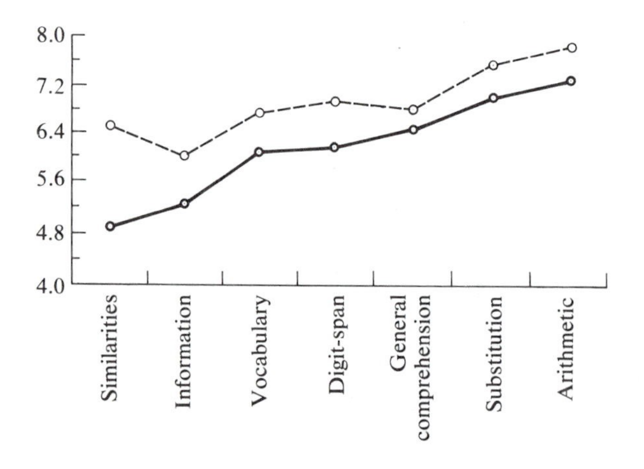

FIGURE 8.5

Scores on seven subtests of the WAIS achieved by 38 children of first cousins, plotted on the solid line. Those of 47 matched controls are plotted on the broken line. (After "School attainment in an immigrant village" by R. Cohen et al. In E. Goldschmidt, ed., *The Genetics of Migrant and Isolate Populations*. Copyright © 1963. Used with permission of Foundation for Child Development.)

TABLE 8.6
Effect of consanguinity on Raven's progressive matrices

Degree of Consanguinity	Grade 4		Grade 6	
	Number	Mean	Number	Mean
Children of unrelated marriages	1,054	8.8	1,054	13.1
Children of first-cousin marriages	503	8.6	467	12.3
Children of double-first-cousin marriages	71	7.9	54	10.6

SOURCE: After Bashi, 1977.

Inbreeding can change the variability as well as the mean of a population. However, even though the effects of inbreeding are severe in individual cases, inbreeding probably does not have an appreciable effect on the means or variances of human traits in the population as a whole. In present-day human populations, the inbreeding coefficient is almost always less than 0.04, even for very small breeding isolates. Thus, changes in population means and variances resulting from inbreeding should be negligible.

Assortative Mating

Old adages are sometimes contradictory. Do "birds of a feather flock together," or do "opposites attract"? Studies of assortative mating, or phenotypic similarity between mates, seek answers to this question. It turns out that assortative mating is almost always in a positive direction. "Birds of a feather" do "flock together," in the sense that individuals who mate tend to be similar in certain characteristics. In contrast to inbreeding, assortative mating is much more common and it is character-specific. Thus, individuals sort themselves into mating couples on the basis of certain phenotypic characteristics.

Like inbreeding, assortative mating affects only genotypic frequencies, not frequencies of alleles. If we think about the influence of a single locus on a trait for which positive assortative mating occurs, assortative mating, like inbreeding, reduces heterozygosity. Homozygotes tend to mate with homozygotes, and some heterozygous individuals in each generation have homozygous offspring. However, for characters influenced by genes at many loci, assortative mating will not greatly reduce heterozygosity. On the other hand, assortative mating for such characters may substantially increase genotypic variability. For example, differences in height are mostly the result of genetic differences. If random mating were to prevail for height, tall women would be just as likely to mate with short men as tall men. Offspring of the matings of tall women and short men would be of moderate stature. If, however, there is positive assortative mating for height (as we know there is), children with tall mothers are also likely to have tall fathers, and the offspring themselves are likely to be taller than average. Positive assortative mating thus increases the variance, in the sense that the offspring differ more from the average than they would if mating were random.

In human populations, assortative mating is common. The highest correlation between husband and wife—about 0.75—is for age. Approximately one-third of the 290 correlations for physical characters summarized by James Spuhler (1968) were in the range 0.10 to 0.20. For example, the correlation for height is about 0.25 and for weight it is about 0.20. Few correlations are negative. Thus, it appears that in general there is some positive assortative mating for physical characters; however, the correlations are relatively low.

Among behavioral characters, most personality-rating correlations between mates are found to be in the 0.10 to 0.20 range, comparable to values observed for the physical characters (Vandenberg, 1972). Correlations for cognitive measures, most notably IQ, have been thought to be much higher. However, IQ and age, for which there is much assortative mating, are related to some extent. It has recently been shown that correlations for specific cognitive abilities, as well as for overall cognitive ability, are also in the 0.10 to 0.20 range, when scores are adjusted for age. Of the specific cognitive abilities (such as memory, spatial ability, verbal ability, and perceptual speed), verbal ability seems to show the most assortative mating (Johnson et al., 1976). Although the correlations are not large, assortative mating can still greatly increase genotypic variability in a population because its effects accumulate generation after generation.

GENETIC VARIABILITY

Population genetics provides another perspective for understanding genetic variability. The point of the Hardy–Weinberg–Castle equilibrium is that genetic variability will be maintained generation after generation in the absence of forces that change the frequency of alleles. The forces that change allelic frequencies can also enhance genetic variability. New alleles can be introduced into a population through migration and mutation. Selection can also maintain genetic variation in a population. Selection can be considered a dynamic flux that maintains genetic variability, rather than an orderly process that proceeds in a single direction to make organisms tightly mesh with their environment. Balanced polymorphisms provide a reservoir of genetic variation for future selection, in the face of changing environmental circumstances. Thus, they may have considerable evolutionary significance.

Equilibrium genotypic frequencies can also be changed by certain systems of mating. Even here it seems that the deck is stacked in the direction of genetic variability. While inbreeding can reduce the frequency of heterozygotes, its rarity in the human species makes its effects quite negligible. Assortative mating, on the other hand, increases genotypic variability for many polygenic characters.

SUMMARY

Population genetics considers allelic and genotypic frequencies in groups of breeding organisms. It provides the basis for understanding how genetic variability is maintained in populations. This idea is formalized in the concept of the Hardy–Weinberg–Castle equilibrium, which shows that frequencies of alleles and genotypes remain stable generation after generation in the absence of forces of change. We can use this concept to estimate frequencies of alleles, since genotypic frequencies in a population in equilibrium should correspond to $p^2 + 2pq + q^2$ for a single-locus, two-allele character. Population genetics is also concerned with the forces that change frequencies, such as migration, mutation, and selection. Balanced polymorphisms can be caused by heterozygote advantage and frequency-dependent selection. Inbreeding and assortative mating change genotypic frequencies without affecting frequencies of alleles.

9

Quantitative Genetic Theory

Until now, we have focused on single-gene influences. Characters influenced by only one gene are often called Mendelian because they show the classical segregation ratios described by Mendel. Although there are many examples of the effects of single genes on behavior (see Chapter 4), most of these interrupt the organism's normal course of development. For example, many of the single-gene influences on human behavior cause mental retardation. These examples demonstrate the power of a single gene to throw the organism out of kilter. However, the normal range of behavior variation is more likely to be orchestrated by a system of many genes.

Because behavioral genetics considers polygenic as well as single-gene influences on behavior, we need to study the theory underlying quantitative inheritance. Since quantitative genetic theory is somewhat abstract, some people erroneously believe that a character is really influenced by genes only if the character shows classical Mendelian inheritance. In this chapter, we will show that one can generalize from single-gene, Mendelian theory to the quantitative effects of multiple genes.

Quantitative genetics is more abstract in that it considers variance in a population, rather than specific genotypes. If we just have two or three types of individuals in a population, as is the case for most Mendelian characters, we can simply count the different types. Figure 9.1a illustrates the distribution in a population of a character determined by a single gene with two

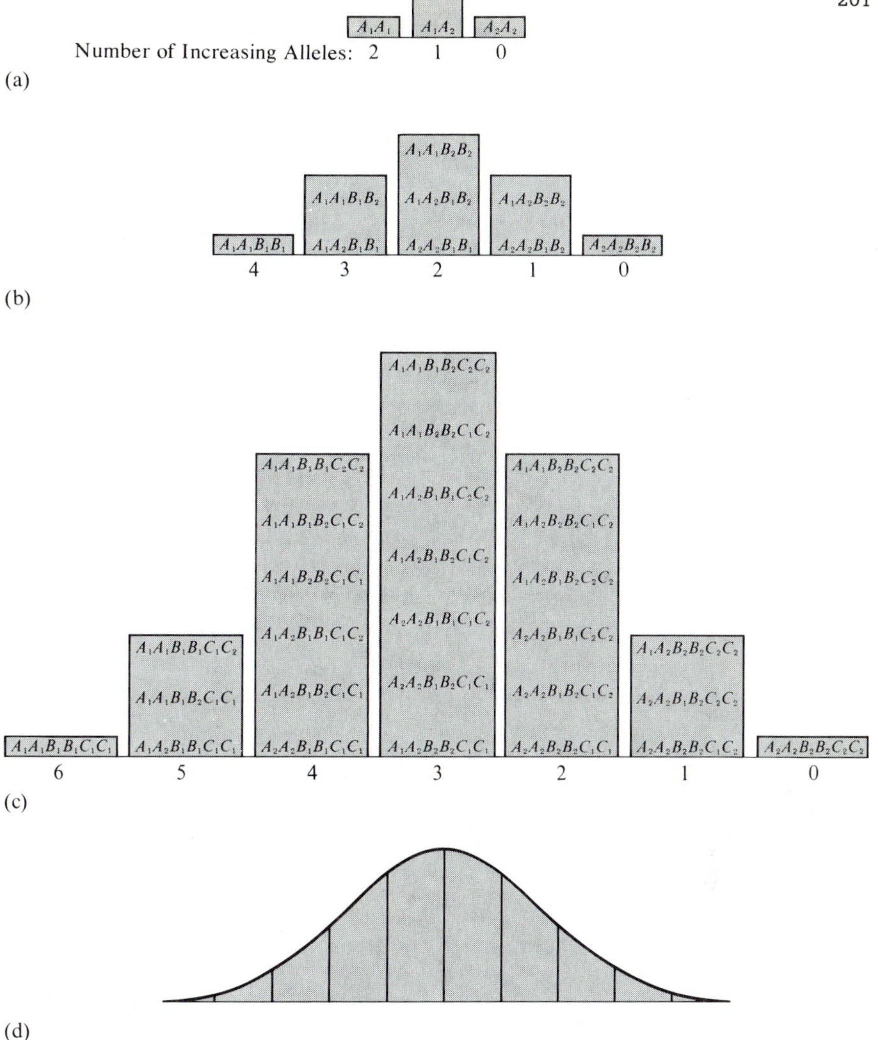

FIGURE 9.1

Single-gene and polygenic distributions for characters with additive gene effects. (a) Distribution of genotypes for a single locus with two alleles. (b) Distribution of genotypes for two loci, each with two alleles. (c) Distribution of genotypes for three loci, each with two alleles. (d) Continuous variation.

alleles. There are three distinct types of individuals. However, there are many characters that show continuous variability similar to the normal bell-shaped curve illustrated in Figure 9.1d. A normal distribution would be approximated if you tossed a handful of 20 coins hundreds of times and each time recorded the number of heads and tails. The average number of heads per toss would be 10, and the other numbers would be evenly distributed around 10, with 0 and 20 being extremely rare. For example, the size of the pea seed is continuously distributed, as Galton discovered when he conducted his early studies of quantitative inheritance. In fact, continuous variation is the rule rather than the exception for behavioral characters. The genetic foundation for such variability has been a recurrent theme in the previous chapters.

Figure 9.1b and c suggest how the qualitative distribution of a single-gene character becomes a quantitative distribution as more loci become involved. For example, when a trait is influenced by two alleles at each of three loci (*A, B, C*), there are 27 different genotypes. Even if we assume that the alleles at the different loci equally affect the trait and that there is no environmental variation, there are still seven different phenotypes, as indicated in Figure 9.1c.

The point is that, even with just three loci and two alleles at each locus, the genotypes begin to approach a normal distribution in the population. When we consider environmental sources of variability and the fact that the effects of alleles at different loci may not be equal (as we assumed in our oversimplified example), it is easy to see that the effects of even a few genes will lead to an approximately normal distribution. Moreover, the complex characters that interest behavioral scientists may be influenced by hundreds of genes. Thus, we should not be surprised to find continuous variation at the phenotypic level.

Because variance is the core concept of quantitative genetic theory, a brief digression is in order.

BRIEF OVERVIEW OF STATISTICS

Statistics Describing Distributions

Figure 9.2 describes the results of testing a small sample of two inbred strains of mice for activity in an open field. The open-field apparatus (see Figure 4.5) is a brightly lit enclosure in which an animal's activity is measured. The activity scores in Table 9.1 were obtained by placing mice one at a time in the open field for 5 minutes. The number of squares entered during this observation period was recorded as each subject's score. How can we describe these two distributions? We first calculate an average score (or some measure of *central tendency*), and then describe the variability of the scores. The average score is not very useful in itself, because it adequately represents the scores only if there is little variability. The distributions in Figure 9.2 show substantial variability, as do distributions for most behavioral characteristics. The *average,* or *mean,* is simply the sum of scores divided by the number of scores:

$$\overline{X} = \frac{\Sigma X}{N}$$

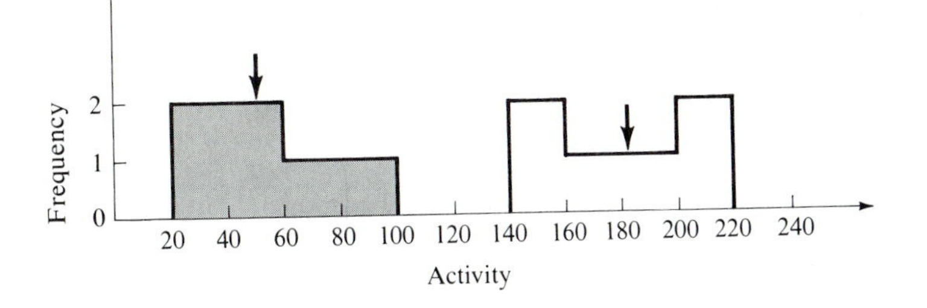

FIGURE 9.2
Frequency histograms of the activity scores of two inbred strains of mice: A (shaded) and C57BL. The means are indicated by arrows.

where $\overline{X}$ refers to mean, ΣX is the sum of scores, and N is the number of scores. The sum of the scores obtained from the six A subjects is 306. Thus,

$$\overline{X}_A = \frac{306}{6} = 51$$

The mean score of C57BL subjects in the sample is:

$$\overline{X}_C = \frac{1092}{6} = 182$$

These means are indicated by the arrows in Figure 9.2, which divide each of the distributions in half.

As the name implies, *variance* is a measure of variability or dispersion. The more spread out the distribution, the greater the variance. Variance is described relative to the mean of the sample. The difference between each subject's score and the mean is computed (i.e., $X - \overline{X}$). Some of these deviations are above the mean and are thus positive numbers; those below the mean are negative numbers. We would like to obtain an average devia-

TABLE 9.1
Activity scores of two inbred strains of mice

A	C57BL
29	155
29	157
44	161
58	199
63	202
83	218

tion from the mean in order to describe the variability in the distribution. However, if we simply summed the deviations from the mean, the positive deviations would balance the negative deviations and the sum would always be zero. The solution is to square the deviations from the mean and then calculate an average squared deviation. This is the definition of variance. However, the sum of the squared deviations from the mean is divided by N − 1 for technical reasons (in order to obtain an unbiased estimate of the variance). In short, the variance of a sample (V) is:

$$V = \frac{\Sigma(X - \overline{X})^2}{N - 1}$$

To illustrate the calculation of V, the data of Table 9.1 are presented again in Table 9.2, along with corresponding deviations from means and squared deviations. As you can see, the variance of activity scores in the C57BL sample is somewhat larger than that in the A sample.

Since variance is the average of the *squared* deviations from the mean, the values obtained are expressed in squared units, rather than in the actual units of measure. In spite of this, as you will see later in this chapter, variance has many important applications in genetics. Nevertheless, a measure of variability expressed in actual units, rather than squared units, is useful. Such a measure is provided by the square root of the variance, the so-called *standard deviation.* If our sample has been drawn at random from a population with a normal distribution of a trait (see Figure 9.3), the sample standard deviation (s) provides a precise estimate of dispersion of the trait within that population.

Approximately two-thirds of the population (68 percent) fall within one standard deviation above and below the mean, and about 96 percent of the observations fall within two standard deviations. Thus, we can predict that in a large population of mice of the A strain (for which we would expect a normal distribution), approximately two-thirds of their activity scores would fall within the range of 51 ± 21.14, i.e., between 29.86 and 72.14. The precision of such estimates increases along with the sample size.

Statistics Describing the Relationship Between Two Variables

When two variables are measured for each subject, or when the same variable is measured on pairs of subjects (for example, pairs of twins, or parents and their offspring), we can analyze the relationship between the two measures. The question is usually phrased in terms of *covariance,* which literally means "shared variance." It tells us the extent to which the measures relate to one another. If there is substantial covariance between two variables (X and Y), then a subject above the mean on X will also likely be above the mean on Y. Like variance, the covariance statistic is based on

TABLE 9.2

Examples of variance estimation from activity scores of two inbred strains of mice

	A	
X_i	$X_i - \overline{X}$	$(X_i - \overline{X})^2$
29	−22	484
29	−22	484
44	− 7	49
58	+ 7	49
63	+12	144
83	+32	1024
$\Sigma X_i = 306$	$\Sigma(X_i - \overline{X}) = 0$	$\Sigma(X_i - \overline{X}) = 2234$
$\overline{X}_A = 51$		$V_A = \frac{2234}{5} = 446.8$
		$s_A = \sqrt{V_A} = \sqrt{446.8} = 21.14$
	C57BL	
X_i	$X_i - \overline{X}$	$(X_i - \overline{X})^2$
155	−27	729
157	−25	625
161	−21	441
199	+17	289
202	+20	400
218	+36	1296
$\Sigma X_i = 1092$	$\Sigma(X_i - \overline{X}) = 0$	$\Sigma(X_i - \overline{X})^2 = 3780$
$\overline{X}_C = 182$		$V_C = \frac{3780}{5} = 756.0$
		$s_A = \sqrt{V_C} = \sqrt{756.0} = 27.50$

deviations from the mean of each variable. It is computed by multiplying each subject's deviation from the mean of X by the subject's deviation from the mean of Y. Cross products of these deviations are summed across subjects and divided by the size of the sample (actually, N − 1). In short, the sample covariance between two variables (Cov_{XY}) is:

$$Cov_{XY} = \frac{\Sigma[(X - \overline{X})(Y - \overline{Y})]}{N - 1}$$

Consider the hypothetical data presented in Table 9.3 and plotted in Figure 9.4. Note that Y tends to increase as X increases. The variance of X is 2.5, and the variance of Y is 10. The covariance between X and Y is 3.

Covariances are easier to interpret if we divide them by an appropriate variance or a product of standard deviations. The two major statistics are *correlation* and *regression*. Sometimes one of these methods is more appro-

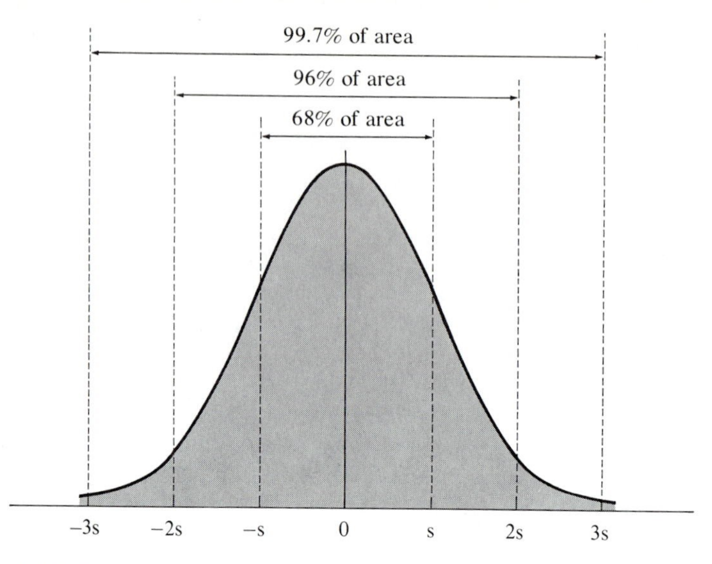

FIGURE 9.3
The normal distribution curve.

priate for certain quantitative genetic analyses, and sometimes the other is more suitable. A correlation coefficient standardizes the covariance by dividing it by the product of the standard deviations of X and Y. This is known as standardization because it results in equal units of X and Y. The regression coefficient divides the covariance by the variance of just one of the variables.

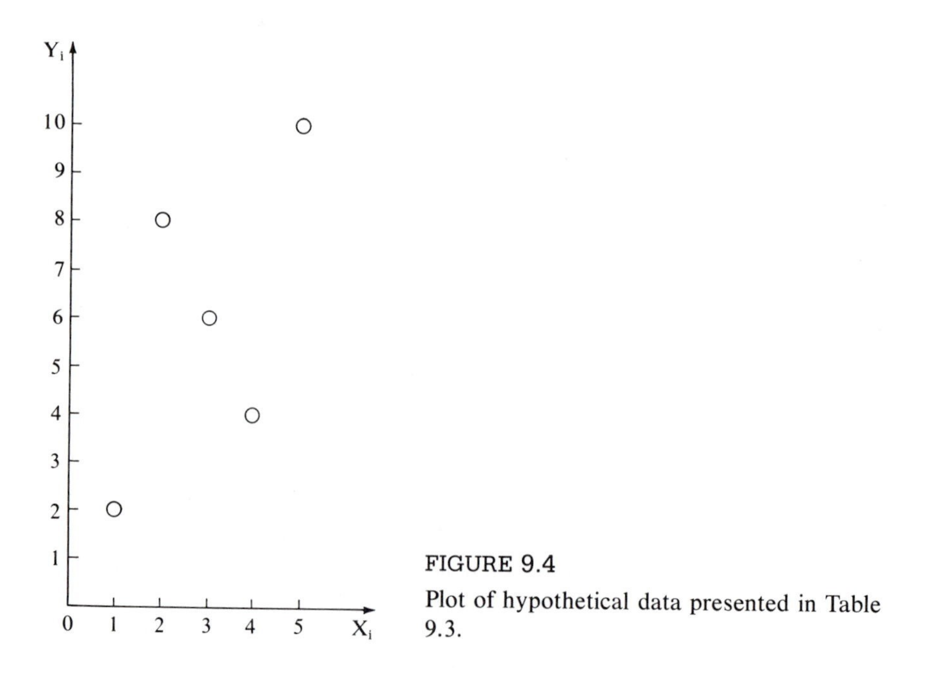

FIGURE 9.4
Plot of hypothetical data presented in Table 9.3.

For example, if we are predicting offspring scores (Y) from parental scores (X), the regression of Y on X (b_{YX}) is the covariance divided by the variance of X. Thus, the regression coefficient is not standardized. It is expressed in terms of observed units of measure. It expresses the number of units, on the average, that Y changes corresponding to each unit change of X.

In summary, the formulas for a correlation coefficient (r_{XY}) and a regression coefficient (b_{YX}) are:

$$r_{XY} = \frac{Cov_{XY}}{\sqrt{(V_X)(V_Y)}} \quad \text{and} \quad b_{YX} = \frac{Cov_{XY}}{V_X}$$

Note that if the standard deviations of X and Y are equal, then the correlation coefficient and the regression coefficient are the same. If $\sqrt{V_X} = \sqrt{V_Y}$, then $\sqrt{(V_X)(V_Y)} = V_X$, so that both the denominator and the numerator are identical for the correlation and the regression.

Table 9.3 illustrates the computation of a correlation. The covariance (3) divided by the product of the standard deviations is 0.6. A correlation of zero (or near zero) indicates that the two variables are independent: scores on one variable tell us nothing about scores on the other. A high positive or negative correlation (close to +1 or −1) indicates a close relationship. Because correlations are standardized, they are easily related to variances. Squaring the correlations yields the percent of variance in one variable related to the variance of the other. The correlation of 0.6 in Table 9.3 indicates that 36 percent of the variance in Y is related to the variance of X (and vice versa). The variance of Y is 10.0. We can express this in terms of variance rather than percent by stating that $0.36 \times 10 = 3.6$ is the variance of Y

TABLE 9.3

Sample calculation of a correlation coefficient, r_{XY}

	X	$X - \bar{X}$	$(X - \bar{X})^2$	Y	$Y - \bar{Y}$	$(Y - \bar{Y})^2$	$(X - \bar{X})(Y - \bar{Y})$
	1	−2	4	2	−4	16	+8
	2	−1	1	8	+2	4	−2
	3	0	0	6	0	0	0
	4	+1	1	4	−2	4	−2
	5	+2	4	10	+4	16	+8
Σ:	15	0	10	30	0	40	12

$$V_X = \frac{10}{4} = 2.5 \qquad V_Y = \frac{40}{4} = 10 \qquad Cov_{XY} = \frac{12}{4} = 3$$

$$r_{XY} = \frac{3}{\sqrt{(2.5)(10)}} = 0.6$$

related to the variance of *X*. This means that the rest of the variance of Y, 6.4, is not related to the variance of X.

We have been using the phrase "related to" rather than "caused by" because correlations do not in themselves prove the existence of a causal relationship. In genetics, however, there are clear causal associations between genotype and phenotype. When a causal relationship between two variables (X and Y) has been established, the correlation coefficient can be used to estimate the variance in Y caused by the variation in X. In the previous example, this would mean that, if X is held constant, 64 percent of the variance in Y will remain.

The regression of Y on X for the same data (Table 9.3) is 1.2:

$$b_{YX} = \frac{Cov_{XY}}{V_X} = \frac{3.0}{2.5} = 1.2$$

Thus, on the average, for every unit of change in X, Y changes 1.2 units. This regression coefficient can be used to show how the variance of Y may be partitioned into two parts—one due to variation in X, and one that is independent of X. In overview, we will use an equation to predict scores on Y, given scores on X. Then we will obtain the variance of the Y scores as they were predicted by X scores. The deviation of the actual Y scores from the predicted Y scores can be squared and averaged to produce the variance of Y, independent of X.

The regression coefficient describes the change in Y predicted by a unit change in X. Such a prediction may seem unnecessary, given that we already have information regarding both variables. However, from the sample regression, we may estimate Y for other members of the population for whom we have information only regarding variable X. More importantly, for our present purpose, the regression can be used to draw a straight line through the observed points, as in Figure 9.5. This line is called a "least squares" regression line because the sum of the squared deviations from the predicted points is at a minimum. This prediction equation is:

$$\hat{Y} = \overline{Y} + b_{YX}(X - \overline{X})$$

where $\hat{Y}$ is the predicted value of Y, given information on X. Thus, the predicted value of Y is derived from the deviation of the X score from its mean, weighted by the regression coefficient. From the data of Table 9.3,

$$\hat{Y} = 6 + 1.2(X - 3) = 6 + 1.2X - 3.6 = 2.4 + 1.2X$$

Using this equation, we can calculate the expected value of Y corresponding to each observed value of X in Table 9.3. These observed and expected values are presented in Table 9.4 and graphed in Figure 9.5. For example, the X score of 2 predicts a Y score of 4.8 because $2.4 + 1.2\ (2) = 4.8$. This predicted value has been entered as a point on the straight line in Figure 9.5.

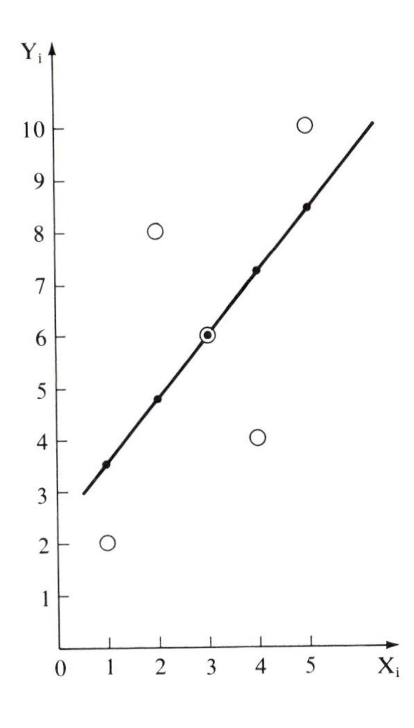

FIGURE 9.5

Plot of observed values (open circles) and expected values (small black dots) of Y, corresponding to observed values of X. Expected values are obtained from the regression equation, $\hat{Y} = 2.4 + 1.2X$.

TABLE 9.4

Observed and expected values of Y

X	Y	$\hat{Y}$
1	2	3.6
2	8	4.8
3	6	6.0
4	4	7.2
5	10	8.4

TABLE 9.5

Calculation of the variance in Y due to both regression and deviations from regression

	Y	$\hat{Y}$	$\hat{Y} - \overline{\hat{Y}}$	$(\hat{Y} - \overline{\hat{Y}})^2$	$Y - \hat{Y}$	$(Y - \hat{Y})^2$
	2	3.6	−2.4	5.76	−1.6	2.56
	8	4.8	−1.2	1.44	+3.2	10.24
	6	6.0	0.0	0.00	0.0	0.00
	4	7.2	+1.2	1.44	−3.2	10.24
	10	8.4	+2.4	5.76	+1.6	2.56
Σ:	30	30.0	0.0	14.40	0.0	25.60

$V_Y = 10 \qquad V_{\hat{Y}} = \frac{14.4}{4} = 3.6 \qquad V_{Y-\hat{Y}} = \frac{25.6}{4} = 6.4$

The variance of these predicted scores of Y is the variance of Y due to variation in X. As calculated in column 4 of Table 9.5, the variance of the predicted Y scores is 3.6, the same answer obtained using the correlation coefficient (r^2V_Y). Of course, the variance of Y *not* predicted by X is the rest of the variance of Y (that is, 10 − 3.6 = 6.4). However, we can directly calculate the variance of Y *not* predicted by X by obtaining the deviation of each Y value from its predicted value, and then deriving the variance of these deviations as in column 6 of Table 9.5.

HISTORICAL NOTE

In 1877, the first regression line was drawn by Galton, the father of behavioral genetics, to describe quantitative inheritance. As an example, he chose the size of the seed in the pea plant. He knew that parental plants with large seeds were likely to have offspring with larger than average seeds. He plotted parent and offspring seed sizes, and drew the regression line, reproduced as in Figure 9.6. Galton noticed that the slope of the line described the following relationship: As parental size increases one unit, the offspring size increases one-third unit. Thus, the regression of offspring on parent was 0.33, which is the covariance divided by the variance of the parents. This similarity could be ascribed to inheritance because all plants were raised in similar environments.

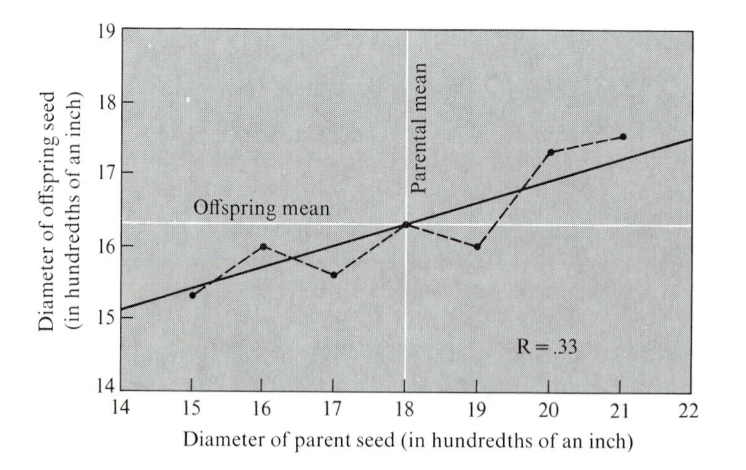

FIGURE 9.6

First regression line. Drawn by Galton in 1877 to describe the quantitative relationship between pea seed size in parents and offspring. (Courtesy of the Galton Laboratory.)

Because the size of the pea seed varies continuously, the hereditary mechanism did not appear to operate like Mendelian elements. This led to a controversy between the so-called Mendelians and biometricians during the early twentieth century. After the rediscovery of Mendel's laws, some researchers began to equate heredity with Mendelian segregation ratios. It was difficult for Mendelians to reconcile continuous variation with the type of qualitative, discrete difference, mediated by single genes, with which they had worked. The biometricians, on the other hand, vigorously pursued Galton's approach to problems of the inheritance of continuously varying characteristics. However, they supported the blending hypothesis of inheritance, and generally regarded Mendelian inheritance as an unimportant exception to the general rule. With justification, they pointed to the importance of the smoothly continuous, quantitative characteristics, such as height, weight, and intelligence. The biometricians thought that the characteristics investigated by Mendelians, resulting in qualitative differences and usually abnormalities, could not account for such continuous distributions.

The groundwork for the resolution of this conflict was actually provided by Mendel himself, with his suggestion that a certain characteristic might be due to two or three genes. General acceptance of this idea, however, was not forthcoming until the work with plants by H. Nilsson-Ehle (1908) and E. M. East and collaborators (East and Hayes, 1911; Emerson and East, 1913). These researchers showed that, if it was assumed that a number of gene pairs, rather than just one pair, each exerted a small, cumulative effect on the same characters, and if the effects of environment were taken into consideration, the final outcome would appear to be a continuous distribution of the character instead of discrete categories such as those featured in the typical Mendelian researches. This was quite different from the blending hypothesis. In this multiple-factor hypothesis, it was not presumed that the hereditary determiners vary continuously in nature from one individual to another, creating a continuous distribution in the population. Rather, the genes were acknowledged to occur in discrete alternate states (typically two, sometimes more). But when a number of such discrete units bear on the same character, the final outcome approximates a continuous distribution, as discussed in the beginning of this chapter. Elaborate statistical development of this notion was provided by R. A. Fisher (1918) and by Sewall Wright (1921). Their work presented convincing demonstrations that the biometrical results, in fact, follow logically from a multiple-factor extension of Mendel's theory.

THE SINGLE-GENE MODEL

Although quantitative genetics was developed for application to characters influenced by genes at many loci, the underlying model is based on segregation at only a single locus. Once we have described gene action at a single locus, we can generalize to the polygenic case.

Genotypic Value

Genotypic values are expressed as deviations from the mid-homozygote point, as indicated in Figure 9.7. The homozygote with the higher value will be referred to as A_1A_1. The genotypic value for A_1A_1 will be +a. The genotypic value of other homozygote, A_2A_2, is −a. The values +a and −a are equidistant from the mid-homozygote point. However, the genotypic value of the heterozygote, A_1A_2 (symbolized by d), is dependent on the gene action at the locus. If there is no dominance, d will equal zero and will fall at the mid-homozygote point. If A_1 is partially dominant to A_2, d will be closer to A_1A_1, as in the example in Figure 9.7. If dominance is complete, that is, if the observed value for A_1A_2 equals that of A_1A_1, then d = +a.

Additive Genetic Value

The additive effect of genes is merely the extent to which they "add up" or sum according to gene dosage. More specifically, the additive genetic value is the genotypic value expected from gene dosage, as illustrated in Figure 9.8. Gene dosage is the number of a particular allele (say the A_1 allele) present in a genotype. As gene dosage increases by one (for example, from the A_2A_2 genotype to A_1A_2), the expected genotypic value increases by a constant unit. If the frequencies of the two alleles were not equal, we would need to weight each allele according to their respective frequencies in the population. But this will not affect our example. If there is no dominance, these expected genotypic values will be the same as the actual genotypic values. However, dominance can cause the actual genotypic values to deviate from expected values.

Another way of thinking about additive genetic values is to consider that every allele in the genotype has some average effect. In this sense, the additive genetic value is the sum of these average effects of alleles across the genotype. Additive genetic value is a fundamental component of genetic influence because it represents the extent to which genotypes "breed true" from parents to offspring. If a parent has one "dose" of a particular allele,

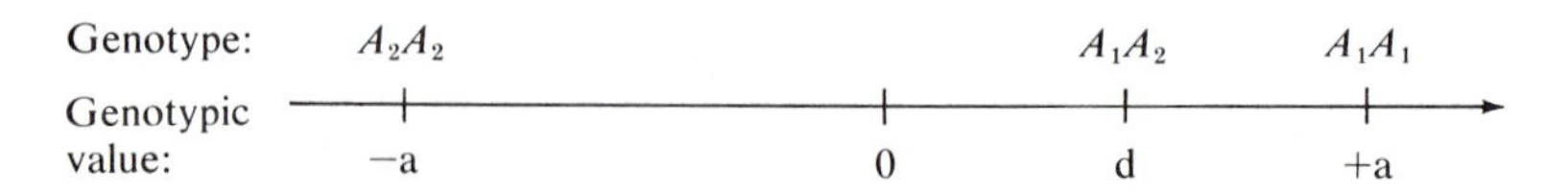

FIGURE 9.7

Assigned genotypic values. (After *Introduction to Quantitative Genetics* by D. S. Falconer. Copyright © D. S. Falconer, 1960, p. 113, Longman Group, Ltd., London and New York.)

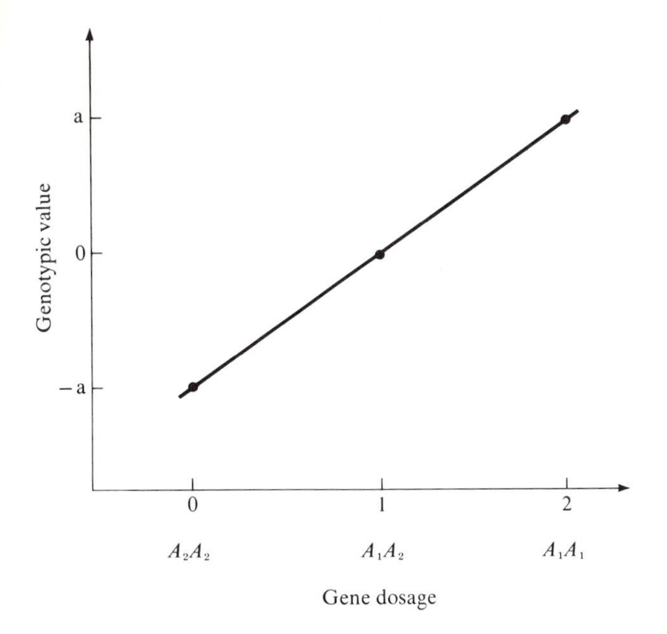

FIGURE 9.8
Additive genetic values predicted by gene dosage, when d = 0. Because there is no dominance, the genotypic values are the same as the additive genetic values.

the offspring of that parent each have a 50 percent chance of receiving that allele. If the offspring receive that allele, its effect will be added in to the same extent it was added in the total genotype of the parent. It does not matter how many alleles are involved at a locus (or, as we shall see in the next section, how many loci are involved). Additive genetic values are simply the extent to which the effects of the alleles add up according to gene dosage.

Dominance Deviation

If there is dominance, genetic values do not simply add up according to gene dosage as in Figure 9.8. Dominance deviations are the difference between the expected (or additive) genotypic value and the actual genotypic value. Dominance allows for the fact that alleles at a given locus can interact with one another, rather than simply adding up in a linear fashion. For example, if there is complete dominance, genotypic values will fall on points as plotted in Figure 9.9. In Figure 9.8, the genotypic values were the same as the additive genetic values, and they fell on a straight line. In Figure 9.9, however, the genotypic values are not on a straight line. For this reason, we use a regression equation to fit the best straight line through these points. Regression of genotypic value on gene dosage yields the genotypic values

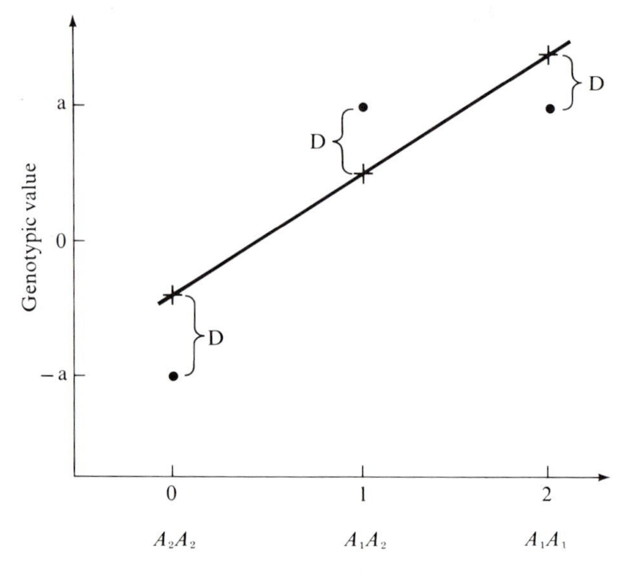

FIGURE 9.9

Genotypic values (black dots) when dominance is complete. Regression line predicts additive genetic values (crosses) based on gene dosage. Dominance deviations (D) are the difference between the additive genetic values and the actual genotypic values.

predicted by gene dosage. This, of course, is our definition of additive genetic values. Thus, the crosses on the regression line in Figure 9.9 are additive genetic values. If there is dominance, this prediction of genotypic values from gene dosage will be slightly off. Dominance, as represented by the Ds in Figure 9.9, is thus the deviation of the genotypic value from the regression line, which represents the predicted genotypic values based on gene dosage.

Dominance is important because it represents genetic influence that does not "breed true." If dominance occurs, a parent's genotypic value is due to some particular combination of alleles at a locus. Offspring cannot receive both of those alleles from the parent. Therefore, they will be genetically different from the parent to some extent if alleles do not add up in their effect. In summary, we have partitioned the genotypic value into two parts—one predicted by gene dosage, and one that is not. Additive genetic values are the extent to which genotypic values add up or sum according to gene dosage; dominance is the extent to which they do not add up.

THE POLYGENIC MODEL

Not only can we consider the additive and nonadditive effects of alleles at a single locus, we can also sum these effects across loci. This is the essence of the polygenic extension of the single-gene model. Just as additive genetic values are the summation of the average effects of two alleles at a single locus, they may also be summed across the many loci that may influence a particular phenotypic character. Similarly, dominance deviations from additive genetic values may also be summed for all the loci influencing a character. Thus, it is relatively easy to generalize the single-gene model to a polygenic one with many loci, each with its own additive and nonadditive effects. However, we need to introduce one more concept, *epistatic interaction*.

Epistatic Interaction Deviation

Dominance is the nonadditive interaction of alleles at a single locus. When we consider several loci, we need to consider the possibility that a particular allele interacts not only with the allele at the same locus on the homologous chromosome, but also with alleles at other loci. This type of interaction is called *epistasis*. In other words, dominance is *intralocus* interaction, and epistasis is *interlocus* interaction. For example, consider two loci (*A* and *B*) that affect a phenotypic character. Both the additive genetic values and the dominance deviations are summed across the two loci. However, a particular combination of a certain allele at locus *A* and another allele at locus *B* may influence the phenotype in ways not explainable by the additive and dominance effects. *Epistasis* refers to this sort of effect.

In summary, we may partition genetic effects into three components: additive, dominance, and epistatic. At a single locus, the genotypic value includes additive and dominance effects. When we consider the effects across two or more loci, the additive and dominance effects are summed, but may not yield the joint genotypic value, due to epistatic interaction among alleles at different loci. In symbolic terms,

$$G = A + D + I$$

where G is the genotypic value due to all loci, A is the sum of the additive genetic values across all loci, D is the sum of the dominance deviations across all loci, and I symbolizes the deviations due to epistatic interactions. Epistatic interactions may be of several types. They may involve interactions between additive genetic values at different loci, between dominance deviations at different loci, between additive genetic values at one locus and dominance deviations at another locus, and so on.

Variance

Up to this point, we have considered only genetic influences on a phenotype. Although that is a workable approach for traits such as those that

Mendel studied in pea plants in a controlled environment, environmental influences are so important for other traits that analyses must consider both the genetic and environmental factors. The basic model of quantitative genetic theory simply says that the phenotype of an individual is due to a genotypic value (including A, D, and I), and an environmental effect due to all nongenetic causes. However, science seldom studies a single individual. Our focus is on phenotypic differences in a population and on the genetic and environmental differences that create those differences. So, instead of thinking of P as an individual's phenotypic value, we will consider it as the individual's deviation from the population mean.

Thus, quantitative genetic theory begins with a model in which observed (phenotypic) deviations from the mean for some character in a population are a function of environmental (E) and genetic (G) deviations, which combine in an additive (linear) manner. However, this model may also include a nonadditive, or interaction, term (G × E) to deal with possible nonadditive combinations of genetic and environmental effects, just as dominance and epistasis allow for the possibility of nonadditive effects for single and multiple loci. Symbolically,

$$P = G + E + (G \times E)$$

The symbol G × E does not necessarily refer to multiplication of G and E. It designates the contribution of some nonadditive function of G and E to the phenotype, independent of the main effects of G and E. That is, an environmental factor may have a greater effect on some genotypes than on others, and a genotype may be expressed differently in some environments than others. We shall consider G × E interactions in greater detail in Chapter 14.

Each of the components is expressed as a deviation from the mean, but we want to express them in terms of variance. As described earlier in this chapter, variance is the sum of individuals' squared deviations from the mean, divided by the number of individuals. Let us take the G deviation and express it as variance. The variance of G simply involves squaring the genetic deviations, summing the squared deviations, and dividing by the sample size. Let us also obtain the variance of each of the components of G = A + D + I. The variance of G (V_G) can be expressed as the covariance of G with itself. (The covariance of a variable with itself is the same as its variance.) Therefore,

$$\begin{aligned} V_G &= \text{Cov(G)(G)} \\ &= \text{Cov(A + D + I)(A + D + I)} \\ &= V_A + V_D + V_I + 2\text{Cov(A)(D)} + 2\text{Cov(A)(I)} + 2\text{Cov(D)(I)} \end{aligned}$$

Because A, D, and I are not correlated, we are left with:

$$V_G = V_A + V_D + V_I$$

In other words, genetic variance is due to additive genetic variance, dominance variance, and variance resulting from epistatic interactions. Additive genetic values are equivalent to genotypic values expected from gene dosage. Thus, additive genetic variance may be thought of as genetic variance due to variation in gene dosage. In the same way, dominance and epistatic variance (or nonadditive genetic variance) is the genetic variance that is not predicted by gene dosage. It should be noted that, even if dominance is complete (that is, $d = +a$), genetic variance may still have a substantial component due to additive genetic variance.

In a similar manner, we can determine the variance for the general model $P = G + E + (G \times E)$. The symbol $G \times E$ is defined as being uncorrelated with either G or E; however, G and E may themselves be correlated. The variance of the phenotypic deviations (V_P) is a function of the squared deviations for the other components, as follows:

$$\begin{aligned} V_P &= \mathrm{Cov}(P)(P) \\ &= \mathrm{Cov}[G + E + (G \times E)][G + E + (G \times E)] \\ &= V_G + V_E + 2\mathrm{Cov}(G)(E) + V_{G \times E} \end{aligned}$$

In other words, observed variance in a population includes components due to genetic variance (V_G) and those due to environmental variance (V_E). Phenotypic variance also contains components added by the correlation between genetic and environmental effects [2Cov(GE)], as well as by the interaction between G and E. Although error of measurement is also likely in the variance of a phenotype, we will ignore it for now.

An Example of the Polygenic Model

An example illustrating this model may be helpful. The example is hypothetical because we cannot often measure genotypic values, and we do not know the environmental values. We measure the phenotypes. Behavioral genetics employs methods to estimate genetic and environmental variance from observed phenotypic values, as discussed in the next section. For now, we will use a hypothetical example to clarify the underlying model.

Suppose that we knew the genetic, environmental, and phenotypic deviations from the mean for a number of individuals, as indicated in Table 9.6. Because these values are expressed as deviations from the mean, the mean in all cases is zero. In this example, the genetic variance is 2.0, the environmental variance is 2.0, and the phenotypic variance is 4.0. Thus, $V_P = V_G + V_E$. There is no variance added by the covariance between G and E because there is no covariance between G and E in this example. (Satisfy yourself that this is true by multiplying the deviations of G by the deviations of E and then summing the cross products.)

TABLE 9.6

Hypothetical genetic, environmental, and phenotypic deviations from the mean for five individuals

Individual	G	+	E	=	P
1	−2		+1		−1
2	−1		−2		−3
3	0		0		0
4	+1		+2		+3
5	+2		−1		+1
	$V_G = 2.0$		$V_E = 2.0$		$V_P = 4.0$

NOTE: To keep the example as simple as possible, we will consider these individuals as constituting a population rather than a sample, thus ignoring problems of sampling. As a result, variances are obtained by dividing by N, rather than N − 1.

SOURCE: After Plomin, DeFries, and Loehlin, 1977.

Now let us suppose that genes and environment are perfectly correlated, as in Table 9.7. The genetic and environmental variances remain the same (2.0), but the phenotypic variance is now 8.0 instead of 4.0. The added variance is due to the correlation between genetic and environmental deviations: $V_P = V_G + V_E + 2\text{Cov(G)(E)} = 2 + 2 + 4 = 8$. Although we shall consider the genotype-environmental correlation in greater detail in Chapter 14, it should be noted that even if we somehow removed variance due to the correlation between G and E, V_G and V_E would remain unchanged. In fact, correlation between G and E will contribute substantially to V_P only when both V_G and V_E are substantial (Jensen, 1974). Our example illustrates a positive correlation between G and E, in that large deviations in G correspond to large deviations in the same direction in E. Negative correlation between G and E would decrease rather than increase V_P.

TABLE 9.7

Hypothetical genetic, environmental, and phenotypic deviations from the mean for five individuals when genetic and environmental deviations are perfectly correlated

Individual	G	+	E	=	P
1	−2		−2		−4
2	−1		−1		−2
3	0		0		0
4	+1		+1		+2
5	+2		+2		+4
	$V_G = 2.0$		$V_E = 2.0$		$V_P = 8.0$

SOURCE: After Plomin, DeFries, and Loehlin, 1977.

TABLE 9.8

Hypothetical genetic and environmental deviations from the mean and phenotypic values for five individuals when genetic and environmental deviations are perfectly correlated and when there is an interaction between G and E

Individual	G	+	E	+	G × E	=	P
1	−2		−2		+4		0
2	−1		−1		+1		−1
3	0		0		0		0
4	+1		+1		+1		+3
5	+2		+2		+4		+8
	$V_G = 2.0$		$V_E = 2.0$		$V_{G \times E} = 6.8$		$V_P = 14.8$

SOURCE: After Plomin, DeFries, and Loehlin, 1977.

Let us now add the G × E interaction to the example, retaining the positive correlation between G and E. We said that G × E refers to any nonadditive effect of G and E. In our example, however, we will assume that the nonadditive function is, in fact, G multiplied by E. (See Table 9.8.) The variance of the G × E values around their mean of 2.0 is 6.8. Genetic variance, environmental variance, and variance due to the correlation between G and E [2Cov(G)(E)] remain 2.0, 2.0, and 4.0, respectively. Adding the $V_{G \times E}$ term yields 14.8, which is the phenotypic variance.

Although we cannot often measure genetic variance, environmental variance, or genotype-environment interaction directly, this hypothetical example indicates that all four components can contribute to phenotypic variance for a character. Because we cannot measure these components directly, we estimate them indirectly from the resemblance of relatives.

COVARIANCE OF RELATIVES

If we could measure genetic and environmental effects for individual subjects, we could directly estimate V_G and V_E in populations. Instead, our analyses proceed indirectly, estimating the various genetic and environmental components of variance from relationships that differ in genetic or environmental relatedness. For example, full siblings who have both parents in common are twice as similar genetically as half-siblings with only one parent in common. If genes influence a particular behavior, then the double genetic similarity of full siblings should make them more similar for that behavior than half-siblings. Quantitative behavioral genetic methods involve comparisons of several such relationships, in which genetic similarity is varied while environmental similarity is held constant or vice versa. The purpose of this section is to provide the theoretical background for behavioral genetics studies of familial resemblance.

Covariance

Earlier in this chapter, we discussed covariance, correlation, and regression. Covariance between X and Y is the sum of the cross products of the deviations from the mean of X and the corresponding deviations from the mean of Y, divided by $N - 1$. Correlation and regression express covariance as a proportion of variance. For now, we will focus on covariance. Previously, we considered the covariance between two variables, X and Y, for many individuals—that is, the extent to which individuals' scores on X covaried with scores on Y. Now, we will consider covariance between relatives rather than between variables. For example, instead of considering the covariance between the two traits, X and Y, for many individuals measured on both traits, we will consider the covariance between twins or between parents and their offspring for a single variable. If members of a family are more similar than individuals picked at random from the population (i.e., if their deviations from the mean are in the same direction), there is covariance.

Both genetic and environmental hypotheses predict similarities between relatives. Relatives share genes to some extent, and thus should be similar if genes affect the particular behavior under study. Environmental hypotheses also predict that members of the same family should be similar because they are subject to much the same environmental influences. For example, if certain child-rearing practices in human families are thought to be important influences on the development of personality, then children in the same family subjected to similar child-rearing practices should be similar in those aspects of personality. Later, we shall see how the knowledge that certain family relationships are not as similar genetically or environmentally as others provides the basis for untangling genetic and environmental influences. However, the point here is that both genetic and environmental hypotheses predict covariance among relatives living together.

There is zero covariance between pairs of unrelated individuals picked at random. Because such individuals share neither genes nor environment, their scores do not covary. Other relationships, however, share both genes and environment. We can describe covariance between relatives as $\mathrm{Cov}(P_1)(P_2)$, where P_1 is the phenotype of one relative and P_2 is the phenotype of the other. In the previous section, we noted that $P = G + E$, and we can substitute that for $\mathrm{Cov}(P_1)(P_2)$:

$$\mathrm{Cov}(P_1)(P_2) = \mathrm{Cov}(G_1 + E_1)(G_2 + E_2)$$

Shared and Independent Influences

Not all genetic, nor environmental, influences for a particular behavior make family members similar to one another. Identical twins are, of course, identical genetically, and thus share all genetic influences. However, for other family relationships there are both shared genetic influences and those that are not shared (due to segregation). Genetic theory predicts differences between genetically related individuals other than identical twins. In contrast, environmental theories rarely predict differences for members of the same family.

Although V_A, V_D, and V_I contribute in various ways to different familial relationships, for the moment we shall consider only genetic variance that the relatives have in common (V_{G_c}). Parents and offspring are first-degree relatives, as are full siblings. Consider a single locus with two alleles. An offspring has a fifty-fifty chance of inheriting one particular allele rather than the other from the parent. For this reason, first-degree relatives are 50 percent similar genetically; in other words, half of the genetic variance is shared between them. The other half of the genetic variance does not covary between them, so, it makes them different from one another. Such reshuffling of genes is the consequence of meiosis and the source of genetic variability. Thus, we can divide the genetic contribution to the phenotype of an individual into two parts—that part which the individual shares, or has in *common,* with the relative (G_c); and the part that is not shared with the relative (G_w). Influences not shared by family members have traditionally been labeled with a w to indicate differences *within* families.

Similarly, some environmental influences are shared by relatives, while other aspects of the environment make family members different from one another. Some parents are physically punitive toward their children. If punitiveness affects some aspect of personality development (such as aggression), then it will make the children in the family more similar to each other in aggressiveness. There are very few known examples of systematic environmental factors that make family members different from one another, although such influences can be important. For example, the order of birth may well cause behavioral differences among full siblings. If earlier-born children are different from later-born children in the same family, some environmental factor (perhaps prenatal influences, child-rearing practices, or interactions with siblings) operating within the family makes them different from one another. It is clearly not a genetic factor because full siblings are equally similar genetically to each other and to their parents. Other environmental influences of this type include those factors that are independent of the family relationship, such as interactions in school and with peers, not shared by family members. Thus, we can also divide the environmental contribution to the phenotype into influences shared with the relative (E_c) and those independent of the relative (E_w). In the previous equation, by definition, only G_c and E_c can contribute to the phenotypic covariance between relatives:

$$\mathrm{Cov}(P_1)(P_2) = \mathrm{Cov}(G_c + E_c)(G_c + E_c)$$

The covariance of G_c with G_c is equivalent to the variance of G_c (that is, V_{G_c}). As we indicated earlier, a variable completely covaries with itself, meaning that the covariance of a variable with itself is the same as its variance. In the same way, $Cov(E_c)(E_c) = V_{E_c}$.

Now we can express the phenotypic covariance between relatives in terms of components of variance:

$$Cov(P_1)(P_2) = V_{G_c} + V_{E_c}$$

In other words, for a particular character, the covariance between relatives includes the genetic variance and the environmental variance resulting from shared genetic and environmental influences.

Genotype-Environment Correlation and Interaction

The model we have used up to this point is oversimplified. Earlier, we mentioned the correlation and interaction between genetic and environmental factors. These components of variance also enter the picture when we consider the covariance among relatives. $Cov(G_c + E_c)(G_c + E_c)$ also includes the covariance between G_c and E_c. Covariance between genetic and environmental deviations can add to phenotypic variance. It can also add to the covariance between relatives. In addition, when we substituted G + E for P, we did not consider the G × E interaction. The G × E interaction shared by relatives will also contribute to their phenotypic covariance. In Chapter 14, we shall consider the genotype-environment correlation and interaction in more detail, as well as their effects on behavioral genetic analyses.

Genetic Covariance Among Relatives

Our general model for the covariance of relatives is also too simple because it treats only shared genetic variance, rather than distinguishing between V_A, V_D, and V_I. These components of genetic variance contribute variously to different types of family relationships. (See Table 9.9). Parents and their offspring share one-half of their additive genetic variance, as discussed in the previous section. (For this reason, additive genetic variance provides a measure of the extent to which characters "breed true.") However, parents and offspring do not share genetic variance due to dominance. Remember that dominance is the result of nonadditive combinations of alleles at loci. Offspring cannot obtain a chromosome pair from one parent. Thus, although dominance may contribute to the phenotypes of parent and offspring, this genetic factor will not be shared by them.

TABLE 9.9

Contribution of additive genetic (V_A), dominance (V_D), and common environmental (V_{E_C}) influences to the phenotypic covariance of relatives

Phenotypic Covariance Between:	V_A		V_D		V_{E_c}
Parents and offspring (PO)	½	+	0	+	$V_{E_{c(PO)}}$
Half-siblings (HS)	¼	+	0	+	$V_{E_{c(HS)}}$
Full siblings (FS)	½	+	¼	+	$V_{E_{c(FS)}}$
Fraternal twins (DZ)	½	+	¼	+	$V_{E_{c(DZ)}}$
Identical twins (MZ)	1	+	1	+	$V_{E_{c(MZ)}}$

Another factor that contributes to genetic covariance among relatives is assortative mating (discussed in Chapter 8). Assortative mating adds to the genetic similarity between parents and their offspring, as well as between siblings (Jensen, 1978). For example, if assortative mating exists, a correlation between mothers and their children will include not only the genetic similarity between the mothers and their children, but also some part of the genetic similarity between the children and their fathers. However, we can get around this problem by using regressions of the offspring on the average parental score (called midparent score), which is mathematically independent of assortative mating, as discussed in the next chapter (Plomin, DeFries, and Roberts, 1977).

Siblings, like parents and their offspring, share half of the additive genetic variance that influences a character. However, siblings also share one-fourth of the dominance variance, since full siblings can be expected to receive the same alleles from both parents one-fourth of the time, and thus have the same dominance deviation.

Fraternal twins are just siblings who happen to be born at the same time. Two eggs are fertilized by different sperm. For this reason they are sometimes referred to as dizygotic (two-zygote) twins. Like other siblings, dizygotic (DZ) twins can be the same sex or of opposite sexes, and they share half of the additive genetic variance and one-fourth of the variance due to dominance. Twins are born about once in every 83 births, and two-thirds of these are fraternal twins. The other third of twin births are identical twins. They are called monozygotic (MZ) twins because they begin life as a single zygote that splits sometime during the first few weeks of life. Because they are genetically identical, identical twins are always of the same sex. They share all genetic variance—V_A, V_D, and V_I.

Finally, half-siblings who share only one parent thus share only one-fourth of the additive genetic variance (half as much as full siblings). However, unlike full siblings, half-siblings do not share any dominance variance. Because half-siblings have only one parent in common, they cannot inherit the same chromosome pairs, and thus cannot share in allelic interactions at a given locus.

Sometimes we need to consider the covariance of behavioral measures for one relative with the average measures for a number of other relatives. For example, we might consider the covariance between offspring and the average parental scores, rather than scores for a single parent. Or, we could turn it around and look at the covariance between a single parent and the average of all of that parent's offspring. In general, the expectations for such averaged relationships are the same as those discussed above for relatives considered one at a time. However, some preconditions must be met (Falconer, 1960).

What about epistasis? We noted earlier that, in addition to additive effects of alleles across loci (V_A), there is also nonadditive genetic variance. Although some of this nonadditive variance is due to interactions between alleles at a locus (V_D), the rest is due to nonadditive interactions between alleles at different loci (V_I). Because identical twins are genetically identical, their phenotypic covariance includes all additive and nonadditive genetic variance. However, phenotypic covariance for other familial relationships (particularly those, such as full siblings and fraternal twins, that share variance due to dominance) includes only some of the variance due to epistatic and dominance interactions (Falconer, 1960). Fortunately, this complexity turns out empirically to be less important than it might seem. We shall see that additive genetic variance accounts for the majority of genetic variance in most behavioral characters for which such information is available.

Table 9.9 summarizes the genetic and environmental components of variance responsible for the phenotypic covariance of relatives. For example, the phenotypic covariance between fraternal twins includes half of the additive genetic variance ($\frac{1}{2}V_A$), one-fourth of the nonadditive genetic variance due to dominance ($\frac{1}{4}V_D$), and environmental influences common to members of fraternal twin pairs ($V_{E_{c(DZ)}}$). In contrast, identical twins' covariance includes all additive and nonadditive genetic variance, as well as environmental influences common to members of identical twin pairs ($V_{E_{c(MZ)}}$). In Chapter 10, such differences in the components of covariance will be used to estimate the various components of genetic and environmental variance.

HERITABILITY

Because the concept tends to be misunderstood, *heritability* has become something of a bad word in recent years. However, if properly defined and employed, heritability is a useful concept. It is simply a statistic that describes the ratio of genetic to phenotypic variance—the proportion of observed variance in a population that can be explained by genetic variance. In other words, heritability describes the extent to which genetic differences among individuals in a population make a difference phenotypically. The environmental contribution to phenotypic variance is directly analogous to heritability. Unfortunately, there is no generally accepted word to express the proportion of individual differences unexplained by genetic factors. Of the various terms that have been proposed, we shall use a word suggested by Fuller and Thompson (1978): *environmentality*.

For any behavior, we are likely to observe a wide range of individual differences. These phenotypic differences may be caused by environmental experiences, as well as genetic differences. One important aspect of behavioral genetics involves partitioning phenotypic variability into parts due to genetic and environmental differences.

What Heritability Is Not

Heritability Is Neither Constant Nor Immutable

Heritability describes a situation involving a particular phenotype in a population with a certain array of genetic and environmental factors at a given time. Heritability does not indicate an eternal truth concerning the phenotype, for it can vary from population to population and from time to time. It is a population parameter, a true character of a population, analogous to the population mean and variance. If the population changes, you can expect its parameters to change accordingly.

If genetic variance or environmental variance change, heritability (and environmentality) can change. A relatively unexplored benefit of the concept of heritability is that it can describe changes in the mix of genetic and environmental factors in various populations, times, or developmental stages.

Heritability Does Not Refer to One Individual

Heritability is a descriptive statistic that applies to a population. If we say that height has a heritability of 0.80, that means that 80 percent of the variation in height observed in this population at this time is due to genetic differences. It obviously does not mean that an individual who is 5 feet tall grew to the height of 4 feet as the result of genes and that the other 12 inches were added by the environment. However, if an individual from this population were 10 inches taller than average, one could estimate (rather imprecisely) that 80 percent of this deviation was due to genetic effects and that 20 percent was due to environmental influence. The same reasoning, of course, applies to behavioral traits.

Heritability Is Not Absolutely Precise

Some people object that heritability implies a high degree of precision. Heritability, as we have said, is a descriptive statistic; like all descriptive statistics, it involves error. Correlations, for example, involve a range of error that is partially a function of the size of the sample from which the estimate is made. As in the case of other descriptive statistics, however, we can estimate the extent of error involved in heritability estimates (Klein et al., 1973).

What Heritability Is

In 1940, Lush defined heritability as "the fraction of the observed variance which was caused by differences in heredity," a useful alternative to the old nature-nurture dichotomy:

> Furthermore, it gradually came to be recognized that the question whether the nature or the nurture, the genotype or the environment, is more important in shaping man's physique and his personality is simply fallacious and misleading. The genotype and the environment are equally important, because both are indispensible. . . . The nature-nurture problem is nevertheless far from meaningless. Asking right questions is, in science, often a large step toward obtaining right answers. The question about the roles of the genotype and the environment in human development must be posed thus: To what extent are the *differences* observed among people conditioned by the differences of their genotypes and by the differences between the environments in which people were born, grew and were brought up? (Dobzhansky, 1964, p. 55)

Or, as R. C. Roberts has stated: "We need to know how much of the total variation (in a population) is due to various genetic causes, for it is axiomatic that the importance of a source of variation is proportional to the contribution it makes to the total variation" (1967a).

Heritability is the proportion of phenotypic variance that is attributable to genotypic variance:

$$\text{heritability} = \frac{V_G}{V_P}$$

In the numerical example presented in Table 9.6, both the genetic and environmental variances are equal to 2.0. In the simplest case, when there is no correlation or interaction between genetic and environmental factors, the phenotypic variance is 4.0. In this case, heritability is 0.5, meaning that 50 percent of the phenotypic variance is explained by genetic variance. The other 50 percent of the phenotypic variance is caused by environmental variance. Thus,

$$\text{environmentality} = \frac{V_E}{V_P} = \frac{2}{4} = 0.5$$

Tables 9.7 and 9.8 indicate that correlations or interactions between genetic and environmental factors will increase phenotypic variance. These effects will be discussed in Chapter 14.

Lush (1949) later distinguished between two types of heritability. *Broad-sense heritability* (h_B^2) is the type of heritability that we have been discussing. It is the proportion of phenotypic differences due to all sources of genetic variance, regardless of whether the genes operate in an additive or nonadditive manner. *Narrow-sense heritability* (h^2), on the other hand, is the proportion of phenotypic variance due solely to additive genetic variance. If V_G refers to all genetic variance and V_A refers to additive genetic variance,

$$h_B^2 = \frac{V_G}{V_P} \quad \text{and} \quad h^2 = \frac{V_A}{V_P}$$

Narrow-sense heritability is particularly interesting in the context of selective breeding studies, where the important question is the extent to which offspring will resemble their parents. As we noted earlier, additive genetic variance involves the extent to which characters "breed true." On the other hand, broad-sense heritability is important in many other contexts. The most important situation involves the relative extent to which individual differences are due to genetic differences and to environmental differences. We can obtain the appropriate answer by assessing broad-sense heritability. In addition to these descriptive functions, broad-sense and narrow-sense heritability can be used in a predictive way. Heritability predicts genotypic values of individuals, given the mix of genetic and environmental variance in a population at a particular time.

Path Analysis

The concept of heritability can also be presented by the analysis of *paths*—the statistical effect of one variable on another independent of other variables. For some, it is easier to understand the concept of heritability visually in a path model rather than strictly in algebraic terms.

We can construct a path model of the effects of genetic and environmental factors on a behavioral phenotype, as in Figure 9.10. This is the same as the statement P = G + E. The "paths" in this case express the extent to which genetic and environmental deviations cause phenotypic deviations. Thus, h_B is the path by which genetic deviations from the population mean

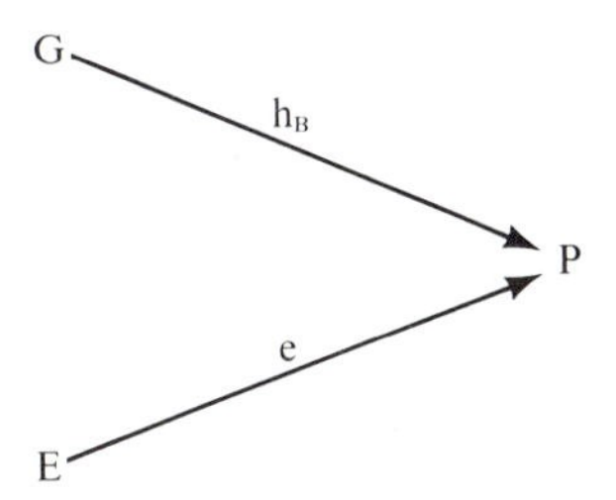

FIGURE 9.10

Path model of the genetic and environmental components of the phenotypic value (P). See text for explanation.

(G) cause phenotypic deviations. In fact, the h_B path is the proportion of the phenotypic standard deviation (s_P) caused by the genetic standard deviation (s_G):

$$h_B = \frac{s_G}{s_P}$$

Remembering that the standard deviation is the square root of variance, you can see that the h_B path is, in fact, the square root of broad-sense heritability. Similarly, e is the square root of environmentality. Path analysis, which was introduced by the geneticist Sewall Wright (1921) over fifty years ago, has recently become popular for describing complexities of multifactorial models (Li, 1975) in both the social and biological sciences.

Multivariate Analyses

We have been focusing on the genetic-environmental analysis of only one behavior of each individual—a univariate (one variable) approach. However, several behaviors can be measured for each individual and subjected to multivariate quantitative genetic analysis (e.g., Plomin and DeFries, 1980*b*). If two characters (X and Y) are measured for each individual in a population and a correlation is observed, this phenotypic correlation may be due to either genetic or environmental factors. Among the genetic causes, pleiotropy is the most interesting, since it results in permanent correlations between characters. Genetic correlations can also result from temporary linkages due to recent admixtures of populations or nonrandom mating. However, these linkages are soon broken up by recombination. Thus, pleiotropy is the most useful way of conceptualizing the genetic correlation between behaviors.

It is easy to visualize how environmental effects may give rise to correlations between characters such as height and weight. A favorable diet, for example, may result in higher height and weight, whereas an unfavorable diet may be accompanied by depressed values for both characters. At the psychological level, the phenotypic correlations among measures of specific cognitive abilities may be due to environmental influences, such as the intellectual environment of the home or the quality of schooling, which affect various specific cognitive abilities in a similar way.

Why would we want to know the extent to which genetic and environmental factors contribute to the phenotypic correlation between two behaviors? When we study behaviors one at a time, many show genetic influence, but it is highly unlikely that each of these is influenced by a completely different set of genes. If the same genes affect different behaviors, we can observe a correlation among the behaviors. The same reasoning applies to environmental influences: they may affect several behaviors, producing correlations among them. Thus, the importance of multivariate genetic-environmental analysis lies in its potential for revealing the genetic and environmental bases of phenotypic covariance.

Using the example of specific cognitive abilities again, the studies discussed in Chapters 11, 12, and 13 suggest substantial genetic influence for each of the specific cognitive abilities. However, it is possible that one set of genes influences all of these mental abilities. The phenotypic correlation among specific cognitive abilities may be largely due either to their genetic correlation or to a single set of environmental influences. Of course, there may also be several independent sets of genetic or environmental influences.

In summary, multivariate genetic-environmental analysis asks some important questions: Are independent gene systems involved, or do genetic influences overlap for some or all of the behaviors? Do various environmental factors make independent contributions, or are broad environmental influences responsible for the behaviors? Although there have been few multivariate behavioral genetic analyses, they suggest that genetic and environmental factors are neither very broad nor very narrow. More surprisingly, genetic and environmental correlations are correlated—that is, the structure of genetic influences seems to be similar to the structure of environmental influences (DeFries, Kuse, and Vandenberg, 1979; Fulker, 1979; Loehlin and Nichols, 1976; Martin and Eaves, 1977). Although most of us would probably predict different patterns of genetic and environmental influence, the possibility of similar genetic and environmental structures is reasonable.

Just as quantitative genetics can be applied to the variance of a single behavior, it can also be applied to the correlation between two behaviors (DeFries, Kuse, and Vandenberg, 1979). In fact, any behavioral genetic method that can partition the variance of a single behavior can also be applied to the partitioning of the covariance between two behaviors. Path analysis provides an easy way to visualize this analysis. Figure 9.11 extends the path analysis of a single behavior (see Figure 9.10) to the analysis of the correlation between two phenotypic characters, P_x and P_y. Just as the variance of a single character (P_x) is due to an environmental path (e_x) and a

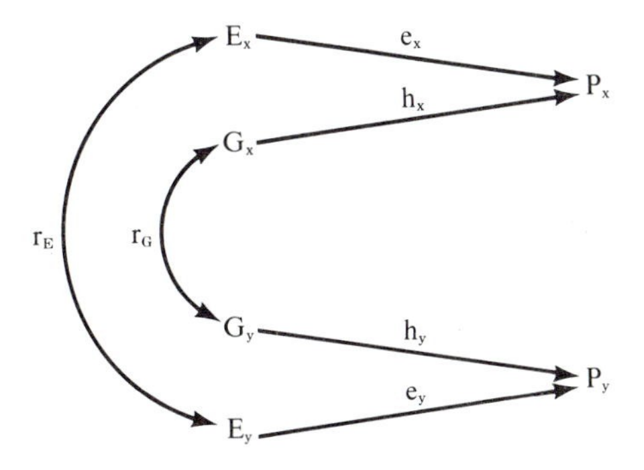

FIGURE 9.11

Path diagram of the phenotypic correlation between two characters (P_x and P_y) measured on an individual as a function of the genetic correlation (r_G) and the environmental correlation (r_E).

genetic path (h_x), the phenotypic correlation between X and Y ($r_{P_xP_y}$) may be caused by an environmental chain of paths ($e_xe_yr_E$) and a genetic chain of paths ($h_xh_yr_G$), where r_E and r_G are environmental and genetic correlations, respectively. These chains of paths are phenotypically standardized, and thus add up to the phenotypic correlation ($r_{P_xP_y}$):

$$r_{P_xP_y} = h_xh_yr_G + e_xe_yr_E$$

Genetic and environmental chains are especially useful for investigating the causes of phenotypic correlations between characters. These chains provide standardized measures of the genetic and environmental contributions to phenotypic resemblance. However, both genetic or environmental correlations (i.e., r_G and r_E by themselves) are also informative. The genetic correlation provides a measure of the extent to which two characters are influenced by the same genes. Likewise, the environmental correlation measures the extent that two characters are affected by the same environmental influences. Whether one estimates genetic chains, genetic correlations, or both, depends on the purpose of the investigator (Plomin and DeFries, 1980*b*).

This merely sums up in more precise terms that the phenotypic correlation between two behaviors may be due to genetic or environmental influences. The phenotypic correlation by itself, however, does not provide a useful index of the importance of the genetic and environmental chains. Even when the phenotypic correlation between two behaviors is negligible, there may be substantial genetic and environmental chains of influence between the two behaviors if the genetic and environmental chains work in opposite directions—that is, if one is positive and the other negative. For example, the same genes may affect specific cognitive abilities, leading to a positive genetic correlation. However, environmentally, one might develop a few abilities to the exclusion of the others, leading to a negative environmental correlation. Moreover, even if two behaviors are both substantially heritable, the phenotypic correlation between them may be environmental in origin. For example, verbal ability and spatial ability are phenotypically correlated, and both show substantial heritability. However, it is possible that completely different sets of genes influence the two abilities. In other words, their genetic correlation could be zero. If this were the case, the environmental chain would be solely responsible for the phenotypic correlation.

Genetic and environmental chains can be estimated by methods analogous to those used to estimate heritability. In Table 9.9, we presented the genetic and environmental components of variance that contribute to the phenotypic covariance between relatives for a single character. When we consider the phenotypic covariance between two characters rather than for one, we need to introduce a new concept, *cross-covariance*. Rather than studying the covariance of character X in parents and character X in offspring, we consider the cross-covariance of character X in parents and character Y in offspring. Phenotypic cross-covariance between parents and offspring may be due to their genetic and environmental similarity. In fact, the components of cross-covariance between relatives are the same as those listed in Table 9.9. This should not be surprising in view of the relationship between the univariate and multivariate analyses as just described. Thus, the phenotypic cross-covariance for characters X and Y, for parents and offspring, involves half of the additive genetic covariance, as well as common environmental influences. Phenotypic cross-covariances for identical twins include all genetic sources of covariance in addition to shared environmental influences.

RESEMBLANCE OF RELATIVES REVISITED

In Table 9.9, we described the genetic and environmental components of covariance for different family relationships. These relationships include both genetic and environmental components of variance. If we could find a relationship or combination of relationships that included only genetic variance, we could easily obtain the heritability statistic by dividing the genetic variance by the phenotypic variance. Determining such relationships is the essence of quantitative genetic methods, which will be discussed in the following chapters.

Univariate Analysis

Regressions and correlations are useful because they are merely covariances divided by variances. If the covariance consists solely of the genetic component of variance, then the correlation between relatives estimates heritability. In this case, the correlation is found by dividing the genetic variance by the phenotypic variance. This is the definition of heritability.

Consider identical twins who have been separated from birth. As shown in Table 9.9, identical twins share all genetic variance, plus common environmental influences. However, if they have been separated from birth, they do not have a common postnatal environment. Thus, their phenotypic covariance estimates V_G, and the correlation between them directly estimates heritability, V_G/V_P. The important thing to remember is that identical twins are genetically identical, whether or not they share environments. If they do not share environments, their correlation estimates heritability. Path analysis presents a picture of this idea.

Each identical co-twin's phenotype is caused by genetic and environmental influences, as shown in Figure 9.12. However, identical co-twins have the same genotype. A useful feature of path analysis is the ability to trace the components of a correlation by following the paths. For identical twins reared together, one chain of paths from the phenotype of one identical

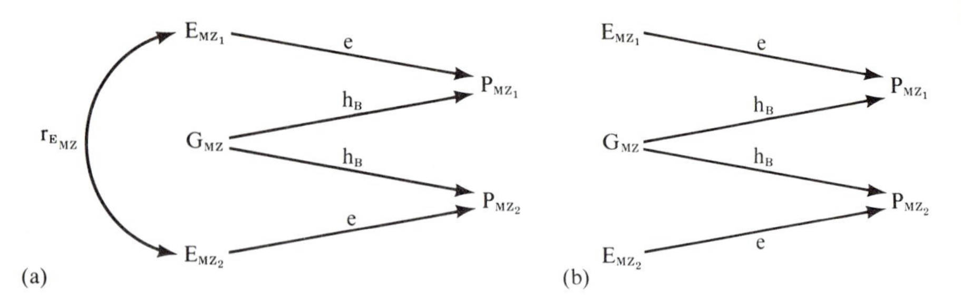

FIGURE 9.12

Path diagram for identical twins reared (a) together and (b) apart in uncorrelated environments.

twin to the phenotype of the co-twin is $(h_B)(h_B)$, or h_B^2. Another chain is (e) $(r_{E_{MZ}})$(e) or $e^2(r_{E_{MZ}})$. The correlation between two phenotypes, given an appropriate path model, is the sum of the chains of paths. Thus, the correlation between identical twins reared together is $h_B^2 + e^2\ (r_{E_{MZ}})$, which means that the correlation includes broad-sense heritability and environmental influences shared by MZ twins. This statement merely reiterates that identical twins reared together have genetic and environmental factors in common. However, as shown on the right side of Figure 9.12, identical twins who do not share environmental influences (that is, where $r_{E_{MZ}} = 0$) share only the genetic paths. Thus, their correlation directly estimates h_B^2 (broad-sense heritability).

Multivariate Analysis

We have just seen in Figure 9.12 that the univariate correlation for pairs of identical twins reared in uncorrelated environments estimates h_B^2. Figure 9.13 extends this relationship to two characters, X and Y. We indicated earlier that the cross-covariance for trait X in one relative and trait Y in another has the same components of covariance as the univariate situation summarized in Table 9.9. Similarly, cross-correlations for two characters for relatives have the same relationship to univariate familial correlations. Thus, as

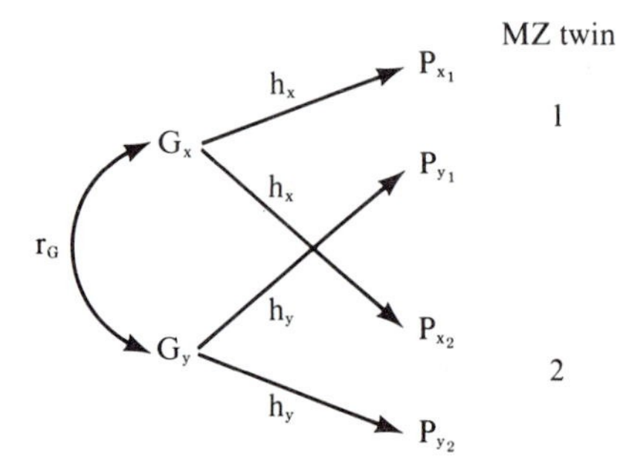

FIGURE 9.13

Path diagram of the phenotypic correlation between two characters (P_x) and (P_y) measured on a pair of identical twins reared apart in uncorrelated environments.

shown in Figure 9.13, the cross-correlation for X and Y for separated identical twins is equivalent to the genetic chain discussed earlier:

$$r_{P_{X_1}P_{Y_2}} = h_X h_Y r_G$$

In other words, if the phenotypic correlation between two traits is due entirely to their genetic correlation, then the cross-correlation for pairs of separated identical twins should be similar to the phenotypic correlation between X and Y observed within individuals. Of course, we do not need to find separated identical twins in order to conduct multivariate quantitative genetic analyses. As we have said, any behavioral genetic analysis of the variance of a single character can be applied to the correlation among characters.

SUMMARY

After a brief overview of statistics (mean, variance, standard deviation, covariance, correlation, and regression), the single-gene model of quantitative genetics is described. Genotypic value, additive genetic value, and dominance deviation are defined. The full quantitative genetic model is a polygenic extension of the single-gene model, which includes additive, dominance, and environmental deviations. We provide a hypothetical example, which also includes genotype-environment correlation and interaction.

Quantitative genetic methods estimate genetic and environmental components of variance from the phenotypic covariance of various types of relatives that differ in genetic relatedness or in environmental relatedness. The covariance of relatives can be used to estimate within-family and between-family environmental influences, as well as heritability. Heritability, either in its narrow sense ($h^2 = V_A/V_P$) or its broad sense ($h_B^2 = V_G/V_P$), is the proportion of phenotypic variance attributable to genotypic variance. Environmentality ($e^2 = V_E/V_P$) is the proportion of phenotypic variance that is attributable to environmental (nongenetic) variance. Quantitative genetic methods are usually applied to the variance of a single character (the univariate approach), but they are equally applicable to the study of the genetic and environmental etiology of correlations among several characters (the multivariate approach). In the remaining chapters, we shall apply these quantitative genetic methods to behavioral examples.

10

Quantitative Genetic Methods: Animal Behavior

In Chapter 4, we discussed methods used to investigate single-gene influences on animal behavior. These methods include analyses of strain distributions, Mendelian crosses, and recombinant inbred strains. In this chapter, we shall consider more general methods of genetic analysis that can begin to untangle genetic and environmental factors, regardless of whether the behavior is influenced by a single gene or by many genes. These methods, like quantitative genetic theory discussed in the previous chapter, can be understood at different levels. They can be viewed simply as experiments, in which genetic factors are manipulated to determine whether genes can influence behavior. Or, they can be coupled with quantitative genetic theory to make estimates of the relative contribution of genetic and environmental factors. The three basic methods are family studies, strain studies, and selection studies.

FAMILY STUDIES

Relatives who share genes ought to be similar for a particular character, assuming that genes influence that character. As indicated in the last chapter (Table 9.9) and summarized in the first column of Table 10.1, genes contrib-

TABLE 10.1

Phenotypic resemblance of relatives, assuming no environmental covariance

Relatives	Genetic Components of Covariance	Regression (b) or Correlation (r)	Relationship to Heritability
One parent and offspring	$\frac{1}{2}V_A$	$b_{op} = \frac{1}{2}V_A/V_P$	$2b_{op} = h^2$
Midparent and offspring	$\frac{1}{2}V_A$	$b_{o\bar{p}} = \frac{1}{2}V_A/\frac{1}{2}V_P$	$b_{o\bar{p}} = h^2$
Half-siblings	$\frac{1}{4}V_A$	$r_{HS} = \frac{1}{4}V_A/V_P$	$4r_{HS} = h^2$
Full siblings	$\frac{1}{2}V_A + \frac{1}{4}V_D$	$r_{FS} = (\frac{1}{2}V_A + \frac{1}{4}V_D)/V_P$	$h^2 \leq 2r_{FS} \leq h^2_R$

ute differentially to various family relationships. We showed that half-siblings share one-fourth of the additive genetic variance and no dominance variance. Parents share half of the additive genetic variance and no dominance variance with their offspring. Full siblings share half of the additive genetic variance and one-fourth of the dominance variance. We can compare these different relationships to determine the extent to which their genetic similarity predicts their phenotypic similarity.

With laboratory animals, environments can be controlled to some extent, so that environmental contributions to phenotypic similarity may often be ignored. If we assume that environments are controlled, the components of covariance include only genetic ones. In the previous chapter, we showed that correlations and regressions are merely covariances divided by variances. We also defined heritability as genetic variance divided by phenotypic variance. The second column of Table 10.1 indicates that the genetic covariances can be divided by an appropriate variance to obtain a regression or correlation. The last column shows the relationship between regression or correlation, and heritability.

For example, the genetic component of covariance between scores of offspring and one parent is half of the additive genetic variance. The regression between parent and offspring divides their covariance (which is $\frac{1}{2}V_A$) by the phenotypic variance of the parent (the variable from which we are predicting offspring scores). Thus, the phenotypic regression of offspring on their parents estimates half of the narrow-sense heritability. Therefore, to estimate narrow heritability, we double this regression:

$$2\left(\frac{\frac{1}{2}V_A}{V_P}\right) = \frac{V_A}{V_P} = h^2$$

As noted in the previous chapter, the covariance between average parental (midparent) and offspring scores has the same components as the covariance between one parent and offspring. However, as indicated in Table 10.1, the regression of offspring on midparent estimates h^2, not $\frac{1}{2}h^2$. The reason is that the variance of midparent scores is half that of single parent scores, when mating is random. The regression takes the covariance component ($\frac{1}{2}V_A$) and divides it by the variance of the midparent scores, which is $\frac{1}{2}V_P$. Thus, this regression directly estimates h^2. The latter method of estimating heritability has several advantages, including the fact that it is unbiased by assortative mating (Falconer, 1960). Remember that these estimates of genetic influence are based on the assumption that environmental influences have been controlled so that they do not contribute to correlations between family members.

For parents and offspring, the preferred statistic is regression (Falconer, 1960). For siblings and twins, however, correlations are preferable. Actually, a special kind of correlation (intraclass) is used, rather than the usual interclass correlation. The intraclass correlation is used so frequently that we shall briefly describe it, although we shall not present details for computing it. (For such details, see Haggard, 1958, or other intermediate level statistics books.) The *interclass* ("between-class") correlation assumes that there are two distinct characters, such as variables X and Y, and this is not the case for the relationship between siblings. The *intraclass* ("within-class") correlation takes into account all possible pairings of siblings within a family. However, if we randomly assign one sib to one arbitrary "class," and the other sibling to another arbitrary "class," and then compute the usual interclass correlation, the answer is much the same as the intraclass correlation. (In fact, if the means and variances are the same for the two "classes," as they would be if we randomly assigned a large number of sibling pairs to two classes, the answer would be exactly the same.) An additional advantage of the intraclass correlation is that it permits the computation of a correlation when there are more than two siblings in a family.

The correlation among half-siblings must be multiplied by 4 to estimate h^2. The consequence of this is that any errors of measurement will also be multiplied by 4. Thus, estimates of h^2, based on half-sibling correlations, tend to be imprecise except when sample sizes are large. In animal breeding research, for example, where records of hundreds of progeny artificially sired by hundreds of males are available, this method has been very useful.

Doubling the correlation of full siblings will overestimate h^2, if dominance variance (V_D) occurs. However, it will underestimate h_B^2 because full siblings share only $\frac{1}{4}V_D$. Doubling the correlation does not yield $(V_A + V_D)/V_P$, which is broad-sense heritability; rather, it yields $(V_A + \frac{1}{2}V_D)/V_P$.

These statistical procedures are the technical basis for the simple point with which we began. The family study method is based simply on the fact that genetically related individuals ought to be similar phenotypically for any behavior that is influenced by genes. Moreover, genes contribute to varying extents for different family relationships (such as parents and their offspring, full siblings, and half-siblings). By comparing such family relationships, we can determine the contribution of genetic similarity to observed phenotypic similarity. An example will help to clarify the family study method.

Open-Field Behavior in Mice

As an example of the use of these methods, we may again consider open-field behavior in mice. DeFries and Joseph P. Hegmann (1970) tested 72 males and 144 females of the F_2 generation derived from C57BL/6 × BALB/c crosses, and then mated each male with two females. In this way, the resulting 128 litters (841 mice) included full siblings, and also half-siblings related through their father.

The heritability of open-field behavior may be estimated in several different ways from these data, as indicated in Table 10.2. The regression of offspring on midparent estimates a narrow-sense heritability of 0.22, while the half-sib correlation estimates a narrow-sense heritability of 0.16. Of these two estimates, which are quite similar, the parent-offspring regression is more accurate because half-sibling correlations are less reliable. However, the full-sibling correlation suggests a much higher heritability. Recall that full-sibling correlations are due both to additive and to nonadditive components of genetic variance. But even this may not tell the whole story. The rest of the answer may lie in the fact that the full siblings were reared in the same litter by the same mother, thus sharing prenatal and postnatal environmental influences. Earlier, we assumed that common environmental sources of covariance among relatives could be safely ignored in laboratory animals in controlled environments. However, in the case of full siblings, the environment is not controlled and evidently contributes substantially to the phenotypic similarity of full siblings.

Although other animal research—such as studies of alcohol preference in mice (Whitney et al., 1970), locomotor activity in *Drosophila* (Connolly, 1968), and avoidance learning in swine (Willham et al., 1963)—has used the family study method, there are surprisingly few such studies, considering the obvious applicability of this technique for estimating heritability. However, there are many examples of family studies of human behavior, and these will be described in the next chapter.

TABLE 10.2
Phenotypic resemblance of relatives for open-field activity

Relatives	Regression or Correlation	Estimate of Heritability
Midparent and offspring	0.22	0.22
Half-siblings	0.04	0.16
Full siblings	0.37	0.74

SOURCE: After DeFries and Hegmann, 1970.

Multivariate Analyses

In the previous chapter, we indicated that quantitative genetic methods can be applied to both multivariate and univariate problems. In other words, we can analyze the phenotypic correlation among characters, as well as the variance of characters taken one at a time. In terms of open-field behavior, a negative correlation (about −0.40) is usually observed between open-field activity and defecation—that is, mice who run around a lot in the open field do not leave many mementos of their travels. We know that both open-field activity and defecation are influenced by genes to some extent, but are any of the same genes involved? In other words, is the phenotypic correlation between activity and defecation due to genetic or environmental factors?

Hegmann and DeFries (1970) addressed this issue using data from the study described above. Cross-correlations for parents and offspring (for example, correlations between activity scores in parents and defecation scores in offspring) in the genetically segregating F_2 and F_3 generations were used to estimate genetic correlations, as discussed in the previous chapter. We indicated that familial cross-correlations include the same components of covariance as univariate familial correlations. Thus, cross-correlations for parents and offspring include additive genetic covariance, as well as environmental influences shared by parents and offspring. However, we can assume that the controlled laboratory setting has weakened such environmental deviations. Environmental correlations were obtained from the genetically invariant parental inbred strains and their F_1 cross. Phenotypic correlations observed between activity and defecation within the inbred or F_1 individuals must be caused by environmental factors because these individuals do not vary genetically.

Table 10.3 shows that three of the four genetic correlations between open-field activity and defecation scores are large and negative. This indicates that many of the same genes that influence open-field activity pleiotropically affect defecation as well. Later in this chapter we shall consider a

TABLE 10.3

Genetic correlations (above diagonal) and environmental correlations (below diagonal) of single-day, open-field behavioral scores of mice

	Day 1 Activity	Day 2 Activity	Day 1 Defecation	Day 2 Defecation
Day 1 activity	—	0.94	−0.51	−0.89
Day 2 activity	0.59	—	−0.10	−0.76
Day 1 defecation	−0.30	−0.25	—	0.20
Day 2 defecation	−0.21	−0.44	0.34	—

SOURCE: After Hegmann and DeFries, 1970.

selective breeding study that substantiates this finding, such that selection for open-field activity also resulted in selection for open-field defecation. This is called a correlated response to selection.

Table 10.3 also shows that the pattern of genetic correlations among the single-day, open-field measures is mirrored by that of the environmental correlations. For example, the highest genetic correlation (0.94) is between day 1 and day 2 activity, and the highest environmental correlation (0.59) is also between these two measures. Such similarity between the genetic and environmental correlations would be rather surprising to those who might expect that some phenotypic correlations are caused solely by genetic factors, and others solely by environmental influences. One might argue that the similarity between the genetic and environmental correlation was caused partially by a confounding of genetic and environmental influences. The genetic correlations contained possible environmental influence shared by parents and offspring. Nonetheless, the results of this study tend to be the rule rather than the exception in multivariate studies of both nonhuman and human behavior and suggest that genetic correlations and environmental correlations are correlated.

In the previous chapter, we suggested that this result is really not surprising, at least in retrospect. Hegmann and DeFries wrote: "From the standpoint of biological efficiency, it would seem most reasonable that correlated characters should respond similarly to both genetic effects and environmental deviations" (1970, p. 285). What could cause the similarity of genetic and environmental correlations? A hypothetical example proposed by DeFries, Allan Kuse, and Vandenberg (1979) involves the metabolic pathway from tyrosine to norepinephrine, as illustrated in Figure 10.1. Dopamine and norepinephrine are neurotransmitters that relate positively to wheel-running activity and negatively to eating. A gene substitution that increases the activity of the tyrosine hydroxylase (TH) or dopa decarboxylase (DD) enzymes will result in more wheel running and less eating. In

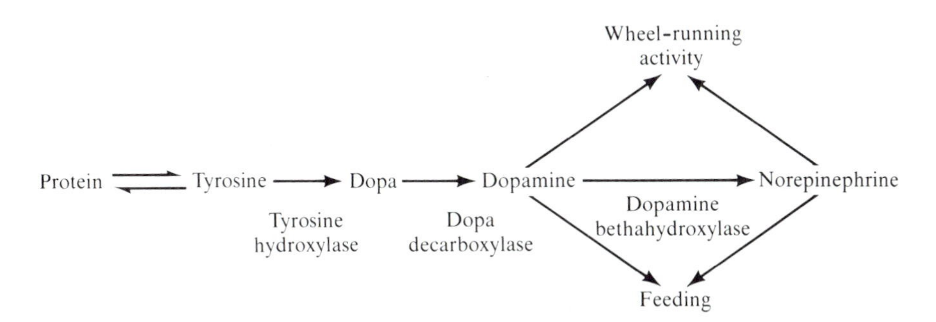

FIGURE 10.1

Metabolic pathway of the relationship between brain amines and rodent behavior. (From DeFries, Kuse, and Vandenberg, 1979.)

other words, mutations that affect TH or DD activity will have pleiotropic effects on wheel running and eating, resulting in a negative correlation between the two behaviors. But how could environmental influences do the same thing? Diet may serve as an example. A diet rich in tyrosine might result in more wheel running and reduced feeding. Changes in room temperature could also do this since lower temperatures might decrease wheel running and increase feeding. Thus, at least at the biochemical level, genetic and environmental correlations may be similar.

STRAIN STUDIES

Animals derived from intense inbreeding, such as brother-sister matings over many generations, eventually become homozygous at all autosomal loci, so that animals of the same sex are genetically identical. The origin of such inbred strains of mice is described in Box 10.1. Because different inbred strains become homozygous for different alleles, the strains are genetically different from one another. This fact can be used to determine whether genetic differences affect behavior. If different strains are reared in similar environments (for example, standard laboratory cages, food, temperature, and lighting), behavioral differences will reflect genetic differences (although prenatal and postnatal parental influences may also affect the behavior). Various numbers of genes may be involved. Many genes are usually responsible for complex behaviors. Inbred strains can often be distinguished on the basis of their coat color (see Figure 10.2), and they also differ behaviorally.

For example, two widely studied inbred strains pictured in Figure 10.2 are BALB/c and C57BL/6. Figure 10.3 shows their average scores for open-field activity and defecation. The C57BL/6J mice are much more active and defecate much less than the BALB/cJ mice. The mean activity and defecation scores of derived F_1, backcross, and F_2 and F_3 generations are also shown. Note that there is a strong relationship between the average behavioral scores and the percent of genes obtained from the C57BL/6 parental strain. From left to right, these percentages are 0 percent (BALB), 25 percent (B_1), 50 percent (F_1, F_2, and F_3), 75 percent (B_2), and 100 percent (C57BL). Such a large strain difference suggests a role for genetic influence. Later in this section we shall discuss more sophisticated uses of data from such strain studies.

Over one thousand behavioral investigations, involving genetically defined mouse strains were published between 1922 and 1973 (Sprott and Staats, 1975). From 1974 through 1978, there were over 650 studies of this type (Sprott and Staats, 1978, 1979). Studies such as these have demonstrated that genetic variance is nearly ubiquitous—almost all behaviors chosen for investigation showed strain differences. Although the strain comparison method now tends to be overshadowed by more sophisticated genetic analyses, it still provides a simple and highly efficient test for the presence of

Box 10.1
Inbred Strains of Mice

The common house mouse (*Mus musculus*) has become the most widely used animal in behavioral genetics research. Part of the reason for its popularity is that although it has the behavioral complexity characteristic of mammals, it has a short breeding time. Another important reason is the diversity of inbred strains that have been developed. As discussed in Chapter 8, brother-sister inbreeding over 20 generations will produce 98 percent homozygosity. Thus, individuals of an inbred strain are nearly identical genetically. Because of the increase in homozygosity, inbred strains are often susceptible to severe inbreeding depression, with a consequent drop in fertility. Many attempts to create inbred strains fail because the strains happen to "lock on" to harmful recessive alleles, which are expressed in the homozygous condition.

The first known inbred strain was created in 1907 by 20 generations of brother-sister matings from a pair of mice, who were homozygous recessive for three genes relating to coat color. The genes were dilute, brown, and nonagouti. (The dominant gene for agouti produces grizzled-looking, banded colors in the hair.) The strain was given the acronym dba (later changed to DBA) to refer to these three recessive genes that make the strain easily identifiable. Animals of the DBA strain and some of the other strains commonly used in behavioral genetic research are shown in Figure 10.2.

In 1913, H. J. Bagg started an albino strain whose name (Bagg albino) was later shortened to BALB/c. In 1921, the BALB/c line was crossed to another albino stock to begin the A strain, which has been particularly useful in cancer research because of its susceptibility to tumors. Crosses between the BALB/c and DBA strains led to several C strains, such as C3H and CBA.

Two other commonly used strains were begun in 1921, when a female identified as C57 gave birth to both black and brown offspring. These were separately inbred to produce the C57BL (black) and C57BR (brown) strains.

There are now over 100 inbred strains available for research. They are designated by the general strain name, followed by a slash and information about the particular laboratory responsible for the strain and other specific designations. DBA/1J, for example, refers to a subline of DBA maintained by the Jackson Laboratory in Bar Harbor, Maine. And DBA/2IBG refers to a different subline of DBA maintained by the Institute for Behavioral Genetics in Boulder, Colorado.

heritable variation. For example, recent strain comparisons have demonstrated considerable genetic variance for such characters as olfaction in mice (Wysocki, Whitney, and Tucker, 1977) and fruit flies (Hay, 1976), taste perception in rats (Tobach, Bellin, and Das, 1974), EEG correlates (Maxson and Cowen, 1976) and developmental patterns of seizure susceptibility (Deckard et al., 1976) in mice, and performance in various learning situations by mice (Anisman, 1975; Padeh, Wahlsten, and DeFries, 1974; Sprott and Stavnes, 1975) and by *Drosophila* (Hay, 1975).

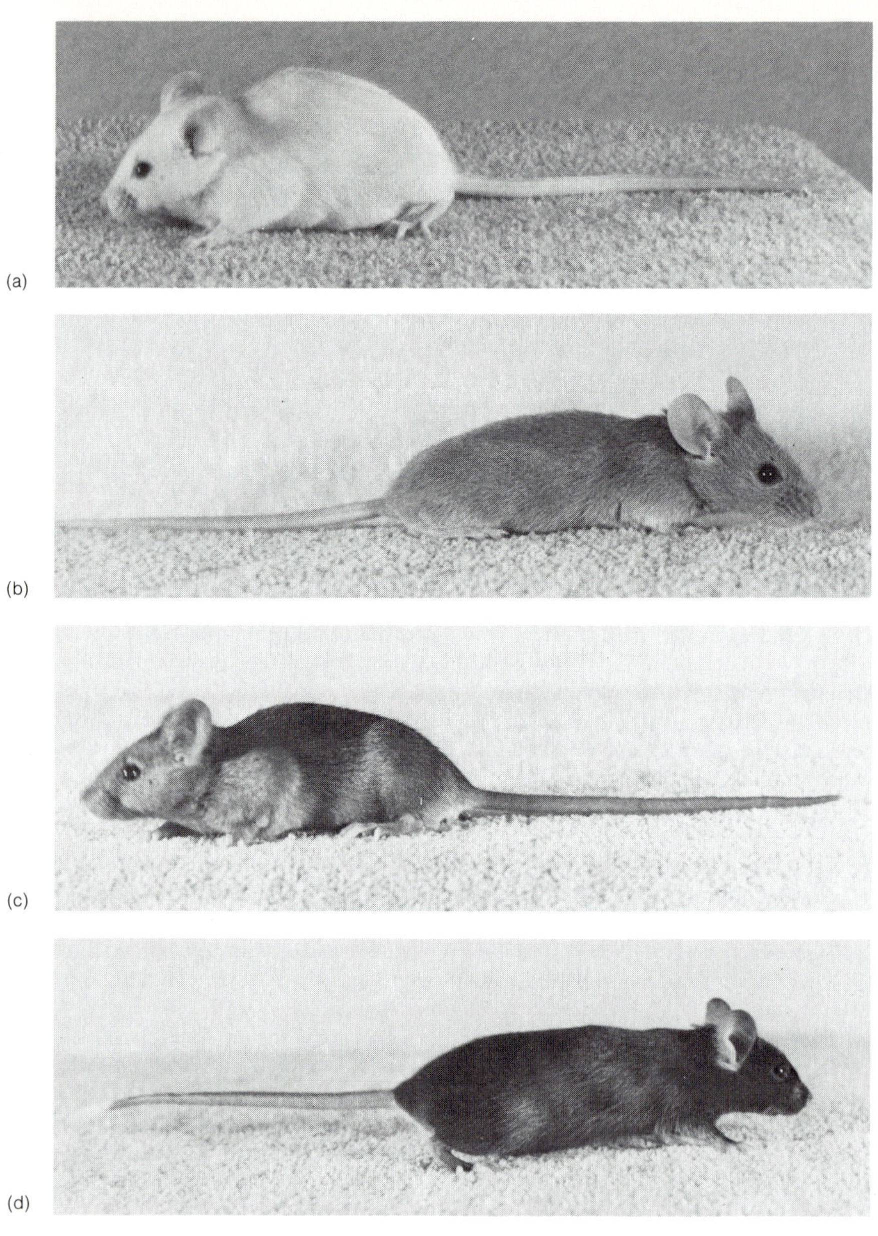

FIGURE 10.2

Four common inbred strains of mice. (a) BALB/cJ. (b) DBA/2Ibg. (c) C3H/2Ibg. (d) C57BL/6Ibg. (Courtesy of E. A. Thomas.)

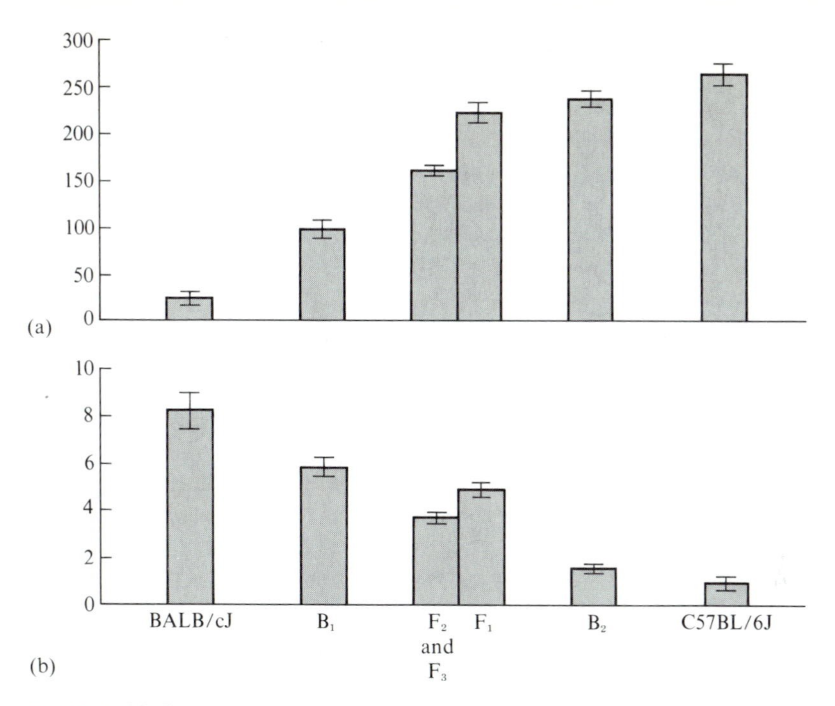

FIGURE 10.3

Mean open-field (a) activity and (b) defecation scores (± twice the standard error) of BALB/cJ and C57BL/6J mice and their derived F_1, backcross (B_1 and B_2), F_2, and F_3 generations. (From "Response to 30 generations of selection for open-field activity in laboratory mice" by J. C. DeFries, M. C. Gervais, and E. A. Thomas. *Behavior Genetics,* 8, 3–13. Copyright © 1978 by Plenum Publishing Corporation. All rights reserved.)

Genetic Effects on Learning

Because behavioral scientists studying animal behavior frequently focus on learning, we shall give a few examples of the widespread differences among mouse strains in learning situations. These data suggest caution in generalizing findings beyond the particular strain studied.

Genetic differences in learning have been found nearly every time they have been studied. (See recent review by Bovet, 1977.) Genetic differences have been shown in mice for active avoidance learning, passive avoidance learning, escape learning, bar pressing, reversal learning, discrimination learning, maze learning, and even heart rate conditioning. Active avoidance learning will serve as an example. This type of learning is usually studied in an apparatus known as a "shuttle box," which has two compartments and an electrified floor. (See Figure 10.4.) An animal is placed in one compartment, a light is flashed on, and the light is followed by a shock (delivered by an electrified grid on the floor) which continues until the animal moves to the

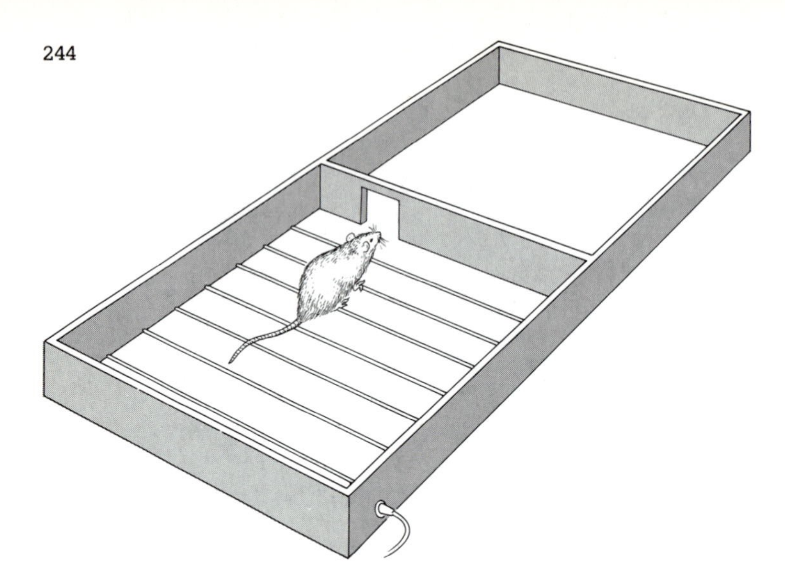

FIGURE 10.4

Shuttle box used to study avoidance learning in mice. (From *The Experimental Analysis of Behavior* by Edmund Fantino and Cheryl A. Logan. W. H. Freeman and Company. Copyright © 1979.)

other compartment. Animals learn to avoid the shock by moving to the other compartment as soon as the light comes on.

Actually, only some animals learn to avoid the shock. Before experimenters became aware of genetic causes of learning differences, they were puzzled by the wide range of differences in their genetically haphazard subjects. They believed that the differences would disappear if they could only measure learning with enough precision. There used to be a joke called "the Harvard law of animal behavior": When stimulation is precisely and repeatedly applied in a highly controlled setting, the animal will react exactly as it pleases (Scott, 1958). The far right side of Figure 10.5 shows that avoidance learning scores for random-bred (heterogeneous stock) Swiss mice range from near zero to greater than 50 percent. Most learning experiments have used rodents that are genetically heterogeneous, but of unknown heritage. The data for the six inbred strains in Figure 10.5 indicate that the inbreds are much more homogeneous than the random-bred mice. Moreover, the substantial differences that exist among the inbred strains point to the influence of genetic differences on active avoidance learning. Four of the inbred strains learn to avoid shock over 40 percent of the time. The C57L strain, however, avoids shock only about 20 percent of the time, and the C57BL/6 strain avoids shock on less than 10 percent of the trials.

Some mice perform even more poorly, as indicated by the day-to-day performances shown in Figure 10.6. The CBA strain avoids fewer than 5 percent of the shocks. This figure also illustrates differences in the rate of learning. By the third day of training, the DBA mice greatly accelerate in

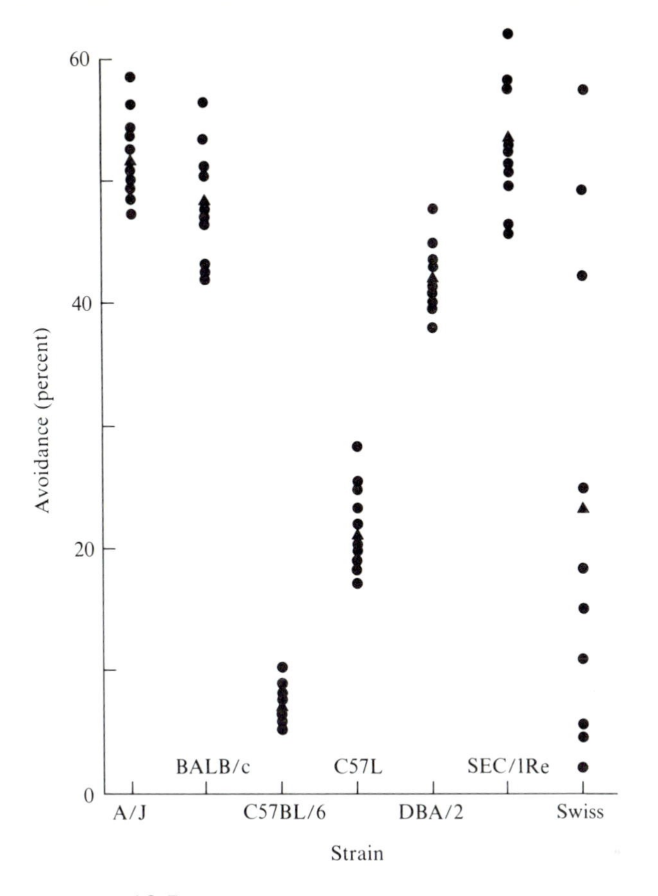

FIGURE 10.5

Avoidance learning in six strains of inbred mice and a random-bred strain (Swiss). Each point represents the performance of a mouse during five sessions. The triangles represent the mean. (From "Strain differences and learning in the mouse" by D. Bovet. In A. Oliverio, ed., *Genetics, Environment, and Intelligence*. Copyright © 1977 by Elsevier/North-Holland Biomedical Press. All rights reserved.)

performance, whereas the BALB/c mice keep plodding along, improving their performance, but at a slower rate. Figure 10.7 shows that such strain differences are not peculiar to aversive learning (that is, learning to avoid an unpleasant event such as shock). Performances of the same three strains on an appetitive task—learning to run through a maze to obtain food—show the same pattern. The DBA/2J strain learned quickly, the CBA animals were slow (although they learned a bit this time), and the BALB/c strain was intermediate. However, strains fastest in one learning situation are not always faster in another (Padeh, Wahlsten, and DeFries, 1974).

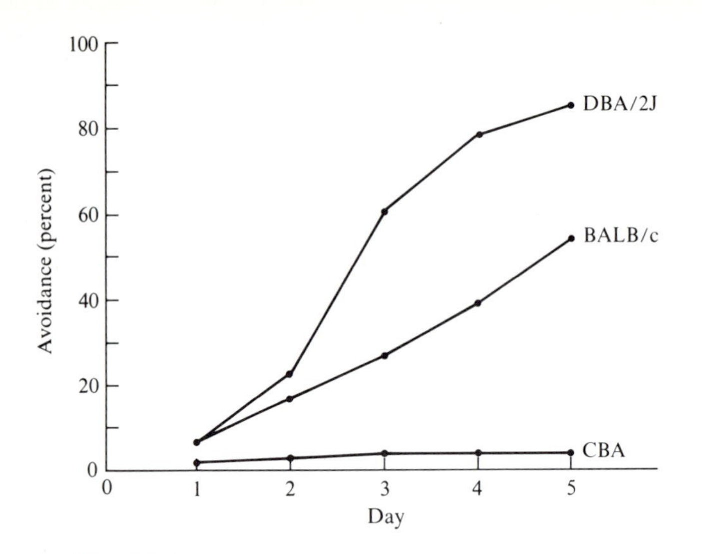

FIGURE 10.6

Avoidance learning (five days, 100 trials per day) for three inbred strains of mice. (From "Genetic aspects of learning and memory in mice" by D. Bovet, F. Bovet-Nitti, and A. Oliverio. *Science,* 163, 139–149. Copyright © 1969 by the American Association for the Advancement of Science.)

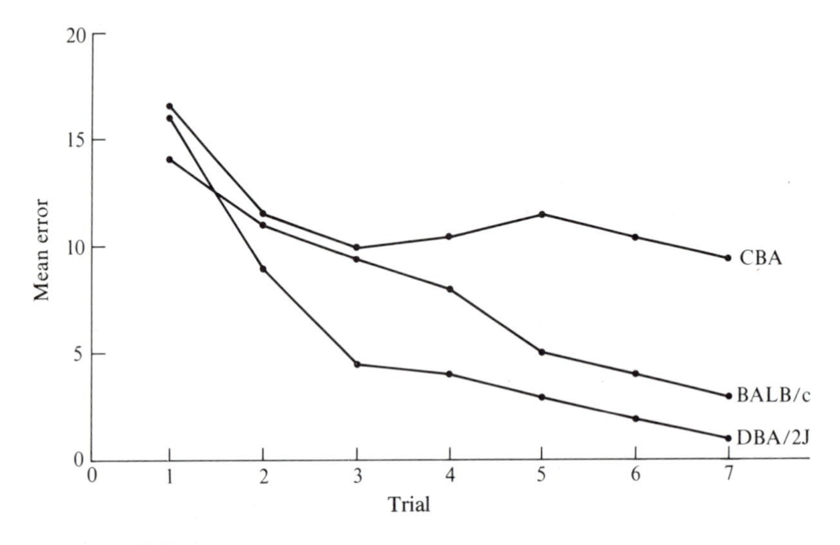

FIGURE 10.7

Maze-learning errors (Lashley III maze) for three inbred strains of mice. Each point represents the mean errors of 16 mice given one daily trial for ten days. (From "Genetic aspects of learning and memory in mice" by D. Bovet, F. Bovet-Nitti, and A. Oliverio. *Science,* 163, 139–149. Copyright © 1969 by the American Association for the Advancement of Science.)

Another classical question in learning concerns the effect of massed trials versus trials distributed over longer periods of time. There is no universal answer to this question. In some situations, C3H inbred mice appear to learn well only when their practice is massed. DBA mice, on the other hand, perform much better when the learning trials are spread out (Bovet et al., 1969).

In summary, inbred strains differ in their performance on various learning tasks. These results suggest that genetic differences affect learning. The next section indicates that inbred strains are also useful in studying the effects of the environment.

Environmental Effects on Behavior

Inbred strains are also useful in demonstrating environmental effects. If mice of a single inbred strain, reared in different environments, differ in behavior, environmental factors are implicated. Studies of this kind have investigated the behavior of inbred strains under various environmental conditions, such as "enriched" environments with playthings and lots of room for running, crowded environments, and environments made stressful by shock. The results of such studies indicate that such environmental circumstances can influence behavior.

A few experiments have studied the effects of genotypic and environmental differences. One of the best-known studies of this type used selectively bred lines rather than inbred strains (Cooper and Zubek, 1958). These researchers worked with rats that had been selectively bred to run through a maze with few errors ("maze bright") or with many errors ("maze dull"), as described in the next section. Rats from these two lines were reared under one of two conditions from weaning at 25 days of age to 65 days of age. One condition was "enriched," in that the cages were brightly colored and contained many movable toys. For the other condition, "impoverished," gray cages without the movable objects were used. Animals reared under these conditions were compared to maze-bright and maze-dull animals reared in a normal laboratory environment (in a different experiment).

The results of testing these animals for maze-running errors (Figure 10.8) showed that there is a large difference between the two lines after rearing in the normal environment. This is not surprising because the lines were selectively bred for maze-running differences in this environment. The enriched condition had no effect on the maze-bright animals, but it substantially improved the performance of the maze-dull rats. On the other hand, an impoverished environment was extremely detrimental to the maze-bright rats, but had little effect on the maze-dull ones. In other words, there is no simple answer concerning the effect of deprived and enriched environments in this study. It depends on the genotype of the animal. This example illustrates genotype-environment interaction, the differential response of genotypes to environments.

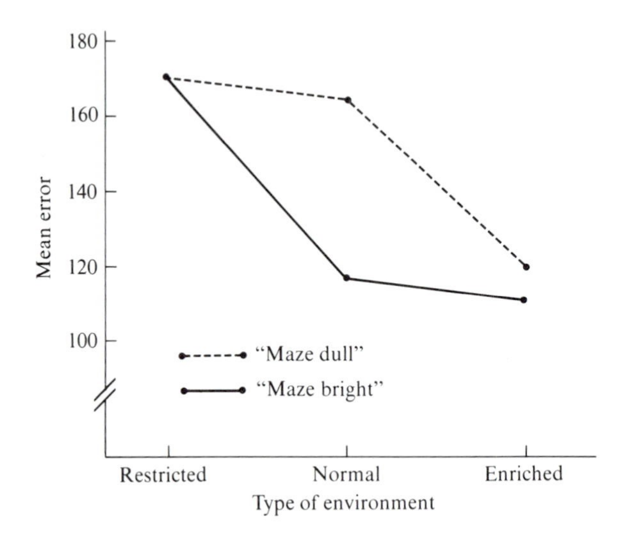

FIGURE 10.8

Genotype-environment interaction. Maze-running errors for maze-bright and maze-dull rats reared in restricted, normal, and enriched environments. (From Cooper and Zubek, 1958.)

Other studies of learning have also found a significant interaction between genotype and environment. However, an important study by Norman Henderson (1972) suggests that the effect of early experience may be mediated by temperamental characteristics, such as curiosity and level of motivation. For example, in the next section we shall see that the maze-bright and maze-dull animals used in the R. M. Cooper and J. P. Zubek study differ more in terms of such temperamental characteristics than they differ in learning ability per se. Henderson studied six inbred strains of mice reared in either standard or ''enriched'' environments. Previous studies of early experience had relied on learning tasks that may be influenced by motivational and exploration differences, such as learning to run through a maze in order to obtain food, and learning to avoid a shock with a certain cue (such as a light). Henderson tested the mice in two life-threatening escape-learning tasks (escape from water and escape from shock), which provided uniformly high motivation and minimized curiosity. As in other studies, learning proved to be substantially influenced by genetic factors, as evidenced by large strain differences. Unlike other studies, rearing environment and genotype-environment interaction appeared to have little effect. Henderson argued that learning, independent of such temperamental characteristics as curiosity and motivation, may not show rearing effects or genotype-environment interaction. We shall return to the important topic of genotype-environment interaction in Chapter 14.

Prenatal and Postnatal Factors

Strain differences may be due to prenatal and postnatal environmental influences, as well as to genetic differences between the strains. For example, differences in activity between BALB/c and C57BL/6 mice may be caused by some prenatal or postnatal rearing difference between the BALB/c and C57BL/6 mothers. Although such "environmental" differences may ultimately be based on maternal genetic differences between the two strains, it is useful to separate such influences from more direct genetic differences.

The most efficient test is simply a *reciprocal cross* between the two strains. This involves crossing BALB/c males with C57BL/6 females, and comparing the offspring to the offspring of BALB/c females and C57BL/6 males. Although their mothers are from different strains, the hybrid offspring in these two groups have the same genotypes. If either prenatal or postnatal maternal effects are important, then the genotype of the mother should make a difference in the pups' behavior.

In the large study of open-field behavior mentioned earlier (DeFries and Hegmann, 1970), hybrid offspring were obtained from reciprocal crosses between BALB/c and C57BL/6 mice. Figure 10.9 depicts mean daily open-field activity of the two inbred strains and their reciprocal-cross offspring. As we saw in Figure 10.3, the C57BL/6 mice are much more active than the

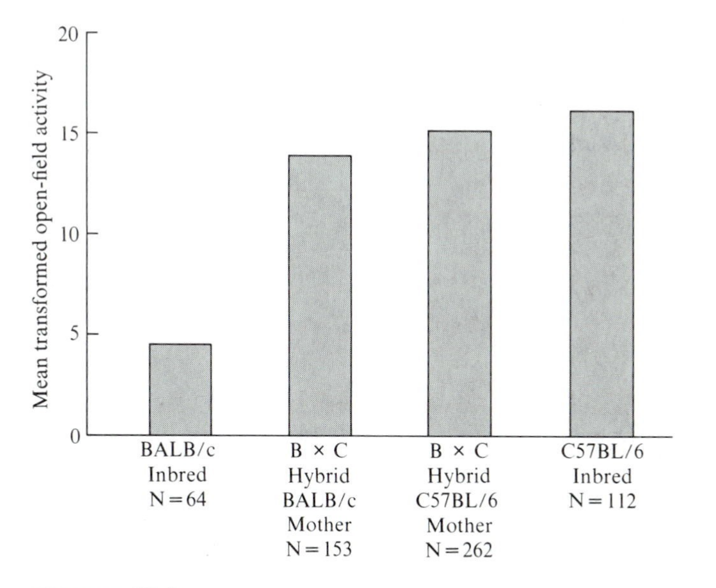

FIGURE 10.9

Mean transformed open-field activity of two inbred strains of mice and their reciprocal crosses. (Data from DeFries and Hegmann, 1970.)

BALB/c mice. The hybrids tend to be more like the C57BL mice, indicating some dominance for open-field activity. Moreover, the hybrids reared by a BALB/c mother are nearly as active as the hybrids reared by C57BL/6 mothers. The slight maternal effect, although statistically significant, is insubstantial compared to the large differences between the two inbred strains. A similarly slight maternal effect was also found for open-field defecation. Thus, although a small maternal effect is indicated, it is clear that it cannot begin to account for the large strain difference in open-field behavior.

Cross-Fostering

The slight maternal effect indicated by the above results of the reciprocal-cross method may be either prenatal or postnatal in origin. Postnatal influences can be isolated by a technique called *cross-fostering*. At birth, pups are transferred (cross-fostered) to mothers of a different strain. If the pups reared by a foster mother behave like the pups of the foster mother's strain, then postnatal influences are implicated. However, if the cross-fostered pups behave true to their own genotype, then the particular strain difference is not influenced by postnatal factors.

Despite the reasonableness of assuming the possibility of a maternal influence, neither reciprocal-cross nor cross-fostering studies have shown many important maternal effects, at least for rodents. However, one study (Reading, 1966) found a slight, but significant, effect of cross-fostering for open-field activity in BALB/c and C57BL/6 mice. Figure 10.10 illustrates the results of this study. C57BL/6 mice were not affected by cross-fostering, but the BALB/c mice were slightly more active when cross-fostered to a C57BL/6 mother. As before, the maternal effect is quite small when compared to the overall difference between the two strains. The C57BL/6 mice cross-fostered to BALB/c mothers were still nearly 40 percent more active than the BALB/c mice cross-fostered to C57BL/6 mothers. Thus, the strain difference between BALB/c and C57BL/6 mice does not seem to be caused to any major extent by postnatal influences.

Ovary Transplants

An elegant technique to determine maternal effects involves transplanting the ovaries from one female to another. Ovaries are the paired female reproductive organs that produce eggs. Although this sounds like science fiction, the operation is relatively easy, and the technique has been used in experimental embryology since 1909 (Palm, 1961). It certainly has elements of science fiction, as evidenced by the title of one of the earliest articles, "Offspring from Unborn Mothers" (Russell and Douglas, 1945).

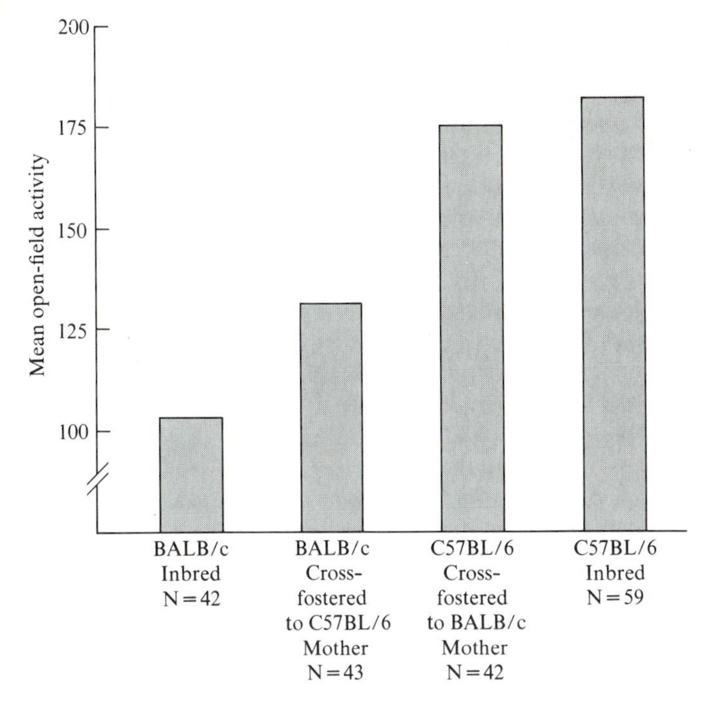

FIGURE 10.10

Mean open-field activity of two inbred strains of mice and reciprocal cross-fosterings. (Data from Reading, 1966.)

Ovary transplants have unique advantages over other techniques for assessing maternal effects. Reciprocal crosses involve hybridization, and the prevalence of hybrid vigor makes it likely that hybrids are less susceptible to environmental influences (Falconer, 1960). Cross-fostering studies require transferring a litter shortly after birth to a different mother. Ovary transplantation avoids these problems. However, just like skin grafts, ovary transplants require histocompatibility between donor and recipient. Thus, transfers between females of the same strain can be made with no difficulty. But this does not help us understand the role of prenatal influences. Fortunately, hybrids are compatible with both of their inbred parents. By transferring ovaries from two inbred strains to genetically identical hybrid mothers, we can determine whether inbred strain differences are due to the genotype of the offspring or that of their mother.

The first behavioral study using ovary transplants was conducted by DeFries et al. (1967). Ovaries of BALB/c and C57BL/6 females were transplanted to F_1 (BALB/c × C57BL/6) females whose own ovaries were removed. The operation is quite simple because the ovary is a self-contained unit. A small incision is made in the abdominal wall and each ovary with its surrounding fat pat is removed. The ovaries, within their transparent capsules, are then simply transferred to the ovarian cavity of the recipient female, whereupon the fat pads graft to the abdominal wall. In just a few days, the donor's ovaries will release ova.

The hybrid females with BALB/c ovaries were mated with BALB/c males, thus producing BALB/c inbred fetuses (from BALB/c sperm and BALB/c eggs) carried by a hybrid mother. The offspring of these females were designated B/H (a BALB/c offspring carried by a hybrid mother). Similarly, hybrid females with C57BL/6 ovaries were mated with C57BL/6 males. These pups were called C/H. Also, inbred pups reared by their own unoperated mother (B/B and C/C) were used for comparison purposes. This procedure is outlined in Figure 10.11.

Open-field activity of these four groups of offspring is illustrated in Figure 10.12. As we have often seen previously, BALB/c (B/B) inbreds are considerably less active and defecate more than the C57BL/6 (C/C) inbreds. More importantly, the activity scores of offspring reared prenatally and postnatally by foster hybrid mothers (B/H and C/H) are essentially the same as those of the inbreds. These results strikingly refute the hypothesis of a maternal effect for open-field activity. Similar results were obtained for open-field defecation. However, the study did demonstrate the effect of ma-

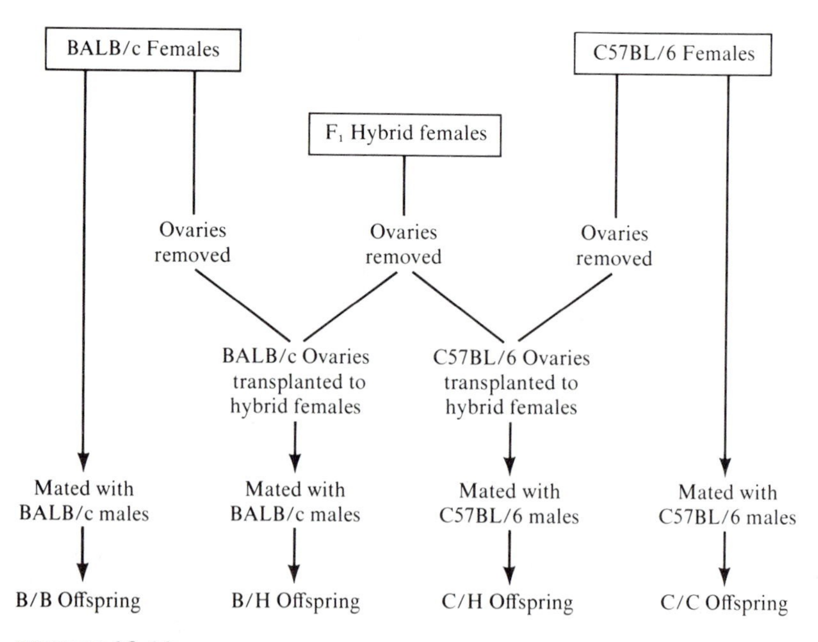

FIGURE 10.11

The ovary-transplant experiment of DeFries and associates. (DeFries et al., 1967.)

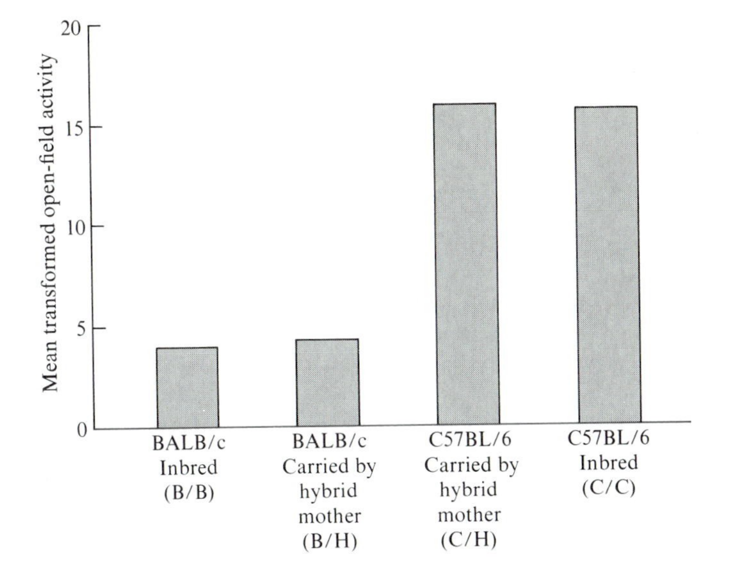

FIGURE 10.12
Mean transformed open-field activity of two inbred strains of mice, and inbreds carried by hybrid mothers. See text for explanation of ovary transplantation. (Data from DeFries et al., 1967.)

ternal environment on body weight, a character previously recognized as being susceptible to maternal effects.

Despite the effectiveness of ovary transplantation, we are aware of no other studies that have used it to separate pre- and postnatal environmental influences from genetic influences on behavior. In the case of open-field behavior, it provides the best case that the maternal effect is not an important source of the difference between the BALB/c and the C57BL/6 strains.

Classical Analysis

In addition to demonstrating that genetic differences between strains produce striking behavioral differences, strain studies can also be used to estimate heritability. Because this method involves the analysis of parental, F_1, and F_2 generations used by Mendel, it has been referred to as the *classical analysis*. When highly inbred strains are crossed to produce F_1 hybrids, individuals within each of the three populations are genetically identical to each other (isogenic). Each hybrid mouse is heterozygous, but all are heterozygous in the same way. Thus, variability within these three populations must be caused by nongenetic factors. The phenotypic variance observed in each of these populations therefore provides an estimate of the environmentally caused variance:

$$V_E = V_{P_1} = V_{P_2} = V_{F_1}$$

where V_{P_1} is the phenotypic variance observed in parental strain 1, etc.

In the F_2 generation, genes assort and the individuals differ from each other genetically as well as environmentally. Because the F_2 individuals differ genetically and environmentally, their phenotypic variance represents all sources of variability:

$$V_{F_2} = V_G + V_E$$

Thus, the variance of the F_2 individuals is equivalent to the total phenotypic variance. Heritability is the proportion of phenotypic variance due to genetic variance. Given that phenotypic variance includes both genetic and environmental variance, we can estimate genetic variance as $V_{F_2} - V_E$. The estimate of V_E can come from the phenotypic variance of the inbred strains or of the F_1 generation. The best estimate is the pooled parental and F_1 variances. To estimate heritability, we divide this estimate of genetic variance by the total phenotypic variance of the F_2 population:

$$h_B^2 = \frac{V_{F_2} - V_E}{V_{F_2}} = \frac{V_G}{V_G + V_E} = \frac{V_G}{V_P}$$

Classical analysis estimates broad-sense heritability because any genetic differences—those caused by nonadditive as well as additive genetic effects—contribute to the genetic variance in the F_2 population.

Open-field behavioral data from BALB/cJ and C57BL/6J mice and their F_1 and F_2 crosses (DeFries and Hegmann, 1970) will again serve as an example. For activity, the variance of the F_2 population was 16.1. The pooled variance for the genetically invariant populations (two parental inbred strains and their F_1) was 9.6. The variance of the F_2, which includes both genetic and environmental variance, is significantly greater than the estimate of V_E, thus suggesting the influence of genes. We can estimate broad-sense heritability as follows:

$$h_B^2 = \frac{V_G}{V_P} = \frac{V_{F_2} - V_E}{V_{F_2}} = \frac{16.1 - 9.6}{16.1} = 0.40$$

This estimate of broad-sense heritability suggests that about 40 percent of the phenotypic variance in the segregating F_2 population is due to genetic differences. This means, of course, that the majority of variance is due to environmental differences. These same data suggest a heritability for open-field defecation of 0.29 ($V_{F_2} = 0.55$, $V_E = 0.39$).

In the previous section, parent-offspring regression suggested that the narrow-sense heritability of open-field activity was 0.22. We have just arrived at a broad-sense heritability estimate of 0.40. The difference between broad- and narrow-sense heritability is due to nonadditive genetic variance, as discussed in Chapter 9. However, it may be that the classical analysis will overestimate genetic variance because, in practice, it usually begins with parental strains that differ markedly. For example, the BALB and C57BL strains differ substantially in open-field activity. Segregation of blocks of genes with positive genotypic values, and blocks of genes with negative genotypic values within the F_2 may result in overestimates of the genetic variance.

Diallel Design

The classical analysis just described considers generations derived by crossing two inbred strains. This analysis can be extended to all possible crosses between several strains. A *diallel design* compares several inbred strains and all possible F_1 hybrid crosses. It was first called "the method of complete intercrossing" (Schmidt, 1919). Like the classical analysis of two inbred strains, diallel analysis provides information concerning genetic variance, heterosis, and maternal effects. Environmental dimensions can also be added (Henderson, 1967; Hyde, 1974). However, the diallel method is more efficient because it can assess most genetic parameters, using only F_1 generations—that is, without waiting for F_2 animals. Moreover, the diallel method includes the genetic variance of several inbred strains and is thus less limited in its conclusions.

As an example, let us once again consider open-field activity in mice. Henderson (1967) conducted a diallel cross involving 1,440 mice of four strains. Table 10.4 summarizes the open-field activity data. Mean scores for the four inbred strains are listed on the diagonal. As we have seen, the C57BL/10 mice are more active than the BALB/c mice, and the C3H/He strain is the least active. The other scores are those obtained by the reciprocal crosses between the strains. Comparing the scores above the diagonal to those below, it is apparent that maternal effects are not important. For

TABLE 10.4
Diallel analysis of four inbred mouse strains for open-field activity

	Paternal Strain			
Maternal Strain	C57BL/10	DBA/1	C3H/He	BALB/c
C57BL/10	56	47	49	50
DBA/1	45	55	29	40
C3H/He	46	22	21	24
BALB/c	49	42	21	35

SOURCE: After Henderson, 1967.

example, there is no maternal effect for the BALB/c × C57BL/10 cross. With a BALB/c mother, the hybrids had an average activity score of 49. With a C57BL/10 mother, their average score was 50. These data generalize the conclusion concerning maternal effects to the other strains and their crosses. Analyses of variance for these four strains suggested broad-sense heritabilities of 0.40 for open-field activity and 0.08 for defecation.

In summary, studies comparing inbred strains are useful, not only for determining the influence of genes on behavior, but also for studying the role of the environment. In combination with manipulated environmental experiences, inbred strain studies can provide information concerning genotype-environment interaction as well as prenatal and postnatal maternal influences.

SELECTIVE BREEDING STUDIES

Both natural and artificial selection are effective only to the extent that the traits selected are under the influence of heredity. Animal breeders have successfully bred for behavioral characters throughout recorded history, long before there was any understanding of why it worked, and selection remains an important part of animal husbandry today. In addition to commercial applications, such as production of poultry, dairy cattle, and animals bred for meat, the results of artificial selection are apparent in our pets. Because dogs are such a familiar example of genetic variability, we shall digress for a moment to discuss breeds of dogs.

Dogs

Despite the tremendous variety of dogs, they are all members of a single species. Dogs have successfully been selected for behavior and morphology for the last ten thousand years. Everyone has seen their vast morphological differences, such as the forty-fold difference in weight between a Chihuahua and a Saint Bernard. However, behavior has clearly been as important in their selection. J. P. Scott and Fuller (1965) describe some very old accounts of breeds. In 1576, the earliest English book on dogs classified breeds primarily on the basis of behavior. Terriers, for example, were bred to creep into burrows to drive out small animals. Another book, published in 1686, described the behavior for which spaniels were originally selected. They were bred to creep up on birds and then spring to frighten the birds into the hunter's net. With the advent of the shotgun, different spaniels were bred to point rather than to crouch, although cocker spaniels still crouch when frightened. However, the author of the 1686 work was more concerned about the personality of spaniels: "*Spaniels* by Nature are very loveing, surpassing all other Creatures, for in *Heat* and *Cold, Wet* and *Dry, Day* and *Night,* they will not forsake their *Master*" (cited by Scott and Fuller, 1965, p. 47).

Behavioral classification of dogs continues today. The American and British kennel clubs still classify dogs on the basis of their behavioral function, rather than their genetic similarity. The selection process can be quite fine-tuned. For example, in France, where dogs are used chiefly for farmwork, there are 17 breeds of shepherd and stock dogs specializing in aspects of this work. In England, dogs have been bred primarily for hunting, and there are 26 recognized breeds of hunting dogs. Breeding, however, has had some mixed blessings. By selecting for certain characteristics, in many cases breeders have accidently selected for defects. For example, many breeds, such as German shepherds, are bred for a "downhill carriage" in which the shoulders are higher than the hips. Though this gives the dogs a powerful appearance, it also makes them more likely to have problems with their hip joints, and may result in lameness.

FIGURE 10.13

J. P. Scott with the five breeds of dogs used in his experiments with J. Fuller. Left to right: wire-haired fox terrier, American cocker spaniel, African basenji, Shetland sheep dog, and beagle. (From *Genetics and the Social Behavior of the Dog* by J. P. Scott and J. L. Fuller. Copyright © 1965 The University of Chicago Press. All rights reserved.)

An extensive behavioral genetics research program on breeds of dogs was conducted over two decades by Scott and Fuller (1965). They studied the development of pure breeds and hybrids of the five representative breeds pictured in Figure 10.13: basenjis, beagles, cocker spaniels, Shetland sheep dogs, and wire-haired fox terriers. The selection history of each of these strains is quite different. Basenjis were recently brought from Africa, where they were bred as general-purpose hunting dogs. They were used to drive small game into a net, to track, and to flush birds. Cocker spaniels are descended from bird dogs, and now are primarily house pets. Terriers are aggressive scrappers, as their selection history described above would suggest. Unlike the other breeds, Shetland sheep dogs have not been bred for hunting, but rather for performing complex tasks under close direction from their masters. They were originally small dogs, which were crossed with the large Scotch collie, and then crossed again with smaller breeds. They are currently being bred as small dogs with a collielike appearance.

These breeds are all about the same size, but they differ markedly in behavior. Before we generalize about the behavior of the breeds, we should note that breeds of dogs are not genetically invariant, like inbred strains of mice. Considerable genetic variability exists within breeds, despite the substantial genetic differences between them. Average behavioral differences among the breeds reflect their breeding history. Spaniels are very people-oriented and nonaggressive. Terriers are considerably more aggressive. Basenjis seldom bark and are very fearful of people until they are a couple of months old, at which time they can be rapidly tamed. Shetland sheep dogs are very responsive to training. In short, Scott and Fuller found behavioral breed differences just about wherever they looked—in the development of social relationships, emotionality, and trainability. They also demonstrated that such differences exist even when the breeds are reared in the same environment.

Heritability

Quantitative genetic considerations can predict the success of selection. At the most basic level, if heritability is zero, selective breeding will completely fail to produce the desired results. If heritability is 1.0, selection will quickly succeed. Intermediate levels of heritability will yield partial success.

More specifically, narrow-sense heritability is used to estimate the response to selection because it involves only additive genetic variance. (See Chapter 9.) Selection will succeed only to the extent that additive genetic variance is present. Consider a behavior such as open-field activity in a genetically heterogeneous group of mice. If we breed mice high in activity in the open-field, and if activity is highly heritable, their offspring will have an average activity score more like their parents' average than that of the average of the base population. However, if open-field activity is not heritable, the offspring of these highly active animals will have an average activity score the same as that of the base population.

We need to introduce two terms to describe this logic. The first is the *selection differential,* the difference between the mean of the selected parents and that of the base population. (See Figure 10.14.) The other is the *response to selection* (or gain), the difference between the mean of the offspring of the selected animals and that of the base population. If heritability is zero, then the response to selection will be zero, no matter how severe the selection differential. In fact, the response to selection (R) is a simple function of heritability (h^2) and the selection differential (S):

$$R = h^2S$$

Thus, as heritability approaches 1.0, the response to selection will approach the selection differential. In other words, the offspring of the selected parents

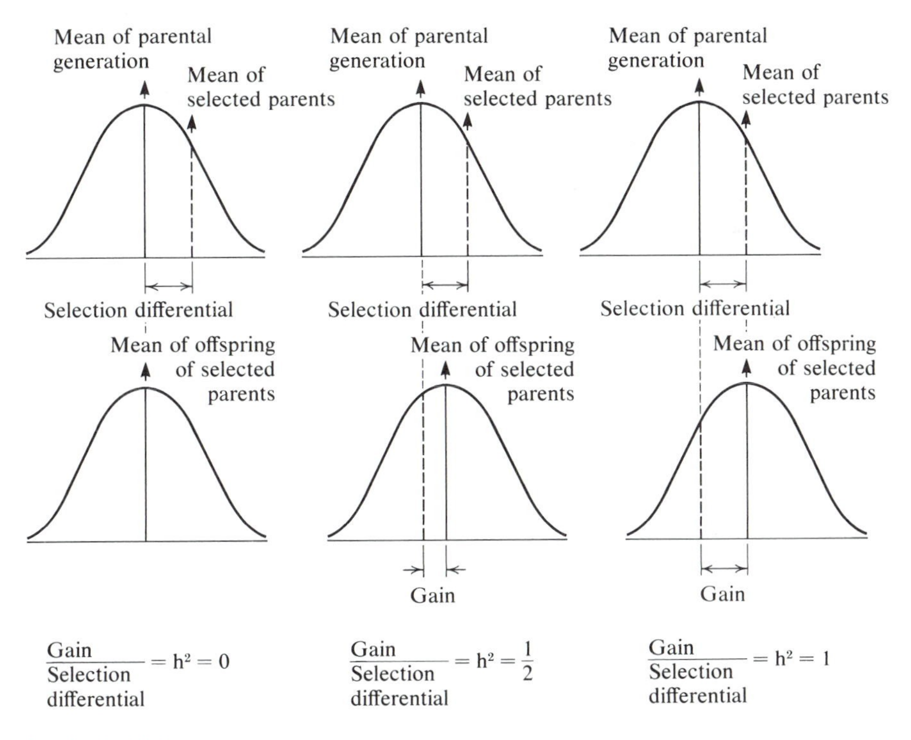

FIGURE 10.14

Relationship between selection differential, response to selection or gain, and realized heritability. (From *Heredity, Evolution, and Society,* 2nd ed., by I. Michael Lerner and William J. Libby. W. H. Freeman and Company. Copyright © 1976.)

Box 10.2
Fundamental Theorem of Natural Selection

Calculations of the selection differential must consider not only the means of the selected parents and the base population, but also the possibility that some of the selected parents may produce more offspring than others. For example, suppose the highest-scoring selected parents did not produce as many offspring as other selected parents. Unless we weighted the parental mean by the number of offspring, the calculated selection differential would overestimate the expected mean of their offspring. Although this may appear to be a small point, it serves to introduce a very important topic, the fundamental theorem of natural selection. This theorem states that changes in evolutionary fitness require additive genetic variance.

The derivation of the fundamental theorem of natural selection can be understood if we refer again to the problem of differential reproduction by selected parents in a selective breeding study. In order to adjust for the problem, a weighted selection differential is used that weights the parental mean by the number of their offspring. A mating pair's relative fitness is the ratio between the number of progeny (f_i) contributed to the next generation by that pair, and the average number of progeny of all selected pairs ($\bar{f}$). Thus, we can obtain the weighted selection differential (S) by multiplying the deviation of each pair's midparent value from the population mean ($X_i - \bar{X}$) by that pair's relative fitness ($f_i/\bar{f}$). We then sum these values for all selected pairs and divide by the number of pairs to get an average. In more concise algebraic terms:

$$S = \left(\frac{1}{n}\right) \Sigma \left[\frac{f_i}{\bar{f}} (X_i - \bar{X})\right]$$

For example, if the population mean for some character were 10.0 and the mean of the selected population was 15.0, the unweighted selection differential would be 5.0. However, if the highest-scoring selected animals reproduced less than the lower-scoring selected animals, then the weighted selection differential (that is, weighted by each pair's relative fitness) would be less than 5.0.

The above equation facilitates an important derivation obtained by Fisher (1930) many years ago. Instead of considering mean values for some trait (X) as above, think about reproductive or relative fitness, the most important characteristic in an evolutionary sense. We described relative fitness as $f_i/\bar{f}$, which we can symbolize as F. If we use the above equation to find a weighted selection differential for relative fitness, (S_F), we substitute as follows:

$$S_F = \left(\frac{1}{n}\right) \Sigma[F_i(F_i - \bar{F})]$$

This is equivalent to:

$$\frac{\Sigma F_i^2 - [(\Sigma F_i)^2]/n}{n} = V_{P_F}$$

This, in turn, is equivalent to the phenotypic variance of F, relative fitness. (Readers who have studied statistics will recognize the form of this equation as the computational shortcut for computing a variance.)

As described in the text, the response to selection is a function of heritability and the selection differential. For relative fitness, the response to selection (R_F) is the change in relative fitness per generation:

$$R_F = h_F^2 S_F$$

Because narrow-sense heritability is equivalent to the ratio of additive genetic variance to phenotypic variance, and, as seen above, S_F equals the phenotypic variance of F, this equation reduces to:

$$\begin{aligned} R_F &= (V_{A_F}/V_{P_F})(V_{P_F}) \\ &= V_{A_F} \end{aligned}$$

In other words, change in relative fitness per generation is equal to the additive genetic variance. If additive genetic variance for characters related to fitness is present in a species, then there is room for further selection. However, when characters related to fitness have been selected as severely as possible (meaning that there can be no more response to selection), no additive genetic variance will remain. We would, therefore, expect heritability to be low for major components of fitness, such as fertility (Falconer, 1960). However, the converse is not necessarily true. Characters with low heritability are not necessarily important components of fitness. They may be low in heritability because environmental influences are very important.

The fundamental theorem of natural selection also implies that most of the genetic variance of fitness characters should be nonadditive. Because dominance-recessiveness is responsible for inbreeding depression and hybrid vigor, we would also expect to find considerable inbreeding depression and hybrid vigor, as well as low heritabilities, for fitness characters in stable populations (Bruell, 1964, 1967).

will yield an average score similar to their parents' average score. Thus, if we know the heritability of a character, we can predict the response to selection for a given selection differential, although precise predictions are complicated by dominance, pleiotropy, linkages, and differing initial allelic frequencies (Wright, 1977). This relationship is especially important for evolutionary biology, where the character being selected is relative fitness. (See Box 10.2.)

By rearranging terms, we see that

$$h^2 = \frac{R}{S}$$

Thus, a selection study can be used to estimate heritability retrospectively, the so-called realized heritability. The relationship between these variables is illustrated in Figure 10.14. In the first example, heritability is zero because the mean of the offspring is just like the mean of the unselected base population. The middle example indicates a response to selection ("gain" in Figure 10.14) that is half the selection differential; so that heritability is 0.50. In the third example, illustrating a heritability of 1.0, the response to selection is the same as the selection differential.

Maze Running in Rats

Selective breeding experiments were employed early in the history of behavioral genetics. In 1924, E. C. Tolman reported the results of two generations of selection for maze learning by rats. Tolman saw the genetic approach, and selective breeding in particular, as a tool for "dissecting" behavioral characteristics:

> The problem of this investigation might appear to be a matter of concern primarily for the geneticist. Nonetheless, it is also one of very great interest to the psychologist. For could we, as geneticists, discover the complete genetic mechanism of a character such as maze-learning ability—i.e., how many genes it involves, how these segregate, what their linkages are, etc.—we would necessarily, at the same time, be discovering what psychologically, or behavioristically, maze-learning ability may be said to be made up of, what component abilities it contains, whether these vary independently of one another, what their relations are to other measurable abilities, as, say, sensory discrimination, nervousness, etc. The answers to the genetic problem require the answers to the psychological, while at the same time, the answers to the former point the way to those of the latter. (Tolman, 1924, p. 1)

As his own contribution toward this end, Tolman began with a diverse group of 82 rats, which were assessed for learning ability in an enclosed maze. Using as a criterion for selection "a rough pooling of the results as to errors, time, and number of perfect runs,'' 9 male and 9 female ''bright'' rats were selected and mated with each other. Similarly, 9 male and 9 female ''dull'' rats were selected to begin the ''dull'' line. The offspring of these groups constituted the first selected generation. These animals were then tested in the maze, and selection was made of the brightest of the bright and the dullest of the dull. These two groups of selected animals were mated among themselves (brother × sister) to provide the second selected generation of ''brights'' and ''dulls.''

The results were quite clear in the first generation. The bright parents produced bright offspring, and the dull parents produced dull. Due to the completeness of the data presented by Tolman (1924), it is possible to estimate that the realized heritabilities for the three characters selected were as follows: errors, 0.93; time, 0.57; and number of perfect runs, 0.61. However, due to the small sample size, these estimates are subject to large standard errors. The difference between ''brights'' and ''dulls'' decreased in the next generation, primarily because of a drop in the efficiency of performance of the bright line. These second-generation results were, of course, disappointing, and Tolman examined various possible explanations. In the first place, the maze he had used was not a particularly reliable measuring instrument. Secondly, the mating of brother with sister may have led to inbreeding depression.

To facilitate further investigation, an automatic, self-recording maze was developed by Tolman in collaboration with Robert Tryon. With the new maze, which provided superior control of environmental variables and proved to be highly reliable, Tryon began the selection procedure again, starting with a large, heterogeneous ''foundation stock'' of rats collected from several different laboratories. Tolman's energies were spent developing his theory of learning, and he did no further actual experimentation in behavioral genetics. Nevertheless, he made a continuing contribution to the field by insisting on the importance of heredity in his well-known H.A.T.E. (heredity, age, training, endocrine plus drug and vitamin conditions) list of individual-difference variables.

In Tryon's experiment, selection was based on the total number of entrances into blind alleys from days 2 to 19, following a preliminary run of 8 days to acquaint subjects with the maze. As in the Tolman study, deliberate inbreeding was practiced. The results are shown in Figure 10.15. There is a fairly consistent divergence between the bright and dull lines through generation 7, at which time there is practically no overlap between the distributions of the two groups. The dullest bright rats were about equal in performance to the brightest dull rats. Little or no additional response to selection was observed in the later generations. Tryon provided sufficient information for Patrick Tyler (1969) to calculate the realized heritability at 0.21.

A general problem in the interpretation of selective breeding studies is evident in the results of Tryon's experiment. The problem is that we may think that we are selecting for one characteristic, when, in fact, we have

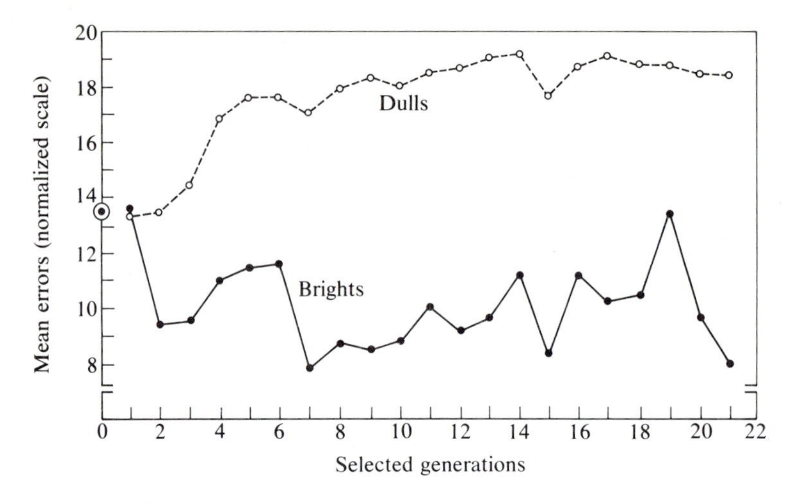

FIGURE 10.15

The results of Tryon's selective breeding for maze brightness and dullness. (From "The inheritance of behavior" by G. E. McClearn. In L. J. Postman, ed., *Psychology in the Making*. Copyright © 1963. Used with permission of Alfred A. Knopf, Inc.)

selected for another. In Tryon's study, we might like to think that he selected for the rat equivalent of intelligence, as the terms "bright" and "dull" connote. But there are many other possibilities. All we know is that in a certain maze, given a certain testing procedure, the bright animals made fewer errors than the dull animals. This difference could be caused by a perceptual or motor defect in the dulls. Although such peripheral hypotheses have not been substantiated by research, a central hypothesis that competes with the "rat intelligence" notion is that the dull animals are more "emotional," or easily frightened, than the bright animals. This hypothesis was suggested by L. V. Searle (1949), who obtained scores for the dull and bright animals on 30 measures of learning and emotionality. He found that the brights performed better than the dulls on only two of five measures of maze learning. His results also indicated that the dulls might be more emotional than the brights. A later selection study (Thompson and Bindra, 1952), using a different maze, also resulted in successful selection for errors in maze running, but the differences between the selected lines in this study did not seem to be caused by emotional or motivational differences.

Although Tryon's selection experiment is a classic in experimental behavioral genetics, the design suffers from several inadequacies that have been perpetuated in some more recent selection research. As indicated, deliberate inbreeding was practiced by both Tolman and Tryon. Although a secondary objective of these studies was to produce highly inbred lines with uniform behavioral differences, inbreeding may impede the response to selection, which was the primary objective. Inbreeding results in a decrease in genetic variance within lines and, thus, a decrease in the potential selection response. In addition, inbreeding is almost always accompanied by a reduction in fertility, resulting in a decrease in the selection differential.

Another inadequacy of this experimental design is the lack of an unselected control group. When such a group is included, it is possible to evaluate the effects of environmental changes across generations. In addition, the response to selection in the high and low lines can each be measured by their deviation from the mean of the control group. In this manner, it is possible to determine the degree of asymmetry of response to selection. Finally, selected and control lines should each be replicated. Since selection experiments involve considerable intergeneration variability, the reliability of the result can be indicated by the inclusion of replicate selected lines. Even more important is the fact that replicate lines are critical in analyzing genetic correlations among characters (see Chapter 9), the so-called correlated response to selection. Fortuitous correlations between the character under selection (for example, maze-running performance) and other characters of interest (such as temperament) may often occur when only one high line and one low line are maintained. However, if similar associations are noted in each of two or more replicates, the correlation is much more likely to indicate a causal (pleiotropic) relationship.

The selection study of open-field behavior described in the next section incorporated the refinements suggested above.

Open-Field Behavior

The largest and longest selection study of mammalian behavior in a laboratory was conducted at the University of Colorado (DeFries et al., 1978). Thirty generations of mice were selected for open-field activity, in a study involving the testing of more than 14,000 mice over a ten-year period. Selection began with an F_3 generation of a cross between BALB/c and C57BL/6 inbred strains. Selection was bidirectional—that is, both high- and low-active lines were selected. Also, a control line was maintained, and each of these three lines was replicated. Selection for the most or least active animals from each generation could lead to inbreeding because the selected animals could come from the same litter. In order to avoid inbreeding, a male and a female were selected from each litter and then mated at random within lines. Ten mating pairs were maintained for each of the six lines in each generation.

Figure 10.16 traces the selection progress over thirty generations. After thirty generations of selective breeding, there was a thirty-fold difference between the high and low lines. Measures on replicated lines indicate the reliability of this result. By the thirtieth generation (S_{30}), the low-active lines had nearly reached the bottom limit of zero activity scores. However, the high lines showed no sign of reaching a "selection limit" (Falconer, 1960). The high-active mice now run the equivalent total distance of the length of a

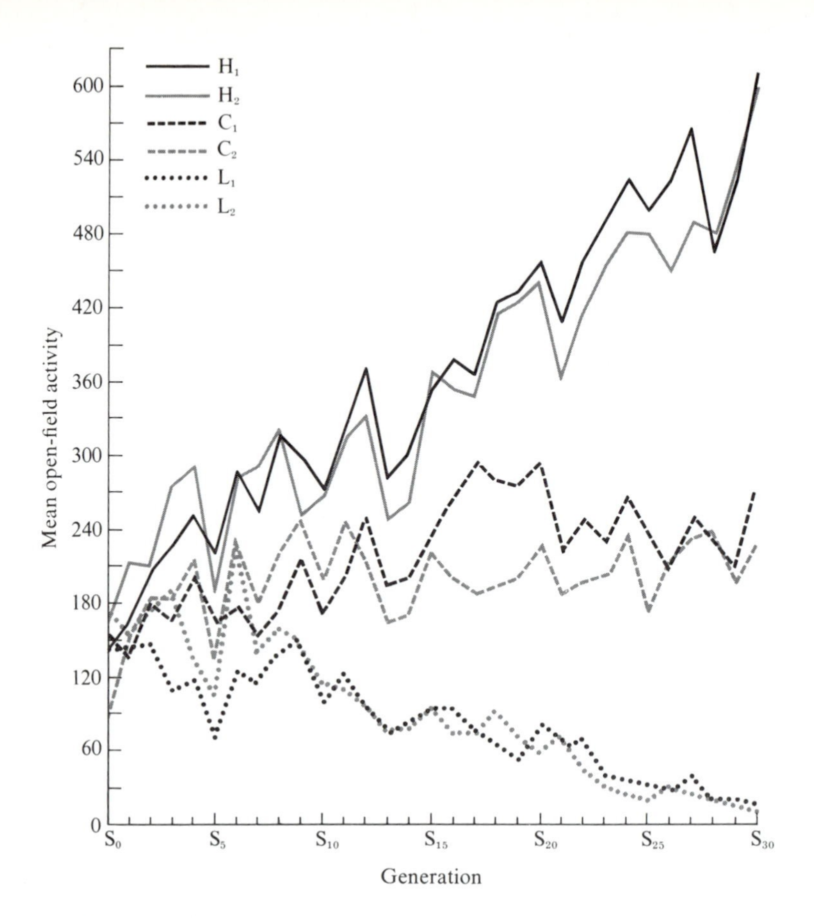

FIGURE 10.16

Mean open-field activity scores of six lines of mice: two selected for high open-field activity (H_1 and H_2), two selected for low open-field activity (L_1 and L_2), and two randomly mated within line to serve as controls (C_1 and C_2). (From "Response to 30 generations of selection for open-field activity in laboratory mice" by J. C. DeFries, M. C. Gervais, and E. A. Thomas. *Behavior Genetics,* 8, 3–13. Copyright © 1978 by Plenum Publishing Corporation. All rights reserved.)

football field during two, 3-minute test periods. As indicated in Figure 10.17, there is no overlap in the distributions of open-field activity for the high and low lines.

Although the mice were selected for open-field activity, there was a correlated response to selection for defecation in the open-field apparatus. (See Figure 10.18.) This suggests that open-field activity and defecation are genetically correlated. (See Chapter 9.) The average defecation scores of the low-active lines were about seven times higher than those of the high-active lines. The study of correlated responses to selection is an example of the multivariate approach discussed in Chapter 9. In this case, open-field activity and defecation seem to be influenced by many of the same genes. Also, as

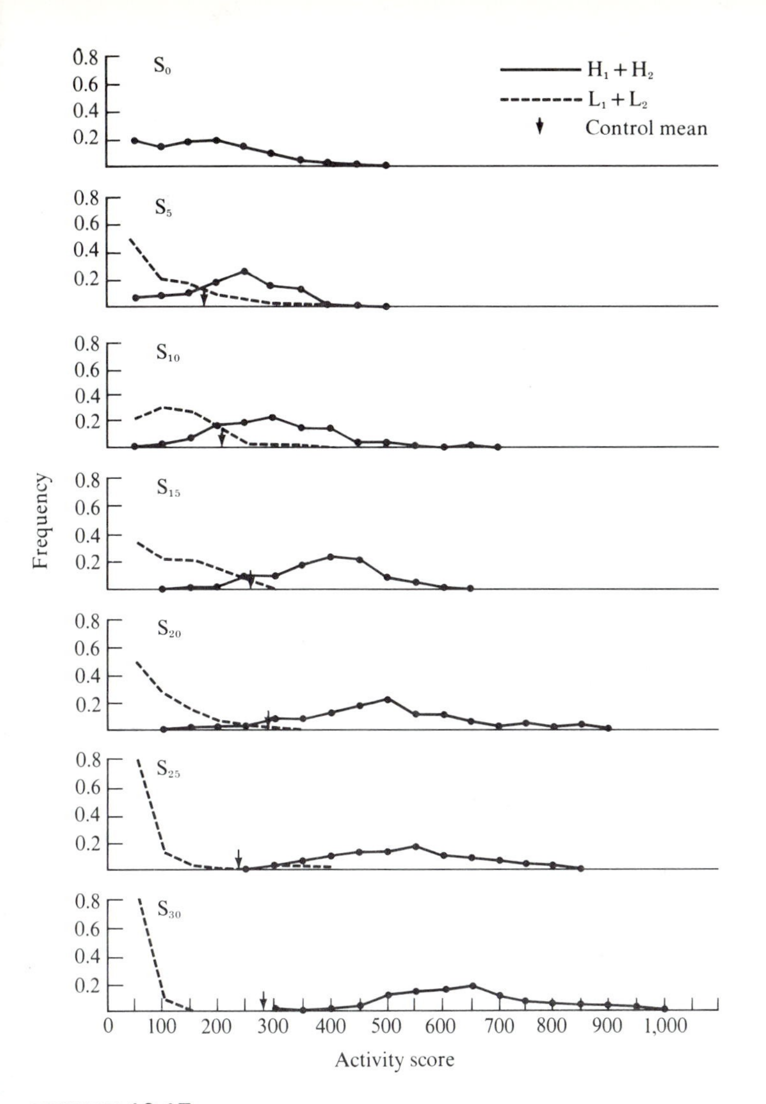

FIGURE 10.17

Distributions of activity scores of lines selected for high and low open-field activity. Average activity of controls in each generation is indicated by an arrow. (From "Response to 30 generations of selection for open-field activity in laboratory mice" by J. C. DeFries, M. C. Gervais, and E. A. Thomas. *Behavior Genetics,* 8, 3–13.)

one would expect from our discussion of albinism and open-field behavior in Chapter 4, there are almost no albinos in the high-active lines, whereas the low-active lines are completely albino. (See Figure 10.19.) It is as if the continued selection pressure has scooped up all those alleles, including the albino one, that make for less activity, and concentrated them in the low-

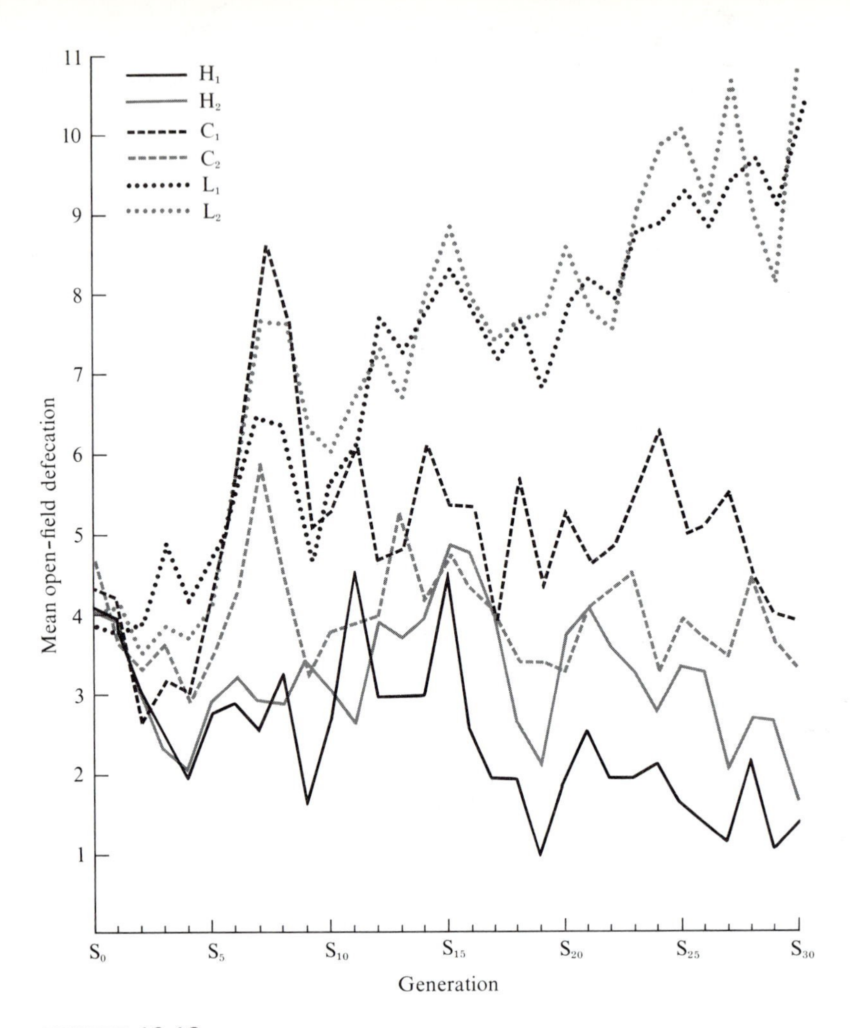

FIGURE 10.18

Mean open-field defecation scores of six lines of mice: two selected for high open-field activity (H_1 and H_2), two selected for low open-field activity (L_1 and L_2), and two randomly mated within line to serve as controls (C_1 and C_2). (From "Response to 30 generations of selection for open-field activity in laboratory mice" by J. C. DeFries, M. C. Gervais, and E. A. Thomas. *Behavior Genetics,* 8, 3–13.)

active lines, while concentrating in the high lines those alleles that make for more activity.

It is possible to use these data to estimate the realized heritability of open-field activity. As previously discussed, selection—either artificial or natural—is based on additive genetic variance. Thus, narrow-sense heritabil-

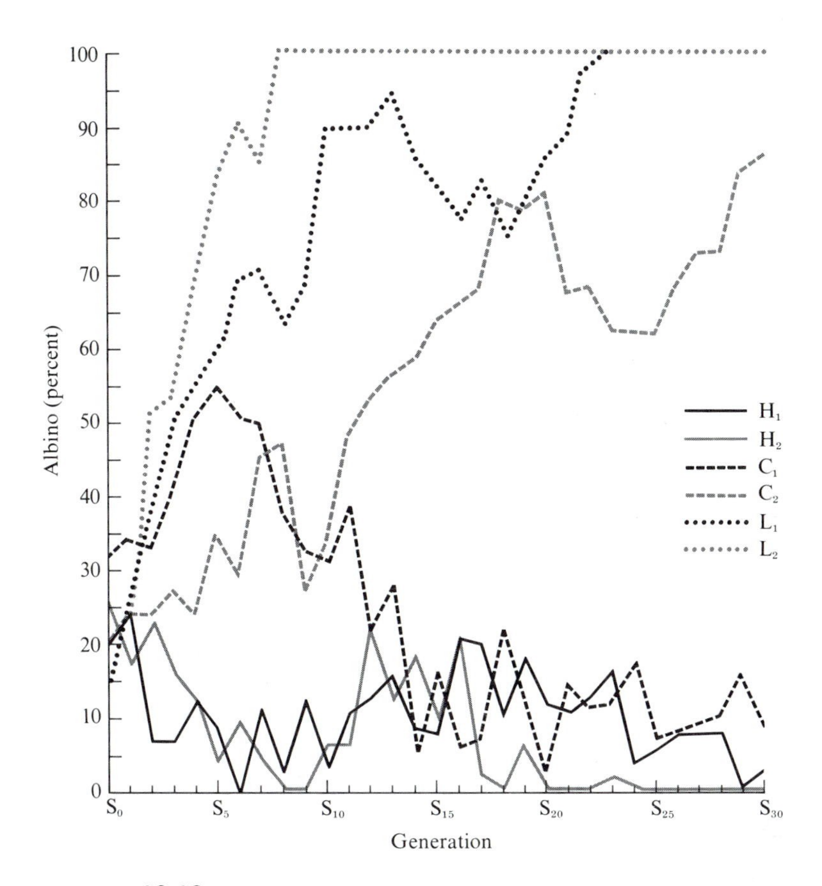

FIGURE 10.19

Frequency of albinism in six lines of mice: two selected for high open-field activity (H_1 and H_2), two selected for low open-field activity (L_1 and L_2), and two randomly mated within line to serve as controls (C_1 and C_2). (From "Response to 30 generations of selection for open-field activity in laboratory mice by J. C. DeFries, M. C. Gervais, and E. A. Thomas. *Behavior Genetics,* 8, 3–13. Copyright © 1978 by Plenum Publishing Corporation. All rights reserved.)

ity estimates from parent-offspring regressions should agree with heritability estimates from selection studies. After five generations of selection for open-field activity, the pooled estimate of narrow-sense heritability was 0.26, which is remarkably close to the parent-offspring estimate of 0.22 discussed earlier. This study demonstrates that marked behavioral changes can occur during selective breeding when heritability is considerably less than 1.0.

Other Behavioral Selection Studies

Selection has been successful for many other behaviors in the laboratory. One study focused on a socially relevant behavior, sensitivity to the effects of alcohol (McClearn, 1976). Mice, like humans, differ in their reaction to alcohol. When mice are injected with the mouse equivalent of several martinis, they will "sleep it off" for varying lengths of time. Sleep time in response to ethanol injections was measured by the length of time it took an injected mouse to right itself in a cradle like that pictured in Figure 10.20. The selection progress is pictured in Figure 10.21. After 15 generations of selective breeding, the long-sleep animals sleep for an average of two hours. Many of the short-sleep mice are not even knocked out, and their average sleep time is only about 10 minutes. Figure 10.22 indicates that there is no overlap between the long-sleep and short-sleep lines. In fact, although one would hesitate to say it while inebriated, the shortest of the

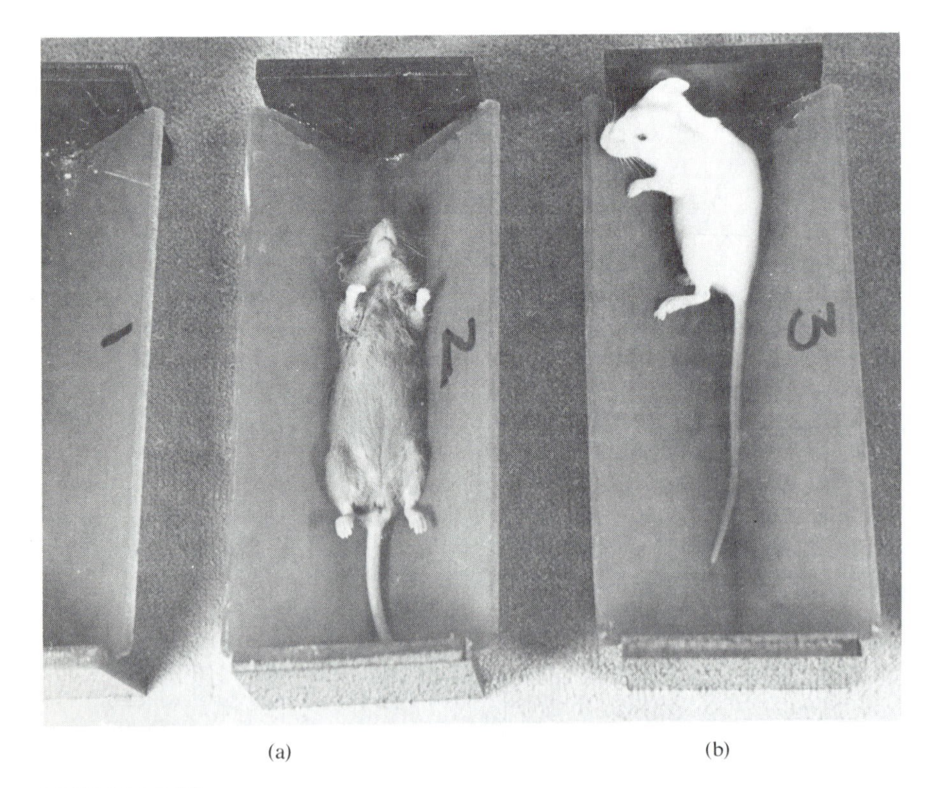

FIGURE 10.20

The "sleep cradle" for measuring loss of righting response after ethanol injections in mice. (a) A long-sleep mouse still on its back sleeping off the ethanol injection. (b) A short-sleep mouse that is just about to right itself. (Courtesy of E. A. Thomas.)

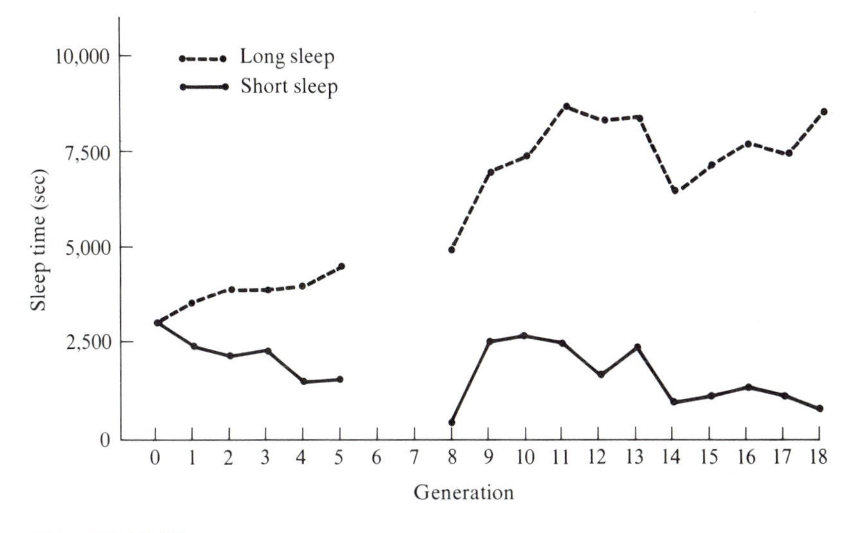

FIGURE 10.21

Results of ethanol sleep-time selection study. Selection was suspended during generations 6 through 8. (From McClearn, unpublished.)

long sleepers sleep about a half-hour longer than the longest of the short sleepers.

Selection has been successful for other mouse behaviors, such as alcohol preference, susceptibility to audiogenic seizures, wildness and tameness, learning, and rearing behavior. Recent selection research on mice includes studies of aggressive behavior (Ebert and Hyde, 1976), nerve conduction velocity (Hegmann, 1975), and seizure susceptibility (Chen and Fuller, 1976; Deckard et al., 1976). In rats, successful selective breeding has been accomplished for spontaneous activity, open-field behavior, and saccharine preference. Aggressiveness and aspects of mating behavior have been selected in chickens. Quail have recently been selectively bred for mating ability (Sefton and Siegel, 1975). Geotaxis, phototaxis, mating speed, and activity have been selected in *Drosophila* (including recent studies by Polivanov, 1975; Watanabe and Anderson, 1976; and Kekic and Marinkovic, 1974).

This success in selective breeding is another indicator of the considerable genetic variation that exists for behavior. More importantly, such studies generate groups of animals ideal for research on the physiological mechanisms underlying behavior. Peter Broadhurst (1975, 1976), for example, has reviewed the large number of experiments conducted from 1964 to 1974 with his Maudsley Reactive and Nonreactive rats selected for open-field defecation. Broadhurst (1978) has also reviewed selective breeding studies that involved selection for pharmacological differences. These studies are a part of the fast-developing area *psychopharmacogenetics*.

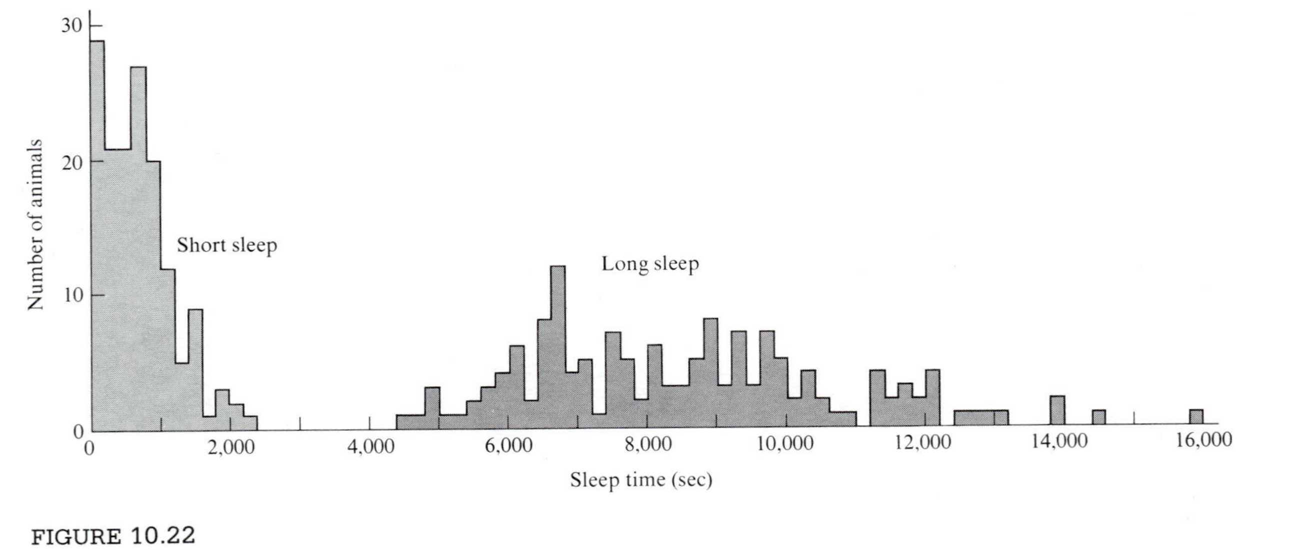

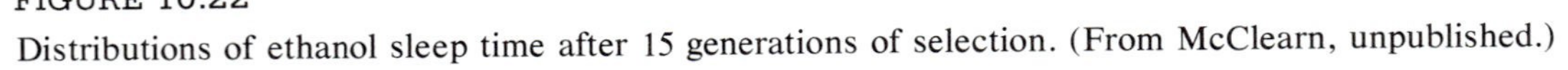

FIGURE 10.22
Distributions of ethanol sleep time after 15 generations of selection. (From McClearn, unpublished.)

SUMMARY

Three basic methods are used to assess polygenic influences on animal behavior. *Family studies* determine whether genetic similarity is evidenced phenotypically for different family relationships, such as parents and their offspring, full siblings, and half-siblings. There are surprisingly few family studies of animal behavior. *Strain studies* compare the behavior of inbred strains. Prenatal and postnatal environmental effects (especially maternal effects) can be studied using reciprocal crosses, cross-fostering, and ovary transplantation. Classical analysis of inbred strains and their F_1 and F_2 crosses can be used to estimate heritability. The diallel method compares several strains mated in all possible combinations. The third method is *selection*. Open-field behavior in mice has been used throughout this chapter as an example of the utility of the various methods. All of them converge on a narrow-sense heritability estimate of about 0.25.

11

Family Studies of Human Behavior

Genes influence human behavior in the same way that they affect any phenotype. They control the production of proteins, which interact in physiological systems, thus affecting behavior indirectly. Obviously, we cannot (nor do we wish to) selectively breed people, assign them to controlled environments, or produce genetically identical groups analogous to inbred strains of mice. We can, however, find naturally occurring situations in which (1) genetic influences are controlled or randomized so that the effects of the environment can be studied, and (2) environmental influences are controlled so that the effects of genes can be studied. Although studies of complex human behavior lack the experimental control of studies of nonhuman animals, there is a silver lining in this particular cloud. From the beginning, human behavior has been studied in the context of genetic and environmental variation as it exists naturally in populations. Thus, human behavioral geneticists have provided information concerning the relative contributions of genetic and environmental factors to observed variation, given the mix of genetic and environmental influences in the real world. In contrast, little is known about the full range of environmental and genetic variation in mice in natural populations because inbred strains are typically studied in the controlled environment of the laboratory.

In this and the next two chapters, we shall consider three basic methods of studying the relative influences of genetic and environmental factors on human behavior: family studies, twin studies, and adoption studies. Throughout the chapters, cognitive behavior and psychopathology will be used as examples of complex behaviors influenced by both genetic and environmental factors. Far more behavioral genetic data are available for cognition and psychopathology than for other human behaviors. In Chapter 1, we discussed three perspectives on behavior. The first recognizes no important differences between or within species, the second accepts differences between but not within species, and the third recognizes differences between and within species. The third view not only represents that of behavioral genetics, but it also characterizes the orientation of much research in the areas of cognitive behavior (particularly IQ and specific cognitive abilities), personality, and psychopathology. Other aspects of behavioral science have been slower to take advantage of behavioral genetics methods because their predominant orientation has been closer to the first or second perspectives. There are, however, clear signs of change.

COMPONENTS OF COVARIANCE IN FAMILY STUDIES

Genetic similarity applies to human family relationships as it does to the nonhuman ones discussed in the previous chapter. (See Table 10.1.) Genetically related individuals should be similar phenotypically to the extent that genes influence a particular behavior. Thus, in addition to testing specific genetic models, as described in Chapter 5, family studies have been useful in demonstrating familial resemblance or familiality, the sine qua non for establishing genetic influence. With laboratory animals, we could assume that environments are controlled, and we could test the adequacy of this assumption by reciprocal crosses, cross-fostering, and ovary transplantation. For human families, this assumption cannot be made. Thus, for human behavior, phenotypic similarity for different family relationships can only be viewed as compatible with a genetic hypothesis, not as conclusive proof of genetic influence.

In this sense, family studies provide an upper-limit estimate of heritability. (See Chapter 9.) That is, genetic influence is usually no greater than the degree of familiality (except when environmental factors are negatively correlated for family members). If family studies show no familial resemblance, the genetic hypothesis is disconfirmed. The finding of no familiality would also suggest that there are no familial environmental influences making family members similar to one another. Such environmental influences shared by family members are known as between-family, or common-family, environment, as discussed in Chapter 9. When behavioral scientists talk about environment, they usually mean between-family influences that make family members similar to one another and different from other families. For example, if families differ in child-rearing practice or income, these environmental factors may underlie observed phenotypic similarity among family members. All other environmental influences are called within-family environment. These factors create differences among members of the same family. Although within-family environmental influences have been given much less attention than between-family influences, human behavioral genetics studies point to a major role for this type of environmental effect, particularly for personality and psychopathology. Heredity, of course, creates differences between, as well as within, families. This is one explanation for the fact that like sometimes, but not always, begets like. Although hereditary factors predict differences within families, we know next to nothing about possible within-family environmental factors.

As described in our discussion of nonhuman research in the previous chapter, we can study various family relationships such as parents and their offspring, full siblings, and half-siblings. Half-siblings are individuals with only one parent in common. This situation occurs, for example, when divorced individuals have children from their previous marriage, as well as from their current marriage. In human research, we can also consider twin relationships. The genetic relationship between fraternal twins is just like that between any full siblings—as shown in Table 9.9, they share $\frac{1}{2}V_A + \frac{1}{4}V_D$. They also share between-family environmental influences. Identical twins have the same genes because they come from the same fertilized egg. Thus, their genetic covariance includes all additive and nonadditive genetic variance ($V_A + V_D$) plus between-family environmental influences.

In the following sections, we shall review family studies of cognition and psychopathology. Until recently, family studies of general cognitive ability have dominated the research scene. As explained below, attention has now turned to specific cognitive abilities.

COGNITION

There is much more to cognition than IQ, but more familial data are available for IQ than for any other behavior. For this reason, we shall present a summary of family studies of IQ, in spite of the recent controversies concerning IQ testing. This is clearly not the place for an extended discussion of the nature of IQ or its application. (See, for example, Caroll and Maxwell, 1979; Jensen, 1979; Vernon, 1979; Willerman, 1979.) Behavioral genetics studies have begun to consider specific cognitive abilities, rather than a single measure of general cognitive functioning, and data from this research will also be presented. However, behavioral genetics studies have just begun to open the door on the complexity of cognitive functioning. Future studies are likely to make use of theoretical approaches, such as Jean Piaget's, as well as the theories and instrumentation of experimental psychology, in order to increase our understanding of cognitive processes related to the perception of stimuli, attention, and memory storage and retrieval (Resnick, 1976).

Intelligence Quotient (IQ)

Many studies of familial resemblance for IQ were conducted in the late 1920s and the 1930s. Although these studies individually suffered from various difficulties, such as small sample size or bias in sampling (see McAskie and Clarke, 1976), the combined weight of their evidence demonstrates substantial familial influence. Figure 11.1 summarizes the results of family studies published prior to 1962 (Erlenmeyer-Kimling and Jarvik, 1963). The median correlations are all about the same (about 0.50) for parents and their offspring, full siblings, and fraternal twins. The median correlations for opposite-sex and same-sex fraternal twins are the same. All of these familial relationships involve first-degree relatives who share about half of their segregating genes. The median correlation for identical twins is much higher (0.88), and we shall return to this finding in the next chapter when we consider the twin method.

The median correlation of 0.50 for siblings may be due to genetic or environmental similarity, or both. If the phenotypic similarity for siblings were not influenced by genetic factors, then all of the phenotypic variance would obviously be caused by environmental (more technically, nongenetic) factors. The correlation of 0.50 would indicate that half of this variance is shared by siblings (between-family environment, as discussed in Chapter 9), and the other half is due to environmental influences not shared by siblings (within-family environment). If genetic similarity were responsible for all of the phenotypic similarity, then broad-sense heritability would be 1.0, meaning that all of the phenotypic variance can be accounted for by genetic differences. The twin studies and adoption studies, reviewed in the next two chapters, suggest that the answer is somewhere in the middle. The phenotypic variance for sibs is due both to genetic variance and to environmental variance between and within families.

The summary data in Figure 11.1 also suggest a correlation of 0.50 between parents and their offspring. Taken at face value, this correlation suggests a similar conclusion to those based on the sibling data. However, these dozen studies include both single parent-offspring and midparent-offspring correlations. These have very different consequences for estimating heritability, as discussed in Chapter 10. Midparent-offspring correlations are not very meaningful, but regressions of offspring on midparent estimate the upper limit of heritability. They also have the advantage of being unaffected by assortative mating, which we know occurs for IQ. (See Chapter 8.) On the other hand, single parent-offspring correlations estimate only the upper limit of half of the heritability, and they may also be inflated by assortative mating.

Data from the largest and most recent family study of cognition (DeFries et al., 1976, 1979) provide a better estimate of parent-offspring similarity, and suggest substantially less familial resemblance. The focus of the study, conducted in Hawaii, was specific cognitive abilities. But the test battery included a shortened version of Raven's Progressive Matrices, re-

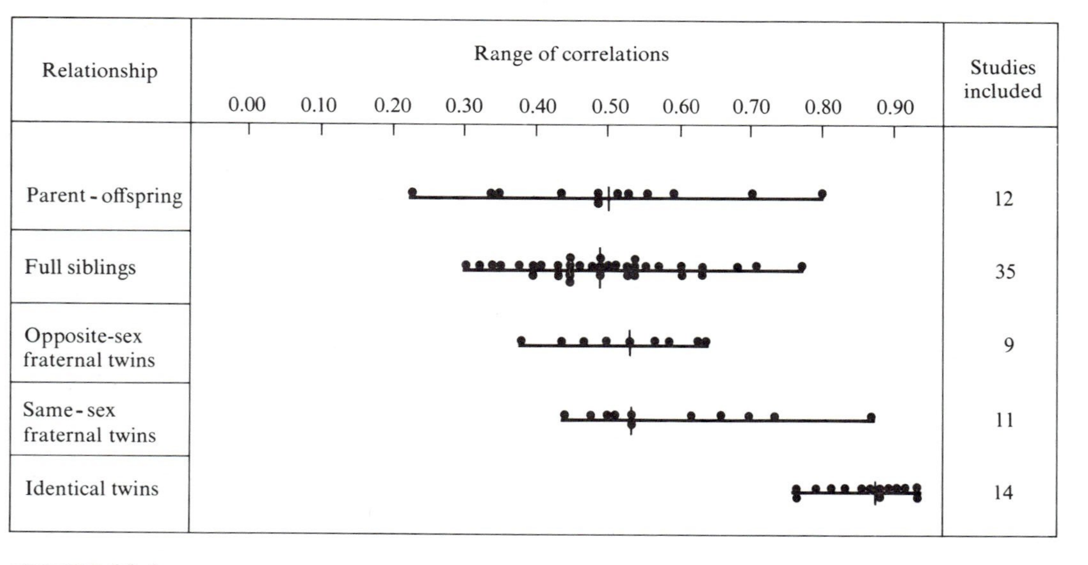

FIGURE 11.1

Family studies of IQ. A summary of correlation coefficients compiled by Erlenmeyer-Kimling and Jarvik from various sources. The horizontal lines show the range of correlation coefficients, and the vertical lines show the median. (From "Genetics and intelligence: a review" by L. Erlenmeyer-Kimling and L. F. Jarvik. *Science,* 142, 1477–1479. Copyright © 1963 by the American Association for the Advancement of Science.)

garded as one of the best single measures of culture-fair abstract reasoning. For 830 Caucasian families, the mean single parent-offspring correlation between parents (aged 35 to 55) and their children (aged 13 to 33) for this test was 0.26. This is lower than the median correlation of 0.50 reported in Figure 11.1. We can estimate the upper limit of heritability by doubling this correlation, thus estimating that narrow-sense heritability should be no greater than 0.52. Bear in mind that this estimate may include genetic variance due to assortative mating. As mentioned above, the best upper-limit heritability estimate is the regression of offspring on midparent (which is not affected by assortative mating). In the Hawaii family study of cognition, this regression was 0.52 (DeFries et al., 1979). The correspondence between this value and the estimate based on single parent-offspring correlations suggests that assortative mating did not affect the single-parent estimate.

Another measure of general cognitive ability from the Hawaii study was a composite score based on the common variance of 15 tests of specific cognitive abilities. This composite, called the first principal component, correlates highly with scores on a standard IQ test. The mean single parent-offspring correlation for this general cognitive factor in the Caucasian families was 0.35. The regression of offspring on midparent, which would probably be the best single estimate of the upper limit of narrow-sense heritability for general cognitive ability, was 0.60.

This same study also provides a good example of the possibility that such estimates can differ in various populations. In addition to the 830 families of European ancestry, the study also included 305 families of Japanese ancestry. For Raven's Progressive Matrices, the offspring-midparent regression was 0.24, compared to 0.52 in the Caucasian families. For the general composite, it was 0.42 rather than 0.60. Thus, there appears to be less familial resemblance (due to genetic or environmental influences) between parents and offspring in these families of Japanese ancestry than in the Caucasian families.

The estimated upper limit for narrow-sense heritability in the Hawaii study of cognition is about 0.60, when the estimate is based on midparent-offspring similarity in Caucasian families. For families of Japanese ancestry, it is lower. Both of these estimates are lower than the estimate based on the median correlation of 0.50 between parents and offspring in Figure 11.1, and we shall return to this issue after describing the sibling data.

The large-scale Hawaii study yielded a mean correlation of 0.31 for 455 pairs of Caucasian siblings. This correlation suggests that the upper limit of broad-sense heritability is 0.62, which is similar to the estimate based on parent-offspring regression. Even more interestingly, the correlation for 147 pairs of siblings of Japanese ancestry was 0.33. This suggests that genetic and environmental influences shared by siblings is about the same for Caucasian and Japanese groups, even though the familial influence shared by parents and offspring may be greater in Caucasian than in Japanese families.

Thus, both the parent-offspring and sibling data from the Hawaii family study of cognition point to less familiality than the older data. Subsequent chapters will show a similar pattern of results. For example, recent twin and adoption IQ data suggest a lower heritability than that suggested by older data. Although it is possible that environmental or genetic changes in the population underlie the differences between newer and older data, methodological variations are a more likely explanation. For example, some of the older studies tested families as a unit, so that any differences in test administration would increase familial correlations. In contrast, tests in the Hawaii family study were administered to many families at the same time in the same testing facility, using highly standardized procedures.

Specific Cognitive Abilities

IQ does not tell the whole story of intelligence. Although there is a general factor of cognitive functioning, there are also specific abilities. We should really speak of intellectual abilities rather than intelligence. The number of these specific abilities depends on the level of analysis. These abilities range from two very general abilities (such as verbal and performance factors), through the 6 to 12 group factors measured by L. L. Thurstone, to the 120 postulated by J. P. Guilford (1967) in his model of intelligence. Although these specific abilities tend to be modestly correlated with one another, lending support to the notion of "general intelligence," they are sufficiently different to permit a more fine-grained analysis of cognitive functioning.

For example, in the Hawaii family study just described (DeFries et al., 1979), 15 tests of specific cognitive abilities were included in the test battery. In addition to the general cognitive factor described in the previous section, application of a technique called *factor analysis* to correlations among scores on the 15 tests yielded four group factors: verbal (including vocabulary and fluency), spatial (visualizing and rotating objects in two- and three-dimensional space), perceptual speed (simple arithmetic and number comparisons), and visual memory (short-term and longer-term recognition of line drawings).

Regressions of offspring on midparent (the upper limit of heritability) for these four factors and the 15 cognitive tests are plotted separately for the two ethnic groups in Figure 11.2. The most obvious fact is that familiality differs for the various tests and factors. These data were corrected for unreliability of the tests to make sure that the differences in familial resemblance were not caused by reliability differences among the tests. For both ethnic groups, the verbal and spatial factors show more familial resemblance than the perceptual speed and memory factors. Other family studies (reviewed in DeFries, Vandenberg, and McClearn, 1976) also indicate that the most familial similarity occurs for verbal ability.

Familial resemblance also varies for tests within each factor. For example, one test of spatial ability, called Paper Form Board, shows high familiality in both ethnic groups. The test involves showing how to cut a

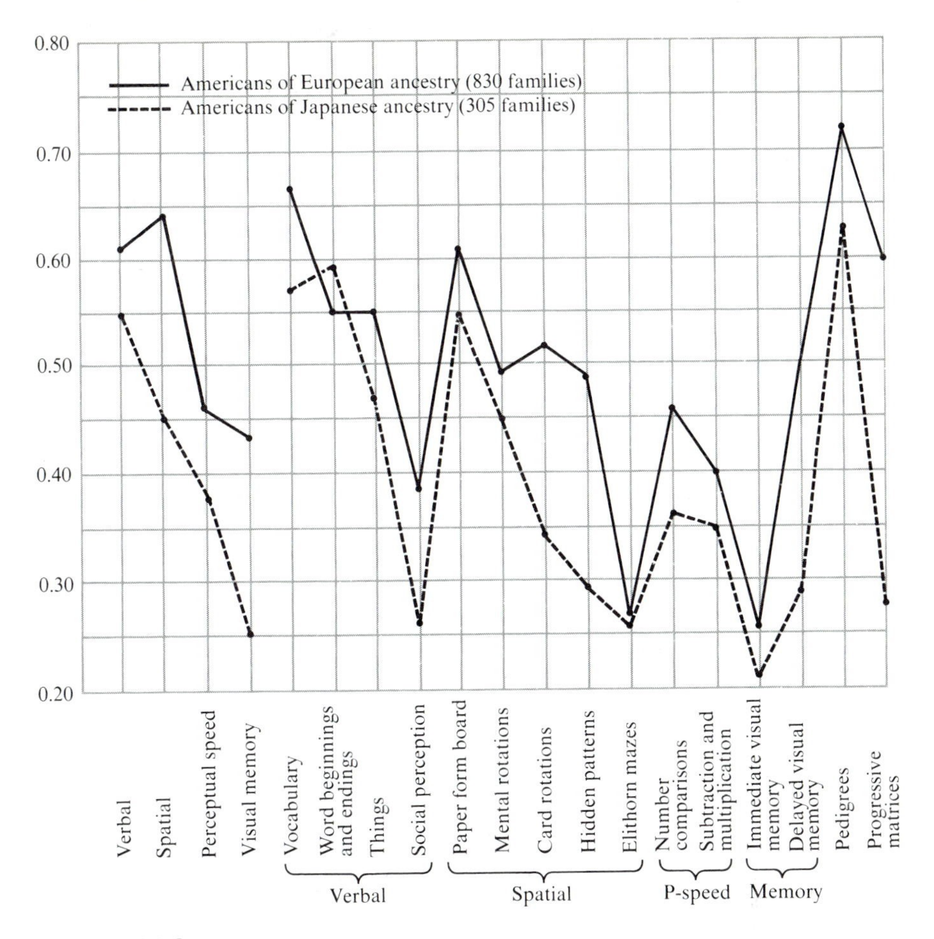

FIGURE 11.2

Family study of specific cognitive abilities. Regression of midchild on midparent for four group factors and 15 cognitive tests in two ethnic groups. (Data from DeFries et al., 1979.)

figure to yield a certain pattern—for example, to cut a circle to yield a triangle and three crescents. Elithorn mazes, a spatial test that involves drawing one line that connects as many dots as possible in a maze of dots, shows the lowest familial resemblance in both ethnic groups. In general, although familial resemblance tends to be slightly lower in the families of Japanese ancestry, the pattern of resemblance is quite similar in the two ethnic groups.

Multivariate Analysis

Do the same genes influence different specific cognitive abilities, or are these abilities genetically unrelated? Are there major environmental factors that have broad, systematic influences across a wide range of specific cognitive abilities, or do different environmental influences affect different abilities? These important questions must be answered in order to understand the nature of genetic and environmental influences on cognition, and they can be answered by multivariate quantitative genetic analyses. (See chapter 9.) Because behavioral geneticists are just beginning to address these issues and the data base is still woefully inadequate, only preliminary information is available at this time.

Data from the Hawaii family study of cognition can be used to illustrate the multivariate approach in a family study of specific cognitive abilities (DeFries, Kuse, and Vandenberg, 1979). In Chapter 9, we emphasized that familial cross-correlations for two characters have the same components of variance as familial correlations for a single character. Familial correlations, or cross-correlations, include both genetic and shared environmental influences. Thus, the cross-correlations among the specific cognitive abilities in the Hawaii study include more than genetic resemblance, although we shall talk about this cross-familiality as genetic. Environmental correlations in this context include within-family environmental factors, as well as nonadditive genetic influences.

Scores on the 15 tests of specific cognitive abilities for parents and offspring in 830 Caucasian families were subjected to multivariate analysis (DeFries, Kuse, and Vandenberg, 1979). Although the average phenotypic correlation among the 15 tests is 0.28, the correlations range from 0.07 to 0.57. As indicated earlier, factor analysis of these correlations yielded four phenotypic group factors (verbal, spatial, perceptual speed, and visual memory). Parent-offspring cross-correlations among the tests were calculated, and genetic and environmental correlations were computed. Because the sheer number of correlations makes interpretation difficult, these genetic and environmental correlations were submitted to factor analysis. The genetic correlations yielded four factors that were virtually identical to the phenotypic factors. The environmental correlations also yielded four factors that were quite similar to the phenotypic factors. These findings suggest that the genetic and environmental influences salient to specific cognitive abilities are neither very broad nor are they idiosyncratic in their effect. Rather, there are several sets of genetic and environmental influences that correspond to the phenotypic factors. As in the multivariate family study of mouse behavior discussed in the previous chapter, these human behavioral data suggest the interesting possibility that genetic and environmental correlations are correlated.

PSYCHOPATHOLOGY

At least 15 percent of the U.S. population is affected by mental disorders during any one year (Regier et al., 1978). The two major kinds of severe mental disorders, called psychoses, are schizophrenia and manic-depressive psychosis, each with an incidence of about 1 percent in the general population. This means that over 4.5 million people in the United States either have a psychosis or will fall victim to one sometime during their lifetimes. Half of the hospital beds in the United States are occupied by patients affected with some form of mental illness, and about half these patients are diagnosed as schizophrenic.

Although early man viewed insanity in demonic terms, the familial nature of insanity was generally recognized by the sixteenth century. At the end of the ninteenth century, E. Kraepelin suggested two major types of insanity—schizophrenia and manic-depressive psychosis. Family studies in the twentieth century have been important in specifying the extent of familial transmission, as well as in studying the heterogeneity of psychopathology. Although schizophrenic individuals are more likely than average to have schizophrenic relatives, they are no more likely to have manic-depressive relatives. The use of family studies to determine the heterogeneity of mental disorders has become even more important in recent years, and we shall consider these advances later.

Psychopathology introduces new problems for quantitative genetic analysis. Mental illness has not been studied in a quantitative way—that is, as the extreme of a normal distribution of mental functioning. "Either-or" sorts of data have been compiled in the past, although that is changing. Because of the qualitative classification of mental illnesses, the usual quantitative analyses are not appropriate. Typically, we compare the incidence of psychopathology in relatives of an affected index case to the incidence in the general population. Familial resemblance is indicated when the incidence in the relatives is greater than the incidence in the general population. However, good incidence figures are not easy to obtain. Incidence is different from prevalence. Prevalence is the number of individuals in a population who are affected at a particular time, regardless of age. Incidence figures, on the other hand, may be viewed as a lifelong prevalence estimate. The age of risk for schizophrenia, for example, is about 15 to 45 years of age. Persons younger than 45 still have a chance of becoming schizophrenic. Also, older individuals may have been hospitalized for a mental disorder, but are no longer. Incidence figures take these cases into account. A special incidence figure used in family studies is called a *morbidity risk estimate,* which is an estimate of the risk of being affected.

Even though mental disorders are often viewed as discontinuous characters, they may well have a quantitative genetic basis. However, the discontinuous method of measurement (qualitative classification) makes it difficult to estimate heritability. We can, however, assume that an all-or-none disorder has an underlying continuous liability. Individuals having a liability above a certain threshold value are assumed to be affected. Those whose liability is below this threshold value are not affected. Liability is thus an unobserved construct that is presumed to be a continuous function of both genetic and environmental deviations. If there is such a threshold for a relatively rare character, concordance even for identical twins can be high only if heritability is very high (Smith, 1970). The heritability of the liability can be estimated, although several assumptions must be made (Falconer, 1965; Reich, Cloninger, and Guze, 1975). Twin data for schizophrenia suggest that the heritability of the liability is 0.85 (Gottesman and Shields, 1972).

We shall now present brief summaries of family studies of schizophrenia and manic-depressive psychosis to illustrate the usefulness of the family study method. It is beyond the scope of our behavioral genetic illustrations to discuss the advances that have been made in studying these disorders biochemically. However, we note in passing that the advances have been considerable and that they indicate the direction of future research on the role of genetic factors in psychopathology (e.g., Motulsky and Omenn, 1975; Kety, 1975).

Schizophrenia

Although there is variability in strictness of diagnosis (Taylor and Abrams, 1978), the risk of schizophrenia in the general population is about 1 percent (Slater and Cowie, 1971). The risks for various relatives of schizophrenics are summarized in Table 11.1 (from Rosenthal, 1970). It is clear that schizophrenia runs in families. In 14 studies of over 4,000 schizophrenics, the risk for parents of schizophrenics is a little over four times that for the general population. It is likely that a lower rate for parents than for other first-degree relatives is found because schizophrenics are less likely to marry and those who do marry have relatively few children. Siblings of schizophrenics provide the most stable data. They are approximately eight times more likely to become schizophrenic than an individual chosen randomly from

TABLE 11.1

Median morbidity risk estimates for the general population and for relatives of schizophrenic index cases

	Percent
General population	0.9
First-degree relatives:	
Parents of schizophrenics	4.2
Siblings of schizophrenics	7.5
Children of schizophrenics	9.7
Second-degree relatives	2.1
Third-degree relatives	1.7

SOURCE: After Rosenthal, 1970.

the population. (See Table 11.1.) This rate almost doubles from those siblings who have an affected parent as well as an affected sibling. These results are quite consistent with six recent European family studies of schizophrenia reported in the 1970s, although one large American study of general psychosis yielded much higher estimates. (See review by Gottesman and Shields, 1976.)

In five studies of children of schizophrenics, their median risk was nearly 10 percent. It is interesting that the risk is no greater when the mother is schizophrenic than when the father is schizophrenic, which suggests that maternal influences are not critical. When both parents are schizophrenic, the risk is four times as great. This introduces an interesting variation of the family study method. *At-risk studies* are those in which the developmental course of schizophrenia is followed in children at risk for the disorder. About twenty studies are in progress (Erlenmeyer-Kimling, 1975), in which the children of schizophrenic parents are being studied longitudinally. The goal of the researchers is to follow the developmental course of schizophrenia before the occurrence of overt disturbances to gain a better idea of the causes of schizophrenia, as well as to find ways to intervene in the developmental process (Erlenmeyer-Kimling, Cornblatt, and Fleiss, 1979). An extensive battery of physiological and behavioral measures has been employed. We shall look forward to the results of the at-risk studies as the target children grow older.

As shown in Table 11.1, the risk for second-degree relatives of schizophrenics, such as grandchildren (who share about one-fourth of the grandparents' segregating genes), is about twice that for the general population. Third-degree relatives (who share about one-eighth of the segregating genes) are slightly more likely than average to be schizophrenic.

Family studies have established that familial influences play a role in schizophrenia. Twin studies and adoption studies are needed to pin down the relative roles of genetic and environmental factors. However, the family studies carry an important message. First-degree relatives are about 50 percent similar genetically, but fewer than 10 percent of the first-degree relatives of schizophrenics become schizophrenic. Thus, the results of family studies indicate that the development of schizophrenia is substantially affected by within-family environmental influences.

Manic-Depressive Psychosis

During the last decade, attention has increasingly turned to other psychopathologies, most notably to manic-depressive psychosis. This syndrome involves affective disorders that come and go, often in regular cycles. Classically, it involves cycles of mania (elation) and depression. Unlike schizophrenics, persons suffering from manic-depressive psychosis are not likely to be chronically hospitalized because they tend to recover from each bout spontaneously—although about 6 percent may commit suicide during depression. Because manic-depressives are often not hospitalized, it is difficult to estimate the incidence of this psychosis in the general population. Estimates fluctuate from 0.07 percent to 7.0 percent. However, the median risk is 0.7 percent, which is close to the estimate for schizophrenia.

Family studies from the 1920s to the 1950s demonstrated familial resemblance for manic-depressive psychosis, and showed that the disease is distinct from schizophrenia. Table 11.2 (from Rosenthal, 1970) documents the increased risk for relatives of manic-depressive psychotics. For example, in nine studies of nearly two thousand parents of manic-depressives, the median risk was 7.6 percent, ten times greater than the risk for the general population. In eight studies of over three thousand siblings, the risk was 8.8 percent. In studies of over one thousand children of manic-depressives, their risk was about 11 percent, and the risk triples when children have two manic-depressive parents.

Clearly, there is a familial basis for manic-depressive psychosis. A variant of the multivariate cross-correlation approach suggests that manic-depressive psychosis is genetically distinct from schizophrenia. Table 11.2 shows that there is no more schizophrenia in relatives of manic-depressive index cases (with the possible exception of their children) than in the general population. This same cross-correlation approach provides indications that there may be distinct subtypes of manic-depressive psychosis.

Some so-called manic-depressives actually alternate between depression and normality, not between depression and mania. This type of behavior has been called *unipolar depression*. In contrast, the classical type, which includes bouts of mania as well as depression, is called *bipolar manic-depressive psychosis*. Four family studies have suggested that the two subtypes tend to be transmitted independently. Unipolar depressive probands tend to have unipolar relatives, and bipolars, to a lesser extent, tend to have bipolar relatives (Perris, 1976; Dunner and Fieve, 1975). One of the largest studies (Perris, 1968) shows little familial overlap between the two types.

TABLE 11.2

Median morbidity risk estimates for manic-depressive psychosis and schizophrenia for the general population and for relatives of manic-depressive index cases

	Manic-Depressive, %	Schizophrenic, %
General population	0.7	0.9
Parents of manic-depressives	7.6	0.6
Siblings of manic-depressives	8.8	0.8
Children of manic-depressives	11.2	2.5

SOURCE: After Rosenthal, 1970.

However, another study (Angst, 1968) yielded less clear-cut results. Although unipolar probands rarely had bipolar relatives, bipolar probands had more unipolar than bipolar relatives (11.2 percent versus 3.7 percent). Recent studies, including a twin study of bipolar and unipolar disorders (Bertelsen, Harvald, and Hauge, 1977—see Chapter 12) and an adoption study of bipolar manic-depressive psychosis (Mendlewicz and Rainer, 1977—see Chapter 13), also suggest that there is considerable overlap between the two subtypes. But these studies provide some support of the usefulness of the distinction. In addition, evidence is mounting that a dominant sex-linked gene is involved in the bipolar, but not in the unipolar, type (Mendlewicz and Fleiss, 1974). Furthermore, the types seem to respond differently to drugs (e.g., Baron et al., 1975). These same studies also provide the foundation for further analysis. For example, there are clear cases of the bipolar type that cannot be due to a sex-linked gene because father-to-son inheritance has been demonstrated. The bipolar type may also be subdivided on the basis of degree of mania (Dunner, Gershon, and Goodwin, 1976). Among unipolar depressives, onset of the disorder occurs in some individuals after the age of forty, while other cases are characterized by early onset (Winokur et al., 1975). Also, there are responders and nonresponders to certain drugs within the unipolar type of depression (e.g., Taylor and Abrams, 1975; Kupfer et al., 1975).

Heterogeneity

Family data on unipolar depression and bipolar manic-depressive psychosis stimulated a burst of research on the heterogeneity of psychopathology. The search for etiologically distinct syndromes has become the major focus of family studies in this area (Winokur, 1975). For example, this is the focus of the Iowa 500 project, a thirty-five-year follow-up study of about five hundred psychotic index cases and their five thousand relatives (Tsuang et al., 1977). This large-scale project, and other ongoing family studies, are considering the heterogeneity of schizophrenia (Tsuang, 1975) and manic-depressive psychosis (Dunner et al., 1976); the possibility of distinct psychoses other than schizophrenia and manic-depressive psychosis (McCabe, 1976); and are studying less severe psychopathology, such as hysteria and antisocial personality (Cloninger, Reich, and Guze, 1975a, 1975b). The search for heterogeneity in psychopathology may initially lead to some confusion. However, we are confident that it is a healthy trend in research and will eventually contribute substantially to our understanding of psychopathology.

Although many researchers favor subdividing various types of psychopathology to find etiologically distinct syndromes (Kidd and Matthysse, 1978), some researchers favor lumping various ones together, particularly for schizophrenia. For example, evidence that will be reviewed in the next two chapters argues for genetic unity of a spectrum of schizophrenic disorders extending as far as *schizoid* characteristics, such as antisocial personality (Rosenthal, 1975; Reich, 1976; Shields, Heston, and Gottesman, 1975).

SUMMARY

Family studies determine whether or not there is familial resemblance for the behaviors being investigated. If familiality is observed, it may be due either to genetic or environmental influences. Family studies do not indicate whether observed familial resemblance is caused by shared environment or shared heredity, although they can provide an upper-limit estimate of heritability (or of between-family environmental influences). Substantial familiality has been found in family studies of IQ, specific cognitive abilities, and psychopathology, particularly schizophrenia and manic-depressive psychosis. Multivariate analyses indicate substantial cross-familiality for specific cognitive abilities, and the absence of cross-familiality between schizophrenia and manic-depressive psychosis or between the unipolar and bipolar subtypes of manic-depressive psychosis. Twin and adoption studies—described in the next two chapters—are needed to determine the relative importance of genetic and environmental influences underlying familial resemblance.

12

Twin Studies

The study of twins has a long history. Galton, the father of behavioral genetics, proposed the method of study in 1876. He studied life histories of two groups of twins. One group consisted of pairs who were similar at birth, and Galton proposed that these twins (now called identical or *monozygotic*) developed from a single fertilized egg. He suggested that the dissimilar pairs of twins in the other group (now called fraternal, or *dizygotic*, twins) derived from two separately fertilized eggs. Galton found that the initial similarity or dissimilarity tended to be maintained throughout life for various aspects of their behavior.

By the beginning of the twentieth century, most biologists were convinced of the existence of two types of human twins (Wilder, 1904). Nonetheless, E. L. Thorndike, who objectively measured specific cognitive abilities in twins in 1905, did not believe there were two types. He used twins to test environmental hypotheses, and found that younger and older twin pairs showed about the same degree of similarity for the cognitive measures. The first behavioral twin study to approach the modern method was conducted in 1924 by C. Merriman, who called the two types of twins "duplicates" and fraternals. He first showed that the similarity in IQ between the opposite-sex twins in his sample was the same as that between non-twin siblings (a correlation of 0.50). He then found that the same-sex twins had a much higher correlation for IQ (about 0.87). He concluded that the greater similarity in IQ between twin pairs in the group of same-sex twins was caused by the inclusion of all the "duplicate" twins in this group. Merriman then tried to separate the two types of same-sex twins using physical similarity as a criterion. This, as we shall see, is a reasonably accurate method for diagnosing zygosity (i.e., number of zygotes). Twenty-two same-sex pairs "resembled each other closely enough to frequently cause confusion of identity." The correlation on the Stanford-Binet IQ test for these twins was 0.99 in this small sample.

While most mammals have large litters, primates, including *Homo sapiens,* tend to have single offspring. However, all types of primates occasionally have multiple births. The embryology of twinning has been thoroughly analyzed for humans (Bulmer, 1970). In the case of single births, the embryo is suspended in a sac made of two membranes formed within the placenta. The innermost layer of the sac, the amnion, is surrounded by an outer layer, the chorion. Fraternal twins have completely separate chambers inside the womb; they develop from two different zygotes (dizygotic) that are fertilized at the same time. When the embryos implant close to one another in the uterus, their placentas sometimes fuse, but they each still have a separate amnion and chorion. (See Figure 12.1.)

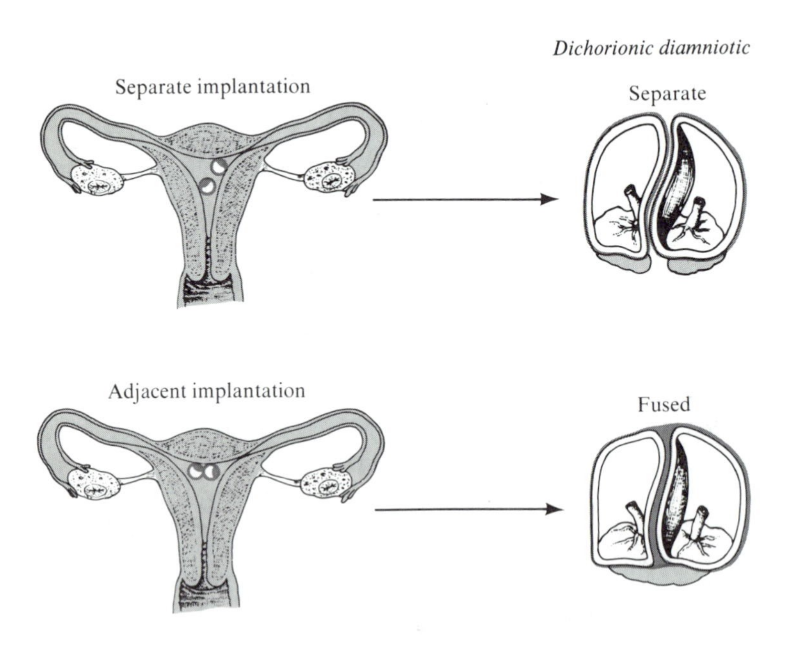

FIGURE 12.1

Fraternal twinning may result in either separate or fused placenta. (From "Twin placentation and some effects on twins of known zygosity" by G. Corney. In W. E. Nance, ed., *Twin Research, Part B: Biology and Epidemiology,* Copyright © 1978 by Alan R. Liss, Inc. All rights reserved.)

Identical twins result from a single zygote (monozygotic) that divides for unknown reasons. They share accommodations in the womb to some extent, depending on when the zygote splits. It used to be thought that identical twins always have the same chorion, but we now know that about a third of them have separate chorions and amnions, as do fraternal twins (Corney, 1978). These identical twins develop from zygotes that split before implantation, which occurs at about five days after fertilization. (See Figure 12.2.) The other two-thirds of identical twins develop from zygotes that split after implantation. These twins develop within the same chorion. When separation occurs five to ten days after fertilization, as it usually does, there are two amnions. About 4 percent separate after ten days, and these share

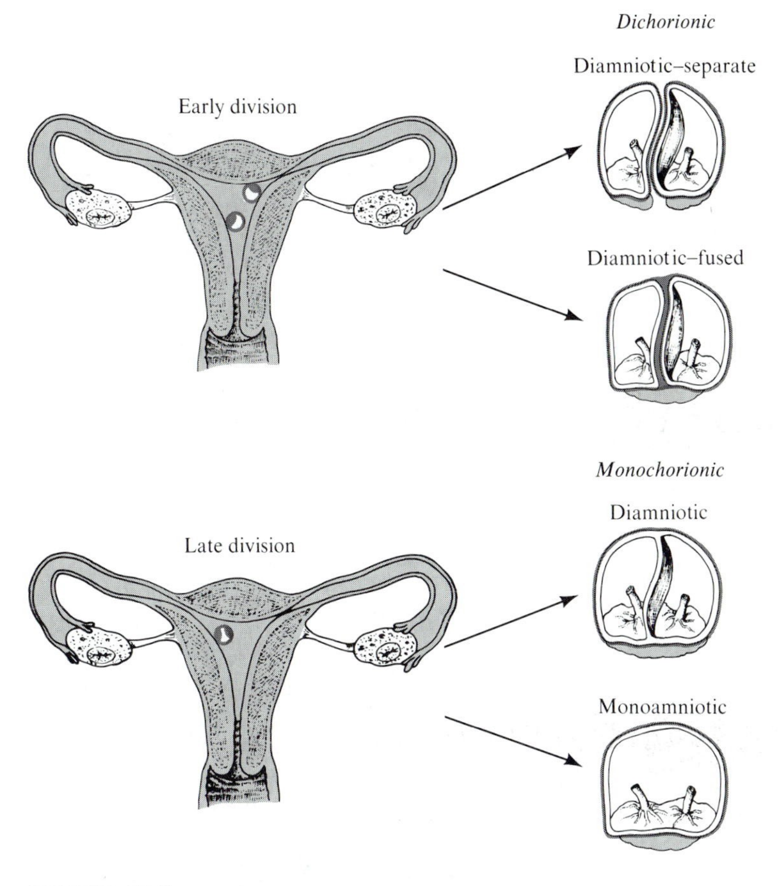

FIGURE 12.2

Identical twinning can result in separate or fused chorions and amnions, depending on the timing of the separation of the zygotes. (From "Twin placentation and some effects on twins of known zygosity" by G. Corney. In W. E. Nance, ed., *Twin Research, Part B: Biology and Epidemiology*, Copyright © 1978 by Alan R. Liss, Inc. All rights reserved.)

their amnion as well as chorion. Extremely late separation results in "Siamese" twins, individuals who are partially fused.

Twin births are not nearly as rare as most people would suppose. One in 83 deliveries in the United States is a twin birth. The rate of twinning, especially fraternal twinning, is affected by maternal age and the number of previous offspring (Rao, 1978), as well as by fertility drugs. In a group of women who take some fertility drugs, the rate of twinning can be as high as 50 percent (Wyshak, 1978). Also, in some countries, such as Nigeria, the rate of fraternal twinning is much higher than in the United States and Great Britain (Nylander, 1978). Finally, the tendency to produce fraternal twins may be inherited in some families (Nance et al., 1978), although heredity does not seem to be an important factor in identical twinning (Segretti, Winter, and Nance, 1978).

About one-third of the conceptions that produce twins result in opposite-sex fraternal twins, one-third in same-sex fraternal twins, and the remaining third in identical twins. After childhood, however, the proportion of identical twins is reduced to about one-quarter due to the decreased viability of this type of twin. Rates of identical twinning are remarkably similar for different maternal ages, but the fraternal twinning rate fluctuates (Bulmer, 1970).

As discussed in the previous chapter, fraternal twins are 50 percent similar genetically on the average, like other first-degree relatives, while identical twins are identical genetically. By comparing the phenotypic similarity of identical twins with that of fraternal twins, we can conduct a natural experiment to investigate the effects of heredity and environment. Despite the twofold differences in genetic similarity, both types of twins share roughly the same environmental influences (an assumption that is discussed in detail later). If many pairs of twins are studied and identical twins are found to be no more similar than fraternal twins for a particular behavior, then genetic factors cannot be important in determining variability for that trait. In that case, the twofold greater genetic similarity of identical twins did not make them more similar phenotypically. On the other hand, if identical twins are significantly more similar than fraternal twins, the particular trait is influenced by genetic factors. Fraternal twins may be of the same sex or opposite sex, while identical twins are always of the same sex. Because of the sex chromosome differences between males and females, and because of the possibility that males and females are exposed to differing environments, twin studies are better natural experiments if we compare only same-sex fraternal twins to identical twins.

ZYGOSITY DETERMINATION

The twin method has often been subject to criticism. We shall discuss in detail two major issues: zygosity determination and the equal environments assumption. With respect to zygosity determination, a problem arose from the fact that some researchers made a subjective judgment of zygosity (whether the twins were identical or fraternal) and a subjective judgment of behavior. This method could have biased the results. It is now possible to perform highly accurate blood analyses by which the members of each pair of twins are compared for genetic markers in the blood. If any of these genetic markers are different, the twins must be fraternal; if all the markers are the same, the twins are identical. In addition, any errors in diagnosing the zygosity of twins will lower estimates of genetic influence based on the comparison of identical and fraternal twin correlations because they will raise the fraternal twin correlation and lower the correlation for identical twins.

We shall consider some new developments in diagnosis in detail because they have important implications for the ease of conducting twin research. As early as 1927, physical traits (such as hair and eye color) were used along with minimal blood analyses to diagnose zygosity (Siemens, 1927). By 1955, the analysis of blood groups began to be accepted as the best method (Smith and Penrose, 1955). However, as early as 1941, blood markers and physical similarity were shown to be highly related (Essen-Moller, 1941).

In 1959, a detailed analysis of this relationship was reported by T. Husen in Sweden. At that time, although blood analysis was much less accurate than it is now, Husen was able to study four markers. One of these was the ABO system, for which six variants were analyzed. Identical twins, of course, must have the same variant. However, about two-thirds of fraternal twin pairs can also be expected to have the same variant by chance. For this reason, Husen could achieve only 60 percent accuracy in classifying twins as identical if they had the same variant for the ABO blood system. However, the probability that a pair of fraternal twins will have the same variant for two blood systems is much less. For example, the accuracy of diagnosis goes up to 0.72 when we consider both the ABO and MNS blood group systems. With all four blood groups that Husen studied, the accuracy of diagnosis was 90 percent. Presently, it is possible to analyze over 20 other genetic markers in the blood, and the accuracy approaches 100 percent (Giblett, 1969).

Husen found that evaluation of physical similarity is a good way of diagnosing zygosity. Even a subjective rating of similarity predicted well. For 25 pairs of identical twins, only 5 were rated as showing some "striking dissimilarities." Only one fraternal twin pair was rated "very similar." As a general rule, identical twins tend to be misclassified as fraternal twins more often than the opposite. Husen also found that fraternal twins are seldom mistaken for one another by family members and friends, while identical twins often are. However, just asking twins whether they are identical or fraternal is not a very accurate method (Husen, 1959; Carter-Saltzman and Scarr, 1977).

Even greater precision in diagnosing zygosity can be obtained by using specific physical characteristics, such as eye and hair color, in the same way that blood markers are used. For example, Husen found that 92 percent of identical twins had exactly the same eye color, as compared to 44 percent of fraternal twins. The comparisons for some other characteristics were: 92 percent versus 38 percent for hair color; 96 percent versus 55 percent for complexion; 96 percent versus 16 percent for hairline pattern; and 93 percent versus 11 percent for position of teeth. For ten such characteristics, over 90 percent accuracy of diagnosis could be achieved.

These results are in accord with more recent analyses of data on adult (Cederlöf et al., 1961) and adolescent (Nichols and Bilbro, 1966) twins, as well as data obtained when children were rated for similarity by their parents (Cohen, et al., 1973, 1975). The results of the study of children are described in Table 12.1, which lists ten physical similarity variables that were evaluated by mothers of identical and fraternal twins. All of these studies found an accuracy of over 90 percent in diagnosing zygosity from physical similarity. Because of the difficulty and expense of obtaining blood samples, particularly from young children, it is indeed fortunate that physical similarity can be used to diagnose zygosity.

TABLE 12.1

Physical similarity and twin zygosity for children

	Percent of Twins "Exactly Similar" (or "Yes" Responses by Mothers)	
	Identical	Fraternal
"Is it hard for strangers to tell them apart?" (Asked of mothers.)	100	8
Eye color	100	30
Hair color	100	10
Facial appearance	49	0
Complexion	99	14
Weight	46	6
"Do they look alike as two peas in a pod?"	48	0
"Does either mother or father ever confuse them?"	79	1
"Are they sometimes confused by other people in family?"	93	1
Height	56	13
Number of pairs	181	84

SOURCE: Adapted from D. J. Cohen et al., "Reliably separating identical from fraternal twins," *Archives of General Psychiatry*, 1975, 32, 1373–1374.

EQUAL ENVIRONMENTS ASSUMPTION

Another issue that arose with respect to the twin method involved the equal environments assumption. The essence of the twin method is to compare the phenotypic similarities of identical twins and fraternal twins, and to be able to ascribe a finding of greater identical-twin similarity to their twofold greater genetic similarity. The method assumes that the degree of environmental similarity is about the same for the two types of twins. If the equal environments assumption is not correct (for example, if identical twins are treated more similarly than fraternal twins), then a finding of greater phenotypic similarity between identical twins might be due partially to greater environmental similarity. It could not be attributed to greater genetic similarity alone.

On the face of it, the equal environments assumption seems quite reasonable for many important variables. Both types of twins share the same womb at the same time. Both are reared in the same family at the same time. They are the same age and the same sex (if opposite-sex fraternal twins are excluded from the sample). Nonetheless, the major criticism of the twin method is that identical twins may be treated more similarly than fraternal twins. It should be noted that one could also argue that identical twins may be treated *less* similarly than fraternal twins, which would result in underestimates of genetic influence. For example, identical twins are so similar in appearance that their parents may accentuate slight behavioral differences. Also, the twins themselves may behave differently from one another in an attempt to forge separate identities.

Speculations such as these have been much more common than research designed to explore the issue. Fortunately, some empirical studies have recently been conducted concerning the question of the equal environments assumption. Such studies have been of three major types: the effects of labeling, direct assessments of environmental differences for identical and fraternal twins, and tests to determine whether such differential environments affect behavior.

The effect of labeling a twin pair as identical or fraternal has been studied using twins who were misclassified by their parents or by themselves. For example, when parents think that their twins are fraternal but they really are identical, will the mislabeled identical twins be as similar behaviorally as correctly labeled identical twins? Two studies (Scarr, 1968; Scarr and Carter-Saltzman, 1979) have suggested that the influence of such labeling is minimal. Twins whose zygosity is mistaken by their parents or by the twins themselves behave according to their true zygosity. For example, in a study of four hundred pairs of adolescent twins in Philadelphia, the twins were simply asked if they were identical or fraternal; 40 percent were wrong about their zygosity. Table 12.2 presents the differences in performance on cognitive and personality tests between correctly and incorrectly classified

TABLE 12.2

The effects of zygosity labeling: average absolute differences on cognitive and personality tests for identical and fraternal twins who were right and for those who were wrong about their zygosity

	Cognitive Tests		Personality Tests	
	Average Difference Within Twin Pairs	Number of Pairs	Average Difference Within Twin Pairs	Number of Pairs
Identical twins who were correct about their zygosity	0.66	89	0.81	98
Identical twins who were incorrect about their zygosity*	0.73	61	0.85	68
Fraternal twins who were correct about their zygosity	0.81	84	0.93	101
Fraternal twins who were incorrect about their zygosity*	0.81	49	0.99	64

* Includes pairs in which one or both twins were in error concerning their zygosity.

SOURCE: After Scarr and Carter-Saltzman, 1979.

twin pairs. Although the fraternal twins showed greater differences than the identical, the identical twins who thought that they were fraternal had only slightly greater differences than those who thought they were identical. The fraternal twins who thought that they were identical were no more similar than those who correctly thought that they were fraternal. We can conclude from this and an earlier study (Scarr, 1968) that labeling twins as identical or fraternal has little effect on their behavioral similarity.

The second approach to the question of the equal environments assumption involves measuring aspects of the environments of identical and fraternal twins to determine whether identical twins, in fact, experience more similar environments. Some differences can be found (Vandenberg, 1976). For example, identical twins tend to be treated more similarly in terms of clothes, they study together more, and they share more friends (Wilson, 1934; Lehtovaara, 1938; Zazzo, 1960; Smith, 1965; Loehlin and Nichols, 1976). Evidence such as this has often been used to reject the twin method as biased. However, it is a mistake to assume that environmental differences such as these make a difference behaviorally. This approach does not go far enough.

The third approach asks the more appropriate question: "Do observed differences in the environment of the two types of twins make a difference behaviorally?" The method was first suggested by John Loehlin and Robert Nichols (1976), who reported data from a large twin study of personality. The study included about 850 pairs of twins from a group of high school students who participated in the National Merit Scholarship Qualifying Tests in 1962 and 1965. The parents of the twins were asked to rate five environmental variables that had been found to differ for identical and fraternal twins. The results are presented in Table 12.3. Consistent with previous studies, the identical twins experienced slightly more similar environments than did the fraternal.

The important question is whether these environmental differences affect behavior. Because identical twins are identical genetically, differences within pairs of identical twins can be caused only by environmental factors. Some identical twin pairs are more similar behaviorally than others; and some identical twin pairs are exposed to more similar environments than others. Loehlin and Nichols tested the adequacy of the equal environments assumption by correlating identical twin differences for cognition and personality with the measures of similarity of environment. If similarity of environment makes a difference behaviorally, then identical twins whose environments are more similar should be more similar behaviorally. Only identical twins were included in this analysis because observed differences between them are clearly environmental. Fraternal twins, on the other hand, differ genetically as well as environmentally, so the investigators could not have been sure whether the genetic differences led to the environmental differences, or vice versa. Table 12.4 presents the correlations between behavioral differences and a composite index of differential environments. A positive correlation means that identical twins who were exposed to more similar environments were more similar in behavior. The low correlations provide strong support for the equal environments assumption of the twin method. Environmental variables that were unequal for identical and fraternal twins simply did not make a difference in cognition, personality, vocational interests, or interpersonal relationships.

What about the greater similarity of appearance of identical twins? The method used by Loehlin and Nichols can also be applied to this question.

TABLE 12.3

Mean scores for identical and fraternal twins on five items and their composite score concerning differential experience

Item	Score Range	Identical		Fraternal	
		M	F	M	F
Dressed alike	1–3	1.7	1.5	2.0	1.8
Played together (age 6–12 years)	1–4	1.3	1.3	1.6	1.6
Spent time together (age 12–18 years)	1–4	1.8	1.7	2.2	2.0
Slept in same room	1–4	1.7	1.7	1.8	1.8
Parents tried to treat alike	1–5	1.9	2.0	2.3	2.2
Composite of above	6–23	9.7	9.6	11.6	10.8
Number of pairs		217	137	297	199

NOTE: A score of 1.0 indicates maximum similarity, and larger numbers indicate less similarity.
SOURCE: After Loehlin and Nichols, 1976.

TABLE 12.4

Test of the equal environments assumption: correlations between absolute differences for identical twins on behavioral measures and a composite measure of differential experience

	Correlation*	Number of Pairs
Cognitive (average of 5 NMSQT subtests)	−0.06	276
Personality (average of 18 CPI scales)	0.06	451
Vocational interests (average of 12 VPI scales)	0.01	276
Interpersonal relationships (average of 6 types of relationships)	0.05	276

NOTE: NMSQT is the National Merit Scholarship Qualifying Tests; CPI is the California Psychological Inventory; and VPI is the Vocational Preference Inventory.

* A positive correlation means that identical twins who were treated more similarly were more similar behaviorally.

SOURCE: After Loehlin and Nichols, 1976.

Although identical twins are usually quite similar in appearance, they vary in the amount they are mistaken for one another, as indicated in Table 12.1 (Cohen et al., 1975). For example, only 48 percent of the mothers of identical twins said that the twins "look alike as two peas in a pod." Thus, we can ask whether twins who are more easily mistaken for one another are more similar in behavior (Matheny, Wilson, and Brown, 1976; Plomin, Willerman, and Loehlin, 1976). Matheny and associates compared absolute differences in IQ scores and personality measures with ratings of physical similarity for blood-diagnosed identical twins between the ages of three and thirteen. Their finding of no significant correlations between behavioral differences and

TABLE 12.5

Test of the equal environments assumption: rank order correlations between absolute differences for identical twins on behavioral measures and physical similarity

	Correlation*	Number of Pairs
Stanford-Binet IQ test	0.05	47
WISC IQ test	−0.19	74
Personality test (average of 16 PF)	0.00	51

NOTE: 16 PF is Raymond Cattell's 16 Personality Factors test.

* A positive correlation means that identical twins who were more similar physically were more similar behaviorally.

SOURCE: Matheny et al., 1976.

physical similarity scores (see Table 12.5) suggests that the greater physical similarity of identical than fraternal twins does not seriously affect the equal environments assumption of the twin method.

Data such as these strongly support the reasonableness of the equal environments assumption. Although other differences between the environments of identical and fraternal twins may yet be found, these results suggest that such differential experiences will not necessarily affect the behavioral similarity of the two types of twins.

HERITABILITY

Some researchers prefer to test the significance of the difference between correlations for identical and fraternal twins and to leave it at that. We know, for example, that identical twins are significantly more similar in height and weight and in the total ridge counts of their fingerprints. Thus, genetic differences are implicated in the etiology of these traits.

It is useful in many contexts to go beyond this statement of statistical significance to estimate broad-sense heritability, although the cautions expressed in Chapter 9 must be kept in mind. Even if one chooses not to estimate heritability, it is important to know whether the pattern of correlations for identical and fraternal twins is reasonable, regardless of the significance of the difference between the correlations. For example, a correlation of 0.90 for identical twins and a correlation of 0.10 for fraternal twins is not consistent with a genetic hypothesis, because identical twins are only twice as similar genetically as are fraternal twins. Twin studies frequently yield reasonable patterns of correlations. For example, in Table 12.6 we have listed the median correlations for height, weight, and total ridge count of fingerprints for adult twins (from Mittler, 1971). These data also agree with other familial information. For example, for total ridge count, the sibling correlation is 0.51 and the single parent-offspring correlation is 0.46.

TABLE 12.6

Median correlations for identical and fraternal twins for height, weight, and total ridge count of fingerprints

	Correlation	
	Identical	Fraternal
Height	0.93	0.48
Weight	0.91	0.58
Total ridge count	0.96	0.49

SOURCE: After Mittler, 1971.

The components of covariance for various familial relationships were described in Chapter 10. Fraternal (DZ) twins have the same genetic components of covariance as full siblings (see Tables 9.9 and 10.1):

$$\text{Cov}_{\text{DZ}} = \tfrac{1}{2}V_{\text{A}} + \tfrac{1}{4}V_{\text{D}} + V_{\text{Ec(DZ)}}$$

Because fraternal twins also share familial environment, we have referred to a component of common environment salient to fraternal twins (which may be greater than V_{E_c} for non-twin siblings). This is only a slightly more precise way of stating that the phenotypic covariance for fraternal twins can include about half of the genetic variance and some environmental variance. Identical (MZ) twins, on the other hand, covary completely genetically and also share a familial environment:

$$\text{Cov}_{\text{MZ}} = V_{\text{A}} + V_{\text{D}} + V_{\text{Ec(MZ)}}$$

Thus, the difference between the phenotypic covariances of identical and fraternal twins is:

$$\text{Cov}_{\text{MZ}} - \text{Cov}_{\text{DZ}} = \tfrac{1}{2}V_{\text{A}} + \tfrac{3}{4}V_{\text{D}}$$

assuming that the common environments are about the same for the two types of twins.

As described in Chapter 9, most researchers prefer to work with standardized covariances, known as correlations. A phenotypic correlation is a covariance divided by a variance. Thus, the fraternal twin correlation reflects the following components:

$$\text{r}_{\text{DZ}} = \frac{\tfrac{1}{2}V_{\text{A}} + \tfrac{1}{4}V_{\text{D}} + V_{\text{Ec(DZ)}}}{V_{\text{P}}}$$

The identical twin correlation is represented as follows:

$$\text{r}_{\text{MZ}} = \frac{V_{\text{A}} + V_{\text{D}} + V_{\text{Ec(MZ)}}}{V_{\text{P}}}$$

The difference between these two expressions is:

$$\text{r}_{\text{MZ}} - \text{r}_{\text{DZ}} = \frac{\tfrac{1}{2}V_{\text{A}} + \tfrac{3}{4}V_{\text{D}}}{V_{\text{P}}}$$

Doubling this difference yields:

$$2(\text{r}_{\text{MZ}} - \text{r}_{\text{DZ}}) = \frac{V_{\text{A}} + \tfrac{3}{2}V_{\text{D}}}{V_{\text{P}}}$$

As described in Chapter 9, broad-sense heritability is the proportion of phenotypic variance accounted for by additive and nonadditive genetic variance:

$$h_B^2 = \frac{V_A + V_D}{V_P}$$

Thus, doubling the difference between the identical and fraternal twin phenotypic correlations can somewhat overestimate broad-sense heritability because this estimate contains 1.5 times the dominance variance. Although we have not expressed epistatic variance, it, too, would lead to an overestimate of heritability. Another possible complexity is that assortative mating by parents will increase additive genetic variance shared by fraternal, but not by identical, twins. A larger additive genetic variance for fraternal twins will increase their phenotypic correlation and reduce the difference between the correlations for fraternal and identical twins. Thus, assortative mating can result in an underestimate of broad-sense heritability, although there are ways to adjust twin heritability estimates for the effect of assortative mating (Jensen, 1967). Yet another complexity is the possibility of genotype-environment interaction and correlation, which will be discussed in Chapter 14. Although there has been considerable research and theorizing about such problems, we can hold them in abeyance because they are likely to have only small and counterbalancing effects on estimates of heritability from twin studies.

Doubling the difference between identical and fraternal twin correlations has been called Falconer's estimate of broad-sense heritability. Analyses of the correlations in Table 12.6 suggest that the broad-sense heritability is 0.90 (i.e., $2(0.93 - 0.48) = 0.90$) for height, 0.66 for weight, and 0.94 for total ridge count. As explained in Chapter 9, this is a shorthand way of saying that, of the individual differences in height for these samples and at this particular time, about 90 percent of the variation in height is due to genetic differences among people and about 10 percent is due to environmental differences.

Although Falconer's formula is most appropriate for our introductory purposes, there are other formulas for computing heritability from twin data. One of the earliest and most widely used formulas (although incorrect, as we shall see) is called Holzinger's H:

$$H = \frac{\text{Identical twin correlation} - \text{fraternal twin correlation}}{1.0 - \text{fraternal twin correlation}}$$

Although it sounds paradoxical, the basic problem with Holzinger's H estimate of heritability is that it always yields a reasonable heritability estimate, even for unreasonable patterns of correlations. For example, suppose that the correlation between identical twins for a particular trait is 0.90 and the correlation for fraternal twins is 0.10. Falconer's formula would yield a heritability of 1.60, which indicates this is an impossible pattern of correlations given our simple model. However, Holzinger's H would yield a deceptively reasonable estimate of 0.89. Another example of a deceptive result of using Holzinger's H would be the finding of a heritability of 1.0, when the identical twin correlation is 1.0 and the fraternal twin correlation is 0.99. Falconer's formula yields the figure of 0.02 for this pattern of correlations. Holzinger's formula, like all heritability estimates, is based on the difference between identical and fraternal twin resemblance. However, it is equivalent to Falconer's formula only under very restrictive conditions.

ENVIRONMENTALITY

The twin method can also be described graphically (Nichols, 1965), as in Figure 12.3, which emphasizes the assumptions of the method. The rectangle represents the phenotypic variance of individuals for a particular trait. The first assumption is that the total variance of the identical and fraternal twins is the same and that it does not differ from the variance in the general population to which the results will be generalized. Twin studies usually satisfy this assumption. Even in the few instances where there is a mean difference between twins and singletons in the population as a whole (one example is an average twin IQ of about 96), the variance of the two types of twins and the general population is the same. The variance in the rectangle has

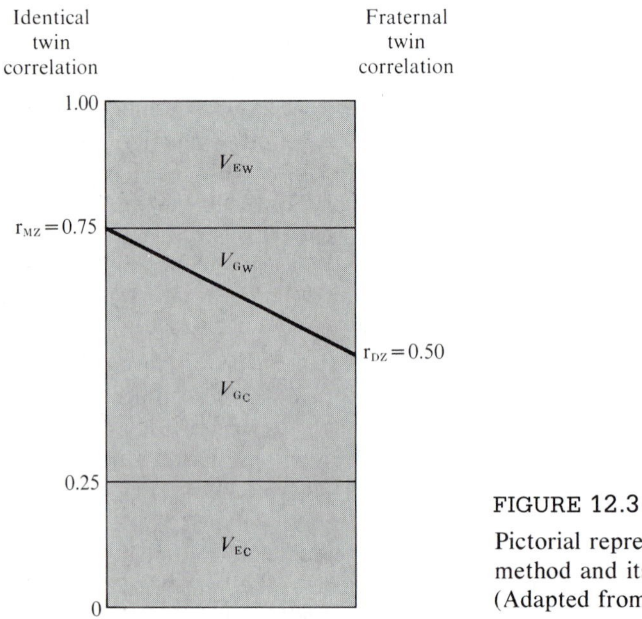

FIGURE 12.3

Pictorial representation of the twin method and its assumptions. (Adapted from Nichols, 1965.)

been broken down into environmental and genetic variance components using an example in which the correlation is 0.75 for identical twins and 0.50 for fraternal twins. On each side of the rectangle, the variance above the correlation represents differences within pairs, while the variance below the correlation stands for variance shared by members of a pair. Examination of the left side of the rectangle shows that identical twins share 75 percent of the variance, which also means that 25 percent of the variance is not shared. The differences within pairs of identical twins can only be environmental (except for somatic mutations and possible cytoplasmic differences at the time of separation). Such environmental influences are called within-family environmental influences—in this case, they are within-pair environmental differences (V_{E_W}). One assumption of the twin method is that these differences are the same for identical and fraternal twins (the "equal environments" assumption discussed above).

The right side of the rectangle shows that the difference between the correlation for identical and fraternal twins is genetic. This variance is sometimes referred to as within-family genetic variance (V_{G_W}) because it is due to genetic influences that cause family members (in this case, fraternal twins) to differ genetically. Comparing the two sides of the rectangle reveals that the genetic covariance (V_{G_c}) for fraternal twins is half that for identical twins. Together, V_{G_W} and V_{G_c} types of genetic variance represent the extent to which phenotypic variance is due to genetic variance (i.e., $V_G = V_{G_c} + V_{G_W}$). In this example, the total genetic variance is 50 percent—double the difference between the correlations for identical and fraternal twins. Of course, twins and other family members can be similar environmentally as well as genetically. This is represented by the component labeled V_{E_c}. Between-family environmental influences were defined in Chapter 9 as environmental influences that make family members similar to one another and different from other families. We need to keep in mind that between-family environmental influence may be greater for twins than for non-twin siblings.

In this example, genetic variance accounts for half of the phenotypic variance. (Broad-sense heritability is 0.50.) Environmentality, the phenotypic variance due to environmental differences, is also 0.50. Half of this environmental variance, in turn, is due to differences within families and half to differences between families. Environmental variance is not always distributed evenly within and between families. As we shall see in the following section, some of the environmental variance relevant to cognition occurs between families. For psychopathology (and personality in general), however, environmental variance is due primarily to within-family influences.

OTHER TWIN DESIGNS

Three other twin study designs deserve attention: studies of families of identical twins, identical co-twin control studies, and studies of genetic and phenotypic similarity within pairs of fraternal twins. The study of adopted identical and fraternal twins is particularly powerful, and will be described in our discussion of adoption studies in Chapter 13.

Families of Identical Twins

An interesting set of family relationships revealed in families of identical twins has recently been subjected to behavioral genetics analysis. This type of analysis is called the families-of-identical-twins method (or the monozygotic half-siblings method) (Corey and Nance, 1978; Nance and Corey, 1976; Nance, 1976). Figure 12.4 illustrates the family configurations that occur when both members of an adult identical twin pair marry and produce offspring. These two families include relatives that share: all of their genes (relationship A in Figure 12.4); half of their genes (relationships such as B, C, and E); one-fourth of their genes (relationship F); and none of their genes (relationship D, if there is no assortative mating). This type of family study compares parents and offspring, as well as siblings. As in other family studies, both heredity and family environment are shared in these relationships. The unique feature of this methodology involves the comparisons across the two families. In the families of male identical twins, for example, uncles and nephews/nieces (relationship C) are just as similar genetically as fathers and sons/daughters (relationship B). Also, cousins (relationship F)

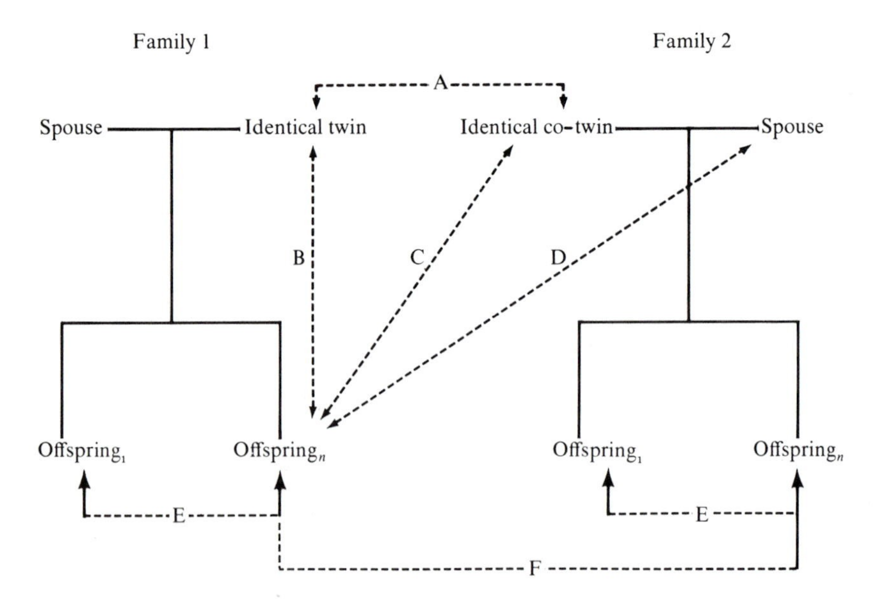

FIGURE 12.4

The families-of-identical-twins method. See text for explanation.

are half-siblings genetically because they have, in effect, the same parent from a genetic standpoint. These half-siblings are particularly interesting because they are reared in different families, thus avoiding the problems of other studies where half-siblings are separated by divorce, death, or desertion of one of their parents.

These comparisons can be fitted to a model to ascertain genetic and environmental parameters (Nance, 1976). Although the model is complicated mathematically, the idea is simple. The most critical comparison is between the identical twin parents and their children versus their nephews/nieces. The similarity between identical twin parents and their children can be compared to the similarity between the same individuals and their nieces/nephews. Genetically, the degree of similarity is the same. But because the nieces/nephews are reared in a somewhat different home environment, fewer environmental influences are shared between the identical twin uncles and their nephews/nieces than between the identical twin fathers and their children. The comparison between the siblings (relationship E) and the half-siblings (relationship F) is also interesting. The half-siblings are less similar than the full siblings genetically and environmentally because they are reared in different families. Another useful feature of this method is the opportunity to compare families of female identical twins and families of male identical twins to detect maternal effects and sex linkage. If there are maternal effects, half-siblings who are children of female identical twins should be more similar than half-siblings produced by male twins. For sex-linked characters, paternal half-brothers of male identical twins should be less alike than maternal half-brothers of female identical twins. Also, maternal half-sisters should be less similar than paternal half-sisters.

The basic problem with the families-of-identical-twins method is that the environments of the two families will be similar to the extent that genetically identical parents set up similar home environments. For example, suppose that the homes of the identical twin parents are essentially the same. Then identical twin parents and their nephews/nieces would be just as similar genetically and environmentally as parents and their offspring. It is possible to use path analyses to clarify not only this problem, but also the possible effects of assortative mating.

Studies have begun to apply the families-of-identical-twins method to behavioral traits. The earliest reports considered specific cognitive abilities such as spatial ability (Rose et al., 1979a) and perceptual speed (Rose et al., 1979b).

Identical Co-twin Control Studies

Co-twin control studies test the ability of environmental variables to alter phenotypes by exposing the members of identical twin pairs (co-twins) to differential environmental influences. The few (less than a dozen) co-twin control studies in the behavioral genetics literature fall into three categories.

The earliest and largest category includes studies of the interaction between maturation and training. The same pair of identical twin girls at different ages was studied by several investigators (Gesell and Thompson, 1929; Strayer, 1930; and Hilgard, 1933). They trained one co-twin and then the other in the same skills, with the major emphasis on the age at which each co-twin was trained. Another co-twin control study of this variety involved reading ability (Fowler, 1965).

A second category compares the effectiveness of different training programs. For example, Vandenberg (1968) reported a study in Stockholm that separated identical twins during their first years in elementary school to test two different instructional programs. The third, and most general, category includes studies of the extent to which environmental influences can produce differences in genetically identical individuals. Two Russian studies (Mirenva, 1935; Levit, reported by Newman, Freeman, and Holzinger, 1937) applied the co-twin method in this way to the study of motoric training. Other studies have applied the method to reading and numerical concepts (Vandenberg, Stafford, and Brown, 1968) and to personality (Plomin and Willerman, 1975).

The co-twin method provides a unique opportunity to control genetic variability in certain kinds of research in order to study environmental influence. The usual experimental method randomizes genetic differences, tests large numbers of subjects, and reports mean differences between the groups. In instances where the intensity and the time requirements of an experimental procedure preclude large numbers of subjects, the co-twin method may be a most useful alternative.

Studies of Genetic Similarity Within Pairs of Fraternal Twins

Even though fraternal twins average half of their segregating genes in common, genetic segregation assures that some fraternal twin pairs are more genetically similar than others, as is the case with non-twin siblings. Attempts have been made to use this fact to test the relationship between genetic similarity, as determined by blood group analyses, and phenotypic similarity. If genes influence a particular behavior, then fraternal twins who are more similar genetically should be more similar phenotypically. The method has been applied to cognitive and personality traits (Carter-Saltzman and Scarr, 1975). However, the usefulness of the method remains in doubt (Loehlin, Willerman, and Vandenberg, 1974).

TWIN STUDIES AND BEHAVIOR

The twin method has been applied to a wide range of behaviors. Many of these studies have been reviewed by Fuller and Thompson (1978), Peter

Mittler (1971), and Vandenberg (1976). Some recent twin studies have considered aging (Jarvik, 1975), scholastic abilities (Martin, 1975), creative abilities (Reznikoff et al., 1973), vocational interests (Roberts and Johansson, 1974), language (Mittler, 1974; Munsinger and Douglass, 1976), reading (Matheny and Brown, 1974), speech (Matheny and Bruggemann, 1973), social attitudes (Eaves and Eysenck, 1974), and sexual behavior (Martin, Eaves, and Eysenck, 1977). In the following sections, we shall continue with our samples of cognitive behavior and psychopathology.

We should note in passing that twins are not as difficult to obtain for research as one might imagine. For example, there are several large registers of twins. In the United States, there is a decade-old register of 16,000 pairs maintained by the National Research Council of the National Academy of Sciences (Hrubec and Neel, 1978), and another register, called the Kaiser-Permanente Twin Registry, begun in California in 1974 (Friedman and Lewis, 1978). There are also large registers in Finland (Kaprio et al., 1978), Sweden (Cederlöf and Lorich, 1978), and Norway (Kringlen, 1978). In addition, there are many mothers of twins clubs in the United States, and many of the clubs are interested in participating in research on twins.

COGNITION

Intelligence Quotient (IQ)

The first behavioral genetics twin study focused on IQ (Merriman, 1924). Since then, IQ tests have been administered to thousands of twins in studies of the heritability of general intelligence. Merriman also first noted the slightly lower average IQ for twins—about 96 compared to the population average of 100. Although the reason for this is not clear, it does not greatly affect the results of twin studies because the variance of twins is the same as that of the general population.

Table 12.7 lists 19 twin studies of general cognitive ability (Loehlin and Nichols, 1976). These are studies in which more than 25 pairs of each twin type were tested. They include 3,454 pairs of identical and 2,885 pairs of fraternal twins. The correlations in all studies except one indicate substantially greater similarity for identical twins.

The median correlation for all identical twins in these studies is 0.86. The median correlation for all fraternal twins is 0.62, which is higher than that reported in other surveys that included small samples. (See Figure 11.1.) As a result, the estimate of broad-sense heritability from these summary data is somewhat lower than usually reported. Falconer's formula for estimating heritability by doubling the difference between the correlations suggests a broad-sense heritability of 0.48 for general cognitive ability. Although this estimate is not adjusted for assortative mating or nonadditive genetic variance, such adjustments would not alter the conclusion that roughly half of observed variation in general cognitive ability is due to genetic differences.

TABLE 12.7

Identical and fraternal twin correlations for measures of general cognitive ability

Test	Correlation		Number of Pairs		Source
	Identical	Fraternal	Identical	Fraternal	
National and Multi-Mental	0.85	0.26	45	57	Wingfield and Sandiford (1928)
Otis	0.84	0.47	65	96	Herrman and Hogben (1933)
Binet	0.88	0.90	34	28	Stocks (1933)
Binet and Otis	0.92	0.63	50	50	Newman, Freeman, and Holzinger (1937)
I-Test	0.87	0.55	36	71	Husen (1947)
Simplex and C-Test	0.88	0.72	128	141	Wictorin (1952)
Intelligence factor	0.76	0.44	26	26	Blewett (1954)
JPQ-12	0.62	0.28	52	32	Cattell, Blewett, and Beloff (1955)
I-Test	0.90	0.70	215	416	Husen (1959)
Otis	0.83	0.59	34	34	Gottesman (1963)
Various group tests	0.94	0.55	95	127	Burt (1966)
PMA IQ	0.79	0.45	33	30	Koch (1966)
Vocabulary composite	0.83	0.66	85	135	Huntley (1966)
PMA total score	0.88	0.67	123	75	Loehlin and Vandenberg (1968)
General ability factor	0.80	0.48	337	156	Schoenfeldt (1968)
ITPA total	0.90	0.62	28	33	Mittler (1969)
Tanaka B	0.81	0.66	81	32	Kamitake (1971)
NMSQT total score:					
1962	0.87	0.63	687	482	Nichols (1965)
1965	0.86	0.62	1,300	864	Loehlin and Nichols (1976)

SOURCE: Adapted from Loehlin and Nichols, 1976.

If about half of the phenotypic variation in IQ is due to genetic differences, then about half must be caused by all other sources of variance. Figure 12.5 shows these median correlations for general cognitive ability in terms of the diagram with which we described the twin method. It shows that about half of the phenotypic variance is genetic in origin. Also, most of the environmental variance operates between families. In other words, twin studies suggest that shared (common) familial influences are the major environmental sources of individual differences in general cognitive ability. However, the correlation for nontwin siblings is lower than the correlation for fraternal twins. (See Chapter 11.) This means that twin studies overestimate the role of the familial environment because twins share more common familial influences than do nontwin siblings.

Developmental Differences

Behavioral genetic analyses can also be applied to developmental data. A common mistake is to think that genes are somehow turned on at the moment of conception and continue to run at full throttle for the rest of our lives. As explained in Chapter 5, genes can be turned on and off. Furthermore, environmental circumstances change as the organism develops. Thus, we might expect to find different patterns of genetic and environmental influences at different stages of development. Although it is reasonable to expect

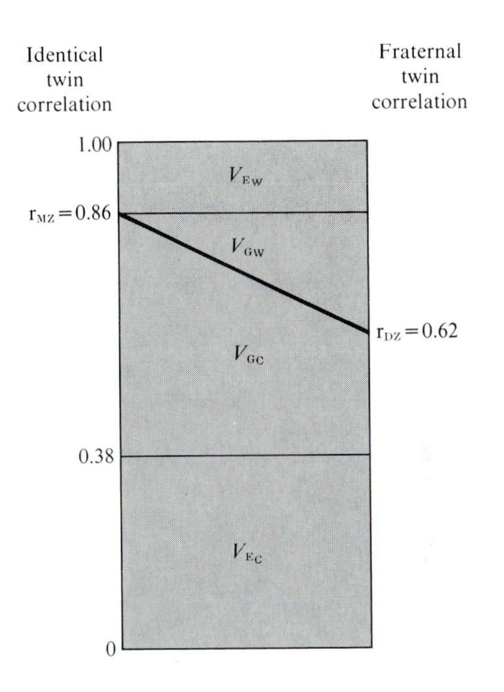

FIGURE 12.5

Pictorial representation of the average results of twin studies of general cognitive ability.

greater similarity as the twins grow up in similar environments, the early twin studies by Galton, Thorndike, and Merriman found little change in twin similarity as a function of age.

These earlier studies were cross-sectional, meaning that they obtained data on twins of different ages at the same time. A longitudinal study follows the same individuals over time. An exemplary longitudinal study of cognitive development in over 400 pairs of twins has been conducted in Louisville, Kentucky, during the last decade (Wilson, 1977a, 1977b, 1978). Twins were tested on the Bayley scales of mental and motor development at 3, 6, 9, 12, 18, and 24 months of age. IQ tests were administered at 2½ and 3 years (the Stanford-Binet), at 4, 5, and 6 years (the Wechsler Preschool and Primary Scale of Intelligence), and at 7 or 8 years (Wechsler Intelligence Scale for Children). The infant twin correlations for the Bayley mental scale are listed in Table 12.8. The average correlation for identical twins during infancy is 0.73; for fraternal twins, it is 0.65. These correlations suggest little genetic influence on individual differences in mental ability early in life. Because both the identical and fraternal twin correlations are high, they also suggest a substantial role for between-family environmental influences. Both conclusions are supported by the results of a smaller study of one-year-old twins in England (Griffiths and Phillips, 1976).

These data pose some important questions concerning cognitive development. Correlations of the same individuals from age to age during infancy are substantially lower than observed for identical or even fraternal twins of the same age. In other words, the score of one member of a twin pair at a given age is a better predictor of the co-twin's score than an earlier score of the co-twin himself. This result suggests that twin pairs follow a similar pattern of mental development across age. The sample profiles for six pairs of identical twins in infancy presented in Figure 12.6 indicate substantial similarities within twin pairs in developmental patterns.

TABLE 12.8

Within-pair correlations of MZ and DZ twins on Bayley mental scale scores

Age (months)	Number of Pairs		Intraclass Correlation		Broad-Sense Heritability
	MZ	DZ	MZ	DZ	
3	73	86	0.66	0.66	0.00
6	83	90	0.74	0.72	0.04
9	75	71	0.67	0.52	0.30
12	81	78	0.67	0.63	0.08
18	90	100	0.83	0.66	0.34
24	86	101	0.80	0.72	0.16

SOURCE: From Wilson, "Synchronies in mental development: An epigenetic perspective," *Science,* 202, 1978. Copyright © 1978 by the American Association for the Advancement of Science.

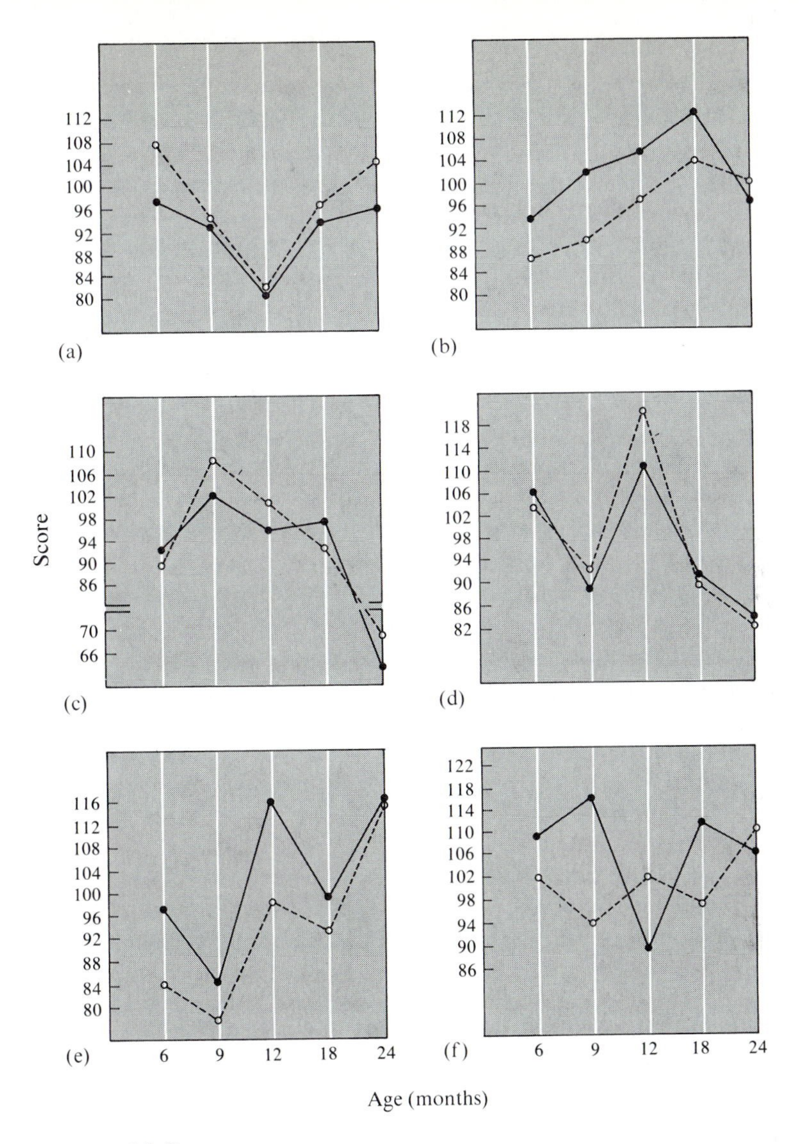

FIGURE 12.6

Profiles of mental development scores for MZ twins at ages 6 through 24 months. The pairs in (a) through (e) exhibit moderate to high profile congruence; the pair in (f) is obviously noncongruent. (From "Twins: early mental development" by R. S. Wilson. *Science*, 175, 914–917. Copyright © 1972 by the American Association for the Advancement of Science.)

The score profile may be described in terms of contour or overall mental level. The contour is a function of age-to-age changes known as *spurts* and *lags*. The overall level is a more accurate reflection of enduring mental development. The similarity between members of identical and fraternal twin pairs with respect to overall level and profile is described by the correlations

TABLE 12.9

Twin concordance for overall mental level and for age-to-age profile changes during the preschool years

Age (years)	Zygosity	Number of Pairs		Correlation	
				Overall Level	Age-to-Age Change
1	MZ	81		0.84	0.40
	DZ	84		0.78	0.15
			h^2_B	0.12	0.50
2 and 3	MZ	74		0.89	0.67
	DZ	95		0.79	0.42
			h^2_B	0.20	0.50
4, 5, and 6	MZ	89		0.90	0.47
	DZ	93		0.75	0.30
			h^2_B	0.30	0.34

SOURCE: From Wilson, "Synchronies in mental development: An epigenetic perspective," *Science*, 202, 1978. Copyright © 1978 by the American Association for the Advancement of Science.

(repeated-measures analysis adapted for use with twin data) presented in Table 12.9. The last column of the table shows that the age-to-age profiles of identical twins are more similar than those of fraternal twins. This indicates that the spurts and lags in development during infancy are influenced by genetic differences, and that genes continue to influence age-to-age changes through the preschool years.

Table 12.9 also shows increasing heritability for overall mental level during development from infancy through six years of age. The identical twin correlations tend to increase and the fraternal twin correlations tend to drop. A recent report from the Louisville twin study (Wilson, 1977a) indicates that the IQ correlation at seven and eight years of age for identical twins is 0.86 (74 pairs) and that for fraternal twins (56 pairs) it is 0.60. These correlations are nearly identical to those for adults.

Specific Cognitive Abilities

The family studies reviewed in Chapter 11 suggest that specific cognitive abilities also have a strong familial component. Twin studies have consistently demonstrated that genetic differences play a major role in these familial influences. However, it is still not clear whether genetic differences influence certain specific cognitive abilities more than others.

Table 12.10 lists the median correlations for measures of specific abilities in eight twin studies (Loehlin and Nichols, 1976). Overall, the me-

TABLE 12.10

Median correlations on measures of specific cognitive abilities in various twin studies

Population	Median Correlation		Number of Pairs		Source
	Identical	Fraternal	Identical	Fraternal	
Schoolchildren (Sweden)	0.67	0.58	128	141	Wictorin (1952)
Military recruits (Sweden)	0.78	0.60	215	415	Husen (1959)
Aged (New York)	0.68	0.44	75	45	Jarvik et al. (1962)
Adult males (Finland)	0.70	0.42	157	189	Partanen, Bruun, and Markkanen (1966)
High school (Michigan, Kentucky)	0.66	0.50	123	75	Loehlin and Vandenberg (1968)
High school (United States)	0.65	0.40	337	156	Schoenfeldt (1968)
High school (Kentucky)	0.64	0.43	137	99	Vandenberg (1969)
High school (United States)	0.74	0.52	509	330	Loehlin and Nichols (1976)

SOURCE: Adapted from Loehlin and Nichols, 1976.

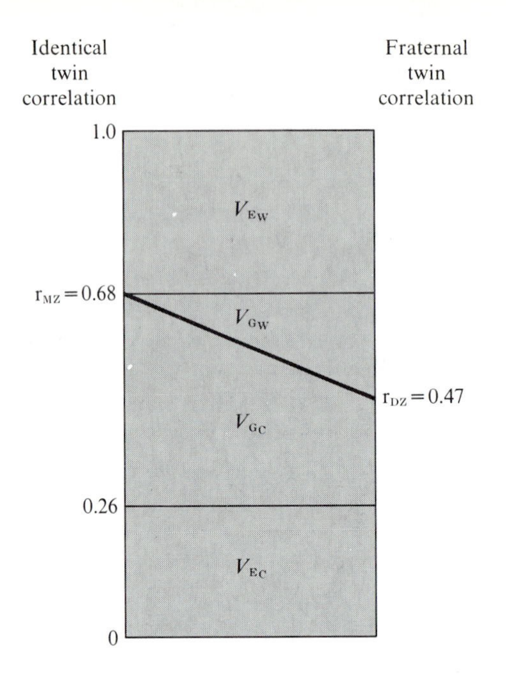

FIGURE 12.7
Pictorial representation of the average results of twin studies of specific cognitive abilities.

dian correlation for identical twins is 0.68; for fraternal twins, it is 0.47. This result suggests that, on the average, the broad-sense heritability for specific cognitive abilities (0.42) is about the same as that for general cognitive ability (0.48). However, as depicted in Figure 12.7, the salient environmental factors that operate within families may be somewhat more important than those that operate between families. A comparison of Figure 12.7 with Figure 12.5 suggests that such within-family influences may have a greater effect on specific cognitive abilities than on general cognitive ability.

As noted earlier, family studies suggest that verbal and spatial abilities show greater familiality than other specific cognitive abilities, such as perceptual speed and visual memory. Although twin studies tend to show greater familiality for verbal ability than memory, this differential familiality does not appear to be genetic in origin. Two of the larger twin studies were conducted in Finland (Bruun, Markkanen, and Partanen, 1966) and in the United States (Schoenfeldt, 1968). Although the studies used different tests, they each included two measures of each of four cognitive ability factors. The correlations for identical and fraternal twins and the heritabilities are presented in Table 12.11, which shows that the results of the twin studies are not consistent. Similarly mixed results have been obtained by other twin studies of specific cognitive abilities (Loehlin and Nichols, 1976).

Multivariate Analysis

In previous chapters, we discussed the application of quantitative genetic analysis to the correlations among characters as a step beyond the

TABLE 12.11

Identical and fraternal twin correlations and twin study heritabilities for four cognitive ability factors in two twin studies

	Bruun et al. Study (1966)			Schoenfeldt Study (1968)		
	Correlation		Heritability	Correlation		Heritability
	Identical (157 pairs)	Fraternal (189 pairs)		Identical (337 pairs)	Fraternal (156 pairs)	
Verbal	0.78	0.53	0.50	0.80	0.48	0.64
Spatial	0.59	0.36	0.46	0.67	0.41	0.52
Perceptual speed	0.73	0.50	0.46	0.67	0.51	0.32
Memory	0.64	0.32	0.64	0.43	0.19	0.48

usual univariate analysis of a single character. In Chapter 11, we presented a multivariate analysis of data from the Hawaii family study of cognition. That indicated that the "genetic" and the "environmental" structures of specific cognitive abilities are quite similar to each other and to the phenotypic structure. Twin studies have been more often used to provide multivariate analyses of human behavior (Loehlin, 1965; Vandenberg, 1965; Bock and Vandenberg, 1968; Loehlin and Vandenberg, 1968; Eaves and Gale, 1974; Martin and Eaves, 1977; Fulker, 1978). We shall describe a multivariate twin analysis of scholastic abilities.

As discussed in Chapter 9, twin cross-correlations for two characters include components of variance similar to twin correlations for a single character. Thus, twin cross-correlations for identical and fraternal twins can be used to assess environmental and genetic contributions to the phenotypic correlation among characters, using the same logic described for univariate analyses earlier in this chapter. For example, doubling the difference between identical and fraternal twin cross-correlations yields an estimate of the genetic contribution to the phenotypic correlation. As implied in Chapter 9 (see Plomin and DeFries, 1980b, for details), we can use twin cross-correlations to solve for genetic and environmental chains and correlations.

As an example of this approach, we shall discuss a large twin study of scholastic abilities conducted by Loehlin and Nichols (1976). They presented data on five scholastic abilities (English, mathematics, social studies, natural sciences, and vocabulary) for high school twins who participated in the National Merit Scholarship Qualifying Tests (NMSQT) in the United States. Phenotypic correlations among the five abilities for a random sample of 1,300 identical and 860 fraternal twin pairs in the NMSQT sample showed that scholastic abilities are much more highly correlated (0.62 on the average) than specific cognitive abilities (0.28 on the average in the Hawaii family study of cognition, as reported in the previous chapter). Loehlin and Nichols computed cross-correlations, from which we computed genetic and environmental correlations. The genetic and environmental correlations help determine the nature of the phenotypic correlations among scholastic abilities. In other words, is each of the scholastic abilities influenced by different sets of genes and different sets of environmental factors, or do the same genes and environmental influences affect several scholastic abilities?

These genetic and environmental correlations from the NMSQT cross-correlations are presented in Table 12.12. The genetic correlations are all substantial, 0.71 on the average. It should be noted that the genetic correlation of 0.99 between social studies and natural sciences seems to be a statistical fluke. In another random sample from the NMSQT study, the genetic correlation was 0.71, which is more in line with the other genetic correlations. These genetic correlations suggest that some, but not all, of the genes that affect each scholastic ability also affect other abilities. They further suggest that there is a general genetic factor that affects the various abilities,

TABLE 12.12

Genetic and environmental correlations for five tests of scholastic abilities from a twin analysis of the National Merit Scholarship Qualifying Tests

	Genetic Correlations			
	Mathematics	Social Studies	Natural Sciences	Vocabulary
English	0.54	0.71	0.86	0.68
Mathematics		0.69	0.58	0.50
Social studies			0.99	0.85
Natural sciences				0.74

	Environmental Correlations			
	Mathematics	Social Studies	Natural Sciences	Vocabulary
English	0.50	0.63	0.40	0.75
Mathematics		0.45	0.59	0.61
Social studies			0.49	0.73
Natural sciences				0.58

SOURCE: Plomin and DeFries, 1980b.

although each ability is also influenced by independent genetic factors. Exactly the same pattern is suggested by the environmental correlations. There appears to be a general environmental factor that affects all of the scholastic abilities, although each ability is also affected by independent environmental influences.

These multivariate findings provide important insights into the way that genes and environments affect scholastic abilities. They suggest that there are genetic reasons why some students who do well in English, for example, also do well in other scholastic areas. They also suggest that there are equally important environmental reasons for general scholastic performance. In other words, there seem to be broad, systematic genetic and environmental factors that affect scholastic abilities, as well as unique genetic and environmental influences on each ability.

Finally, as seen in previous chapters, the structures of the genetic and environmental correlations are quite similar. In this case, because all of the genetic and environmental correlations are of a similar moderate magnitude, one would expect to find no strong correlation between the two kinds of correlations. However, factor analysis (as discussed in the previous chapter in relation to specific cognitive abilities) yields one general factor for the genetic correlations (accounting for 78 percent of the variance) and one general factor for the environmental correlations (accounting for 66 percent of the variance). Thus, this analysis merely reaffirms our earlier conclusion that there is one general genetic factor and one general environmental factor affecting scholastic abilities.

Loehlin and Nichols (1976, p. 80) caution that the similarity between genetic and environmental correlations may be inherent in our measuring instruments. Consider the extreme example in which two moderately heritable subtests happen to include some of the same items. This would, of course, lead to a phenotypic correlation between the subtests, and both genetic and environmental factors would seem to be responsible for the phenotypic correlation. Although this example may seem far-fetched, it is possible that items used to measure scholastic abilities might overlap in a more subtle way. Nonetheless, no such methodological explanation has yet been demonstrated to explain observed similarities between genetic and environmental correlations.

In Chapter 9, we suggested that similarity between genetic and environmental correlations is not so surprising, at least in retrospect. In Chapter 10, we provided an example at a biochemical level. Loehlin and Vandenberg have suggested a possible cultural explanation pertinent to cognitive abilities:

> On general grounds findings such as these are perhaps not unreasonable. Presumably the development of cultural institutions is to some extent influenced by the human biological tendencies they control or exploit. A sex factor, for example, might emerge either from purely sociological or from purely biological data. The case is perhaps less obvious for cognitive traits, but it is at least conceivable that the biological capacities of the human organism have historically had some bearing on what society has tended to recognize, name, and educate as a unit. (Loehlin and Vandenberg, 1968, p. 275)

PSYCHOPATHOLOGY

Twin studies of psychopathology usually report *concordance* values. When a sample of twins has been ascertained for psychopathology, concordance is the percentage of co-twins with the same diagnosis. However, there are several ways of computing concordance (Gottesman and Shields, 1972). The most common method is called *pairwise*. In a sample of twin pairs in which at least one twin is affected, pairwise concordance is the number of concordant pairs divided by the total number of pairs. For example, in Irving Gottesman and James Shields' (1972) study, both members of the pair were schizophrenic in 10 of 24 pairs of identical twins; so the pairwise concordance would be 10/24 = 0.42. The other major method considers index cases rather than pairs. In this *proband* method, the number of affected twins in concordant pairs is divided by the number of index cases (probands). Gottesman and Shields began with 34 identical twin probands and found that 20 of these individuals were in concordant pairs. The probandwise concordance would thus be 20/34 = 0.59. Although there are merits to both approaches, the proband method is more appropriate from a sampling point of view.

Schizophrenia

Family studies of schizophrenia suggested a risk of about 5 to 10 percent for first-degree relatives of schizophrenics. Such familial influence could have either genetic or environmental origins, and twin studies have been useful in investigating these possibilities. Because twins show no more psychopathology than singletons, the results of twin studies can probably be generalized to the rest of the population. Eleven twin studies in seven countries reported concordances for identical and fraternal twins (Kringlen, 1966). The median pairwise concordance for identical twins was 42 percent, while that for fraternal twins was 9 percent. Table 12.13 presents the results of five twin studies of schizophrenia since 1966 that included 210 identical twin pairs and 309 pairs of fraternal twins (Gottesman and Shields, 1976). The weighted average probandwise concordance is 0.46 for the identical twins and 0.14 for the fraternal twins.

Results from the Scandinavian countries of Denmark, Norway, and Finland generally indicate a smaller hereditary contribution to schizophrenia. Attempts to account for these discrepant findings have led to an extensive analysis of differences in diagnostic criteria in different countries, the biases introduced by different types of sampling procedures (resident hospital population, consecutive admission, or twin registry), and the possibility that schizophrenia is really a heterogeneous complex of psychotic conditions (Gottesman and Shields, 1977; Rosenthal, 1970).

The extent to which the data might be influenced by these factors can be illustrated by results of analyses of two of the twin studies. Gottesman and Shields (1977) analyzed separately the concordance of identical co-twins of mild and severe proband cases from their twin study of about forty-five thousand consecutive admissions in a large London hospital. As indicated in

TABLE 12.13
Schizophrenia: probandwise concordance in recent twin studies

Investigator	Year	Country	Identical		Fraternal	
			Concordance %	Number of Pairs	Concordance %	Number of Pairs
Kringlen	1967	Norway	45	55	15	90
Fischer	1973	Denmark	56	21	26	41
Gottesman and Shields	1972	U.K.	58	22	12	33
Tienari	1971	Finland	35	17	13	20
Pollin et al.	1972	U.S.	43	95	9	125

SOURCE: After Gottesman and Shields, 1976.

TABLE 12.14
Schizophrenia: relationship between severity and concordance for identical twins

Degree of Severity of Proband	Concordance
Less than two years in hospital	27% (4/15)
More than two years in hospital	77% (10/13)
Inability to stay out of hospital for at least six months	75% (12/16)

SOURCE: Gottesman and Shields, 1977.

Table 12.14, the concordance for identical twins is only 27 percent for probands in the hospital less than two years. In contrast, the concordance was 77 percent when probands were hospitalized over two years and 75 percent when probands were not able to stay outside the hospital for longer than six months. The relationship between severity and concordance has been documented for most twin studies (Gottesman and Shields, 1977). Clearly, the severity of the cases chosen for study can affect the results obtained. It has been suggested, for example, that the largest twin study of schizophrenia, conducted by F. J. Kallmann (1946), consisted mostly of severe or chronic cases and that this might account for the relatively higher identical twin concordance (69 percent) obtained by Kallmann as compared to subsequent investigators.

Another diagnostic matter of considerable importance concerns the strictness of the definition of schizophrenia. Three of the most recent twin studies can be analyzed using both a strict definition of schizophrenia and a broader definition that includes borderline cases (Fischer et al., 1969). The results are shown in Table 12.15, which shows that concordances for both identical and fraternal twins are higher when the broader definition of schizophrenia is employed.

The issue of diagnostic criteria spills over into the larger question of the heterogeneity of schizophrenia. Although many attempts have been made to break down schizophrenia into more discrete genetic entities, current interest focuses on the possibility that there is a spectrum of schizophrenia that extends as far as antisocial behavior. Generally, researchers talk about a "hard" spectrum, which includes borderline and questionable schizophrenia as well as chronic schizophrenia, as distinct from a "soft" spectrum, which includes personality disorders. Table 12.16 lists the concordances for schizophrenic identical twins when we expand the concept of schizophrenia to include the "hard" and "soft" spectra (Shields, Heston, and Gottesman, 1975). On the average, only about 25 percent of the identical co-twins showed no psychopathology. The median identical twin concordance for the

TABLE 12.15

Concordance for schizophrenia as a function of strictness of diagnostic criteria

Study	MZ		DZ	
	Strict Schizophrenia	Including Borderline Cases	Strict Schizophrenia	Including Borderline Cases
Kringlen (1966)	28% (14/50)	38% (19/50)	6% (6/94)	14% (13/94)
Gottesman and Shields (1966)	42% (10/24)	54% (13/24)	9% (3/33)	18% (6/33)
Fischer et al. (1968)	24% (5/21)	48% (10/21)	10% (4/41)	19% (8/41)

SOURCE: After Fischer et al., 1969.

"hard" spectrum was 61 percent. For the combined categories of "hard" and "soft," the concordance was 77 percent. These twin data suggest that the genetic propensity toward schizophrenia may be broader than we once suspected.

Further research is needed to clarify whether these considerations of diagnostic criteria account fully for the generally lower concordance for schizophrenia of identical twins in the Scandinavian studies. It is possible,

TABLE 12.16

Schizophrenic spectrum: pairwise concordance for schizophrenic identical twins for hard and soft schizophrenic spectra

Study	Number of Pairs	Concordance for Hard Spectrum	Concordance for Hard or Soft Spectrum	Percentage of Pairs with One Co-twin Normal
Luxenburger (1928)	14	72	86	14
Rosanoff et al. (1934)	41	61	68	32
Kallmann (1946)	174	69	95	5
Slater (1953)	37	64	78	22
Kringlen (1967)	45	38	67	33
Fischer et al. (1969)	21	48	57	43
Gottesman and Shields (1972)	22	50	77	23

SOURCE: After Shields, Gottesman, and Heston, 1975.

for example, that populations differ in the frequencies of genes that may modify the expression of the schizophrenic phenotype (Shields, 1968). Another issue that needs to be considered is age correction. None of the concordances that we have mentioned so far has been corrected for age. There is no satisfactory age correction for twin concordances, but it is an important issue because age of onset can vary. One study of Finnish twins (Tienari, 1963) differed from all other twin studies in finding *no* concordance for identical twins. The twins were thirty to forty years of age at the time of that report. In a continuation of the study, one of the identical twin pairs was concordant by 1968. By 1971, 7 of the 20 affected identical twins were in concordant pairs, so that the probandwise concordance was 7/20, or 35 percent, and this figure may increase more as the study continues (Gottesman and Shields, 1976).

In summary, although twin data do not conform to a simple genetic model, they clearly support a genetic hypothesis. The risk of schizophrenia for fraternal twins is about the same as for other siblings (although opposite-sex twins have a consistently lower concordance). Identical twins are much more concordant than fraternal twins. Thus, these studies suggest the influence of genetic differences. Like the family studies, they also suggest that within-family environmental influences play a major role.

Identical Twins Discordant for Schizophrenia

Because differences within pairs of identical twins are environmental in origin, the co-twin control method can be used to study identical twins discordant for schizophrenia. This method is particularly important for studying psychopathology because differences within pairs of identical twins must be caused by within-family environmental influences. These, as we have seen, are important in the etiology of psychopathology. In one study, in which 17 discordant identical twin pairs were studied for a decade (Belmaker et al., 1974), the salient within-family environmental factors have remained elusive. A difference in birth weight was suggested as a possible factor, but other studies have not supported this finding. Another possibility is submissiveness on the part of one twin, but it is difficult to determine the cause and effect of this possible factor. Gottesman and Shields (1976, p. 379) have suggested that "the 'culprits' may be nonspecific, time-limited in their effectiveness, and idiosyncratic." They also suggest that the co-twin control method may be more useful in identifying biochemical "endophenotypes" ("inside" phenotypes), which are not ordinarily measured in clinical examinations. For example, an exciting discovery was that both affected and unaffected identical co-twins had lower peripheral monoamine oxidase levels (Wyatt et al., 1973). However, this finding is currently in dispute because other studies have not found a difference between normals and schizophrenics for this enzyme (Friedman et al., 1974).

Manic-Depressive Psychosis

Although the focus of family studies has shifted to the affective disorders, such as manic-depressive psychosis, twin studies and adoption studies continue to emphasize schizophrenia. Nonetheless, there have been several twin studies of manic-depressive psychosis. Seven studies included a total of 99 pairs of identical twins and 252 fraternal twin pairs (Rosenthal, 1970). The median concordance for identical twins in these studies is 71 percent and that for fraternal twins is 19 percent. Thus, the risk for fraternal twins of being a manic-depressive is slightly greater than the risk for nontwin siblings; the risk for identical twins of a proband is about four times greater than the risk for fraternal twins. Although the concordance for manic-depressive psychosis is higher than for schizophrenia, within-family environmental factors are important for manic-depressive psychosis as well as for schizophrenia.

In the previous chapter, we discussed the possibility that there are two subtypes of manic-depressive psychosis: a unipolar variety and a bipolar manic-depressive type. Although there is clearly overlap between the two types, family data provide some support for the distinction. Twin studies have also been addressed to this issue, which is essentially an ideal candidate for the issue of multivariate genetic analysis. In a review of several small studies totaling 83 identical twin pairs, E. Zerbin-Rudin (1969) found that 27 percent of the identical twin pairs were concordant for unipolar depression and 20 percent were concordant for bipolar manic-depressive psychosis. However, in 6 percent of the pairs, one co-twin was bipolar and the other unipolar. Similar results were obtained in a recent study of 55 pairs of identical twins (Bertelsen, Harvald, Hauge, 1977): 20 percent of the pairs were concordant for unipolar depression, 25 percent were concordant for bipolar manic-depressive psychosis, and one co-twin was bipolar and the other unipolar in 13 percent of the pairs. Thus, the twin data also provide some support for the unipolar-bipolar distinction, although they also suggest overlap between the two types.

SUMMARY

Twin studies take advantage of the natural experimental situation resulting from the fact that identical twins are twice as similar genetically as fraternal twins. If genes make a difference for a particular behavior, identical twins should be more similar than fraternals. Zygosity determination and the reasonableness of the equal environments assumption were discussed, as well as the estimation of broad-sense heritability and between- and within-family environmental influences.

For general cognitive ability (IQ), the average correlations for identical and fraternal twins are 0.86 and 0.62, respectively. This pattern of correlations suggests substantial genetic influence (broad-sense heritability of about 0.50). It also points to a substantial role for between-family environmental influences (E_c) on the behavior of twins. For specific cognitive abilities, the average twin correlations are 0.68 for identical twins and 0.47 for fraternal twins. These correlations suggest about the same degree of genetic influence as for IQ (broad-sense heritability = 0.42). But environmental influences within families (E_W) may be relatively more important for specific cognitive abilities than for general cognitive ability, at least in twins. A multivariate twin analysis of scholastic abilities suggests that some, but not all, of the same genes influence most scholastic abilities. A single set of environmental influences also seems to affect scholastic abilities, although each ability is also influenced by unique environmental factors.

Genetic influences also play an important role in psychoses. The average probandwise concordances for schizophrenia in five recent studies are 0.46 and 0.14 for identical and fraternal twins, respectively. For affective psychoses, the average concordances are 0.71 and 0.19, respectively.

13

Adoption Studies

The first adoption study was conducted in the same year (1924) as the first twin study. However, far more twin studies have been reported since then, no doubt because of the greater ease of conducting twin studies. Nonetheless, adoption studies of complex human behaviors provide the most convincing demonstration of genetic influence. These studies untangle genetic and environmental factors common to members of natural families by studying genetically unrelated individuals living together (to assess environmental influences common to family members), and genetically related individuals living apart (to test genetic influences). In this way, adoption studies can determine the extent to which familial resemblances are due to genetic or environmental similarity.

Adoption studies also provide a powerful tool for evaluating environmental influences as distinct from genetic variables, although this option has not been exercised much in the past. In the behavioral sciences, environmental assessments are often clouded by the effect of genetic influences. For example, parental behavior (such as the quality of parents' responsiveness to their children) has been related to the cognitive development of children (Elardo, Bradley, and Caldwell, 1975). However, the association between parental responsiveness and children's brightness may not be environmental.

If the more responsive parents tend to be more intelligent, their children might inherit their intelligence directly, without any particular influence of their responsive child rearing. It is also possible that an apparently environmental effect may actually result from correlations between genetic and environmental factors. Inherently brighter children may encourage more responsiveness from their parents. The latter possibility is referred to in the developmental literature as the direction of effects in socialization. In terms of generalizing and implementing the results of such research, it is important to know whether such environmental effects are confused with the effects of heredity. Adoption studies in the behavioral sciences facilitate a more refined analysis of environmental influences.

ISSUES IN ADOPTION STUDIES

Because the logic of the adoption design is so straightforward, few criticisms have been leveled at it. If you want to know the extent to which observed resemblance between parents and their offspring is genetic in origin for a particular behavior, you can assess the resemblance between adopted-away children and their biological parents (called "birth parents" by adoption agencies), who gave their children genes but no familial environment. Conversely, you could also study the resemblance between adopted children and their adoptive parents, who provide a familial environment, but no genes. As in any experiment, certain biases must be avoided, or, if that is impossible, they must be assessed. In the case of adoption studies, the representativeness of the sample and the possibility of selective placement must be considered.

Representativeness

Twins, we have seen, are quite representative of the general population. Although many stereotypes of birth parents, adoptive parents, and adoptees have been commonly accepted, such preconceptions tend to fade in the face of data. For example, many people believed that the average IQ of birth parents is lower than that of the general population. However, in the state of Minnesota during the period from 1948 to 1952, when IQ tests were required for all women giving up children for adoption, the average IQ score of 3,600 women was 100 with a standard deviation of 15.4. These values are the same as those for the general population (Scarr, 1977). The most important point is that the representativeness of these groups in terms of variances and means can be measured. If some degree of unrepresentativeness is found, it does not invalidate the results of an adoption study. Rather, it can be taken into account in the interpretation of data.

Selective Placement

An issue that is more specific to adoption studies is *selective placement,* which means that adoptees are placed with adoptive parents who resemble the birth parents in some ways. For example, adoption agencies tend to place children whose birth parents are tall with tall adoptive parents. However, in terms of behavior, adoption agencies have only limited information (usually just education and occupation). Thus, they could not accurately match children for behavioral characteristics even if they wanted to do so. In fact, many adoption agencies now avoid selective placement altogether (even for physical characteristics) because they feel that it causes adoptive parents to have the false expectation of receiving a child similar in many ways to themselves.

Selective placement may increase the resemblance between adoptive parents and their adopted children (if there is genetic influence on the trait being studied). It also may increase the resemblance between birth parents and their adopted-away children (if there is environmental influence). Figure 13.1 presents a simplified path diagram illustrating the influence of selective placement on parent-child resemblance. Path analysis was briefly described in Chapter 9. For now, you need only remember that a path (such as e or g) represents the effect of one variable on another, independent of other influences. In Figure 13.1, e is the path that represents the influence of the adoptive parents (A) on the adopted child (C), independent of the birth parents (B). Similarly, g represents the influence of the birth parents on the adopted child, independent of the adoptive parents. The double-headed path (s) represents the selective placement correlation between adoptive and birth parents. It is possible to solve for these paths to determine parental influences independent of selective placement effects. In this way, we can assess the environmental influence of adoptive parents independent of selective placement (path e), and the genetic influence of birth parents independent of selective placement (path g).

This path diagram leads to an interesting conclusion concerning the effects of selective placement. As indicated in Chapter 9, we can use a path model to visualize relationships. For example, the correlation between adop-

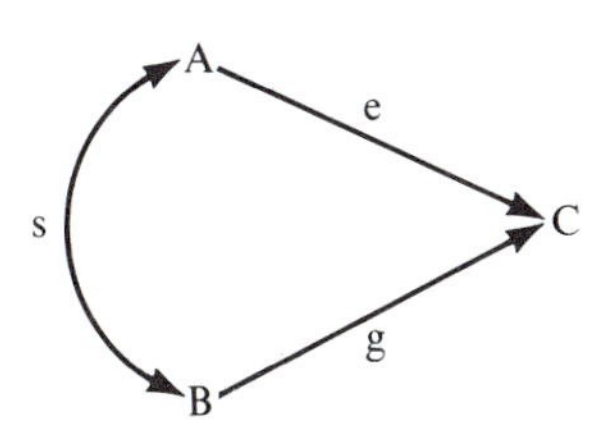

FIGURE 13.1

Path diagram illustrating the influence of selective placement on parent-child resemblance in an adoption study. The letter A symbolizes the adoptive parents, B birth parents, and C the adopted child. (From "Behavioral genetics" by J. C. DeFries and R. Plomin. Reproduced, with permission, from *Annual Review of Psychology,* 29, 473–515. Copyright © 1978 by Annual Reviews, Inc. All rights reserved.)

tive parents and their adopted children includes not only path e, but also the g and s chain of influences. Thus,

$$r_{AC} = e + gs$$

This means that the correlation between adoptive parents and their adopted children will be inflated if there is selective placement for a characteristic, as well as genetic influence on that trait. Similarly, the correlation between birth parents and their adopted-away children will be inflated if there is both selective placement and environmental influence:

$$r_{BC} = g + es$$

Moreover, if the correlation between birth parents and their adopted-away children is greater than the correlation between adoptive parents and their adopted children, then genetic influence must be greater than environmental influence, even if there is substantial selective placement:

$$r_{BC} - r_{AC} = (g + es) - (e + gs) = (g\text{-}e)(l\text{-}s)$$

Thus, if r_{BC} is greater than r_{AC}, g will be greater than e even if selective placement occurs (that is, if s is non-zero).

Other Issues

DeFries and Plomin (1978) have outlined several criteria for an ideal adoption study. These include such general requirements as the inclusion of environmental assessments, studying more than one type of family relationship, measuring many variables, and studying developmental phenomena using a longitudinal design. Although we suggested that an ideal study should meet all these criteria, that is not really essential for the success of an adoption study. One other criterion, the analysis of assortative mating, is critical. We have mentioned several times that the resemblance between parents and offspring is inflated if assortative mating occurs. Assortative marriage is known to occur for cognitive abilities (Johnson et al., 1976) and for psychopathology (Dunner et al., 1976). Unwed birth parents of adopted children also are known to mate assortatively for physical and behavioral characters (Plomin, DeFries, and Roberts, 1977). In short, we can expect that about the same degree of assortative mating occurs for unwed birth parents as for married couples. We need to take this into account when interpreting the resemblance between birth parents and their offspring in adoption studies.

Another important issue involves the effect of prenatal environment. Because birth mothers provide the prenatal environment for their adopted-away children, the phenotypic resemblance between them in an adoption study may reflect prenatal environmental influences. This possibility can be tested by comparing correlations between birth mothers and their adopted-away children to correlations between the birth fathers and the children. This test has been incorporated in recent adoption studies of psychopathology, as we shall see later in this chapter. A particularly valuable asset of adoption studies is their ability to test the influence of prenatal maternal environment, independent of postnatal environment.

HERITABILITY AND ENVIRONMENTALITY

Like the twin studies described in Chapter 9, adoption studies can be used to estimate the extent to which phenotypic variance is due to genetic variance (heritability) and environmental variance (environmentality). In Table 9.9, the phenotypic covariance for various family relationships was divided into genetic and environmental components. A path diagram of the relationship between parents and offspring in natural families (Figure 13.2) shows that the parents share both genetic and environmental influences with their children. The point of an adoption study is to separate these two sets of influence. Figure 13.3 illustrates this separation for parents and offspring. The genetic side of the adoption design involves birth parents who give their adopted-away children genes, but not environment; the environmental side involves adoptive parents who give their children a familial environment, but not genes.

In Chapter 10, Table 10.1 described the relationship between familial correlations and heritability. The genetic side of an adoption study, in the absence of selective placement, estimates genetic influence independent of environment. The regression of adopted-away offspring on the score of one birth parent estimates half of the narrow-sense heritability ($\frac{1}{2}V_A/V_P$). The regression of adopted children on the midparent score of their birth parents directly estimates narrow-sense heritability. The correlation between full

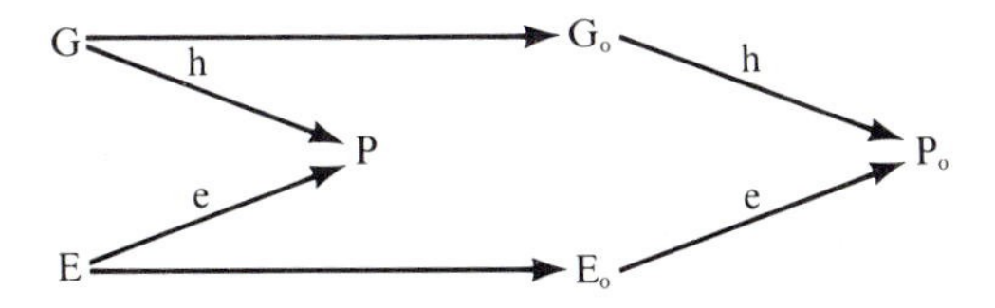

FIGURE 13.2

Path diagram of the relationship between parents and offspring in natural families. The subscript o refers to offspring. See text for explanation.

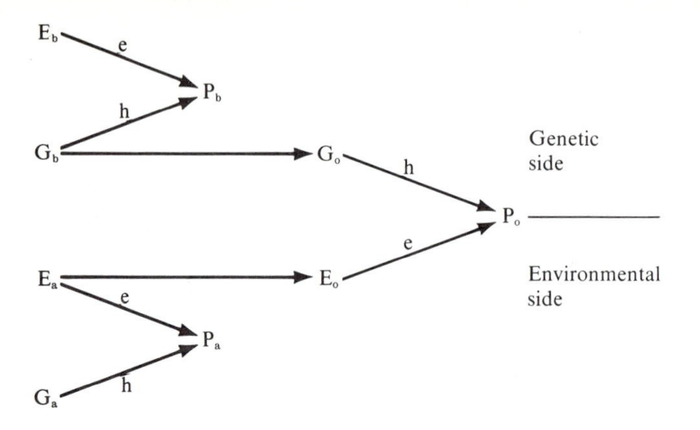

FIGURE 13.3

Path diagram showing the genetic and the environmental sides of the adoption design. The subscript a refers to the adoptive parent, b to the birth parent, and o to the adopted offspring. See text for explanation.

siblings (or fraternal twins) adopted into different families estimates roughly half of the broad-sense heritability, which may also be directly estimated from the correlation between identical twins reared in different families.

Phenotypic variance that cannot be explained by genetic differences is ascribed to the environment. Even better, the environmental side of adoption studies permits the direct assessment of family-shared (between-family) environmental influences. The between-family environmental components described in Table 9.9 can each be assessed by a particular adoptive relationship. For example, common environmental influences shared by parents and their offspring can be measured by the resemblance between adoptive parents and their adopted children. Between-family environmental influences shared by siblings are revealed by the correlation between genetically unrelated children adopted into the same family.

As is true of the results of other behavioral genetics methods, the results of adoption studies can be affected by genotype-environment interaction and correlation. In fact, adoption studies are the only practical tool for isolating interactions and correlations between genetic and environmental influences on human behavior. Both of these issues will be discussed in the next chapter.

Adoption studies have been applied only to two domains of behavior, the same two that we have been using as examples of research in human behavioral genetics: cognition and psychopathology. We shall follow closely a recent review (DeFries and Plomin, 1978), which can be consulted for greater detail. Although we have refrained from being encyclopedic until now, we shall describe all of the major adoption studies relevant to IQ and psychopathology for two reasons. First, adoption studies provide the most powerful and convincing human behavioral genetics analyses and thus merit greater detail. Second, the consistency of results across the various studies, conducted in different countries and using different measures and procedures, leaves an impression of the robustness of these findings. However, readers who prefer to avoid such detail can skip to the summary at the end of each section.

INTELLIGENCE QUOTIENT (IQ)

The first adoption study considered general cognitive ability (Theis, 1924), and every adoption study for about the next 40 years continued to focus on this same trait. There have been two types of adoption studies: full adoption studies, which include both the genetic and environmental sides of the design (see Figure 13.3); and partial adoption studies, which assess either genetic or environmental variables. We shall emphasize full adoption designs because they provide considerably more information. After describing several studies of each type, we shall summmarize their results at the end of this section.

Parent-Offspring Adoption Studies

Full Adoption Designs

S. Theis' study had several methodological problems. For example, mental ability was simply rated on a three-point scale, and only 35 percent of the 910 children were adopted before they were 5 years old. These problems make it difficult to interpret the finding that the adopted children's rated mental ability was affected by the social status of their birth parents more than by that of their adoptive parents.

More than twenty years later, a full adoption study of IQ was reported by Marie Skodak and Harold Skeels (1949). A group of 100 illegitimate children adopted before 6 months of age were administered IQ tests on at least four different occasions at about 2, 4, 7, and 13 years of age. As indicated in Table 13.1, the mean IQ of the adopted children was found to be above average at all ages, even when they were only about 2 years old. In contrast, the mean IQ of the 63 birth mothers who were tested was 86 (with a normal standard deviation of 15.8). However, the lower than average IQ of the birth mothers may be partially attributable to the fact that the test was administered shortly after the birth of the baby, and usually after the mother had decided to relinquish the child for adoption, conditions unconducive to optimal performance. Although studies of average differences between adopted children and their birth parents often suggest possible environmental influences (e.g., Schiff et al., 1978; Willerman, 1979), such studies are particularly prone to problems such as the one just mentioned, as well as to others such as assortative mating (Munsinger, 1975).

Both the birth mothers and the adopted children were similar in variance to the general population. However, there was some selective placement. The correlation between education level of the birth mothers and the adoptive midparent level was 0.30. Correlations between the IQs of the adopted children at various ages and those of their birth mothers are also reported in Table 13.1. The correlation is significant by 4 years of age and increases somewhat thereafter, reaching a level of 0.44 for the 13-year-old test scores on the 1937 revision of the Stanford-Binet. A similar pattern of correlation was observed between the children's IQ and the educational level of the birth mothers. It is interesting to note that the same pattern has also been observed between children's IQ and the educational level of the birth fathers (Honzik, 1957), which suggests that the role of prenatal maternal influences is minimal.

Marjorie Honzik (1957) compared these developmental data to those obtained from her study of parents rearing their own children, as illustrated in Figure 13.4. Clearly, the similarity between the birth mothers and their adopted-away children is much like the similarity between natural mothers and their own children whom they reared. Similar results have been obtained in comparisons between birth fathers and natural fathers.

Thus, the genetic side of the Skodak and Skeels study strongly suggests hereditary influence on IQ. What about the environmental side of the full adoption design? Figure 13.4 also addresses that issue by showing the correlations between the adoptive mothers' education and the IQ of their adopted children as a function of age. The fact that the correlations hover around the 0.05 level suggests little between-family environmental influence of mothers' educational levels on the children's IQs.

There are four other full adoption studies. A small study (Beckwith, 1971) investigated the relationship between development scores of adopted

TABLE 13.1

Mean IQs of adopted children and correlations between IQs of the children and their birth mothers

Test	Mean Age of Adopted Children		Mean	Standard Deviation	Correlation with Birth Mothers
	Years	Months			
Kuhlman revision of Binet	2	2	117	13.6	0.00
1916 Stanford-Binet	4	3	112	13.8	0.28
1916 Stanford-Binet	7	0	115	13.2	0.35
1916 Stanford-Binet	13	6	107	14.4	0.38
1937 revision of Stanford-Binet	13	6	117	15.5	0.44

SOURCE: After Skodak and Skeels, 1949.

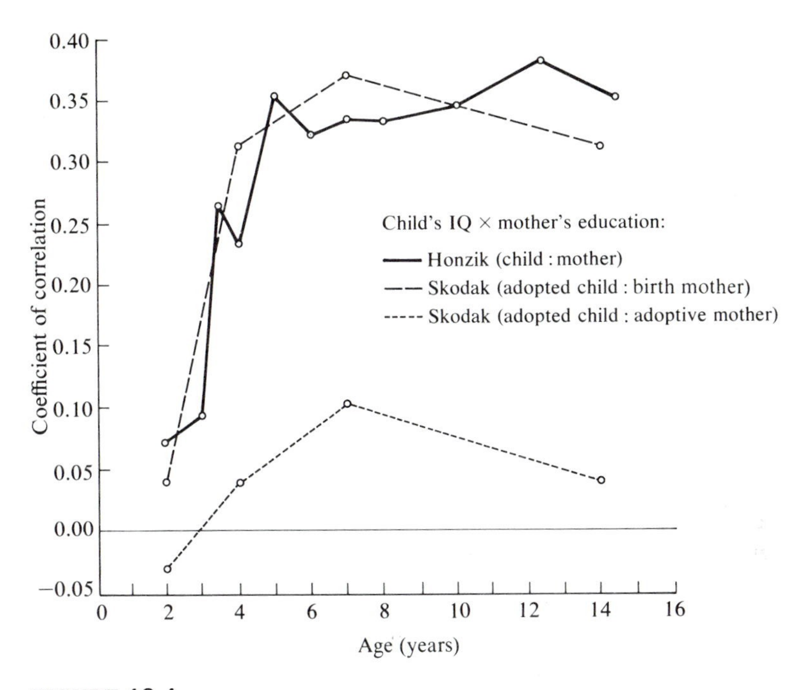

FIGURE 13.4

Coefficients of correlation at different ages between the child's IQ and the educational level of the birth mother or of the adoptive mother. (After "Developmental studies of parent-child resemblance in intelligence" by M. P. Honzik. *Child Development,* 28, 215–228. Copyright © 1957 by the Society for Research in Child Development, Inc.)

infants and socioeconomic status of adoptive and birth mothers. There was no evidence of selective placement, and significant relationships were found on the genetic side, but not the environmental side of the full adoption design. Similar results were obtained by Harry Munsinger (1975) in a study of 41 older (8-year-old) adopted children. Sandra Scarr and Richard Weinberg (1977) have reported an adoption study in which most of the children had one black birth parent, and all were adopted into Caucasian homes. The correlations between their IQ scores and those of their adoptive parents was 0.23 for mothers and 0.15 for fathers. The genetic side of this study compared the birth parents' educational levels with the children's IQs. The correlation was 0.32 for birth mothers and 0.52 for birth fathers. Thus, even though the birth parents' mental ability was estimated crudely by educational level, it correlated more highly with the adopted children's IQ than did the IQ of the adoptive parents. Some selective placement occurred in this study, which means that both the environmental and the genetic estimates are probably a bit on the high side.

The only full adoption study to obtain IQ scores for both birth and adoptive parents is in progress at the University of Texas. IQ tests are being administered to adoptive parents and adopted children whose birth mothers had been tested when they were in residence at a Texas adoption agency. Some slight selective placement occurred, so that both genetic and environmental estimates are likely to be slightly inflated. Preliminary results of this study are summarized in Table 13.2, which shows that the IQ correlation between adopted children and their birth mothers exceeds that between the children and either adoptive parent.

Partial Adoption Studies

There are two types of partial adoption designs. The first type concentrates on either the genetic or the environmental side of the adoption design. The second type compares adoption data on either genetic or environmental influences to data on natural families in which both genes and environment are shared.

One example of the first type of partial adoption design is a Canadian study (Snygg, 1938) that compared the IQ scores of young adopted children to those of their birth mothers. Although this was a large study, the results have been biased by the fact that all 312 birth mothers who were administered an IQ test had not passed the high school entrance examination. The average IQ of the birth mothers was only 78. This restriction of range may account for the low overall IQ correlation of 0.13 between the adopted children and their birth mothers. An even more likely hypothesis results from the fact that preschool tests of mental ability are not highly predictive of an individual's own IQ later in life (e.g., Wilson, 1978). Because the children in Snygg's study were preschoolers, we would not expect to find high correlations between the IQs of the adopted children and their birth mothers. As seen in the study of Skodak and Skeels (1949), significant correlations between birth mothers and their adopted-away children did not occur when the children were younger than 4 years of age. Another study of this type (Casler, 1976) focused on the cognitive abilities of adopted children in the

TABLE 13.2

IQ correlations between adopted children and their adoptive parents and birth mothers from the Texas Adoption Project

	Correlation with Adopted Children's IQs	Number
Adoptive mothers' IQ	0.17	459
Adoptive fathers' IQ	0.14	462
Birth mothers' IQ	0.31	345

SOURCE: After J. Horn, J. C. Loehlin, and L. Willerman, 1979.

first two years of life, and related these scores to IQ scores of their birth mothers. Few significant correlations were found.

Partial adoption designs have also been used to study environmental influences on behavior. For example, W. Claeys (1973) found few significant correlations between primary mental abilities of adopted Flemish children (5 to 7 years old) and the social status of their adoptive parents.

The second major type of partial adoption study compares data from one side of the adoption design to data on natural families (whom we shall call control families), who share both genetic and environmental influences. One study (Lawrence, 1931) reported correlations between the IQ of adopted children and the occupational status of their birth parents, and compared these to similar correlations for control families. The correlations between the adopted children and their birth parents were comparable to those between children and parents in control families, even though the birth parents shared no familial environment with their adopted-away children. The more common type of partial adoption design compares adoptive to control families. In the absence of selective placement (which, unfortunately, is difficult to assess with this design), the adoptive family relationships include only between-family environmental variance, while the control family relationships include variance attributable to both genetic and between-family environmental influences. Comparing these two types of families tests whether the genetic influences in control families lead to greater phenotypic similarity than that observed in the adoptive families.

The classic example of this type of partial adoption design was reported by Barbara Burks in 1928. Nearly two hundred children adopted before they were 1 year old were tested on the Stanford-Binet IQ test when they were between 5 and 15 years old. A control group of about one hundred families was matched to the adoptive families. The control children and the adoptive and control parents were also tested on the Stanford-Binet. Correlations for the adoptive and control families are listed in Table 13.3. The correlations for the adoptive families are clearly lower than the control family correlations,

TABLE 13.3
Parent-child IQ correlations from Burks' adoption study

	Adoptive Family		Control Family	
	r	N	r	N
Father-child	0.07	178	0.45	100
Mother-child	0.19	204	0.46	105

SOURCE: After Burks, 1928.

suggesting that genetic influences are important. However, the adoptive correlations, particularly that between adoptive mothers and children (0.19), provide some indication of between-family environmental influences.

Another study of this type produced conflicting results (Freeman, Holzinger, and Mitchell, 1928). The correlation between the IQs of the adoptive parents and their adopted children at 12 years of age was 0.32, which is much higher than the correlation of 0.19 found in Burks' study. However, substantial selective placement occurred. The children were separated from their biological parents at about 6 years of age on the average, and the correlation between adopted children's IQ at the time of placement, and a rating of the quality of adoptive home, was 0.34. Thus, selective placement could account for much of the observed resemblance between the adoptive parents and their adopted children.

The disparity between correlations for adoptive families in these two studies prompted a third study (Leahy, 1935). A major flaw of the F. N. Freeman study was the late and selective placement of the adopted children. Like Burks, A. M. Leahy studied children adopted early (prior to 6 months of age) and tested them between 5 and 14 years of age. Control families were matched to the adoptive families. As shown in Table 13.4, this study confirmed Burks' results. The control family correlations are about the same as we have become accustomed to finding. As in Burks' study, the correlations for the adoptive families are clearly lower, and also suggest some between-family environmental influence.

Although Leahy's study clearly confirms the findings of Burks, and not those of Freeman and his associates, we shall mention two recent studies of this variety that add weight to the evidence for genetic influence. As part of a larger study in Minnesota, 94 children were found who had been adopted early in life (Fisch et al., 1976). A control group of 50 children was also studied. Correlations between the IQs of the adoptive and control mothers and their children at 4 and 7 years of age are presented in Table 13.5. The correlations between the control mothers and their children are clearly larger than those between the adoptive mothers and their adopted children, which again points to the influence of heredity on IQ.

TABLE 13.4

Parent-child IQ correlation from Leahy's adoption study

	Adoptive Family		Control Family	
	r	N	r	N
Father-child	0.15	178	0.51	175
Mother-child	0.20	186	0.51	191

SOURCE: After Leahy, 1935.

TABLE 13.5

Mother-child IQ correlations obtained in the Minnesota collaborative study

Age of Children (years)	Adoptive Family (N = 94)	Control Family (N = 50)
4	0.07	0.35
7	0.08	0.26

SOURCE: After Fisch et al., 1976.

A second recent investigation with a partial adoption design (Scarr and Weinberg, 1978) studied adoptive and control families in which the children were at least 16 years old. As part of a three-hour battery of tests, four scales of the Wechsler Adult Intelligence Scale (WAIS) were administered to adoptive and control parents and their children. The total score on these four subtests correlates above 0.90 with the full-scale WAIS score. All groups were above average in IQ, but the fact that this was true for both the adoptive and the control group permits comparisons between the two. Table 13.6 presents the results, which again support the conclusion that genetic factors have an important effect on IQ.

Summary of Parent-Offspring Data

Because such a brief presentation of so many complex studies is bound to be overwhelming on first exposure, we shall now summarize and interpret the results. In Table 13.7, we have summarized IQ correlations from the genetic side of the adoption design. For purposes of simplicity, we have not included those studies in which mental ability of the birth mothers was estimated indirectly in terms of educational level or socioeconomic status— even though the results of those studies would not have altered our conclusions. We could discuss the average of the correlations found in the four studies, but it should be noted that two of them tested very young adopted-

TABLE 13.6

IQ correlations for adoptive and control parents and their adolescent children

	Adoptive Family		Control Family	
	r	N	r	N
Father-child	0.16	175	0.40	270
Mother-child	0.09	184	0.41	270

SOURCE: Scarr and Weinberg, 1978.

TABLE 13.7

IQ correlations between birth mothers and their adopted-away children

Study	N	Correlation	Problems
Snygg (1938)	312	0.13	Children only 1–5 years; restriction of range of sample
Skodak and Skeels (1949)	63	0.44	Selective placement
Casler (1976)	150	0.13	Children only 27 months old
Texas Adoption Project (Horn et al., 1979)	345	0.31	Slight selective placement

NOTE: Includes only those adoption studies with IQ data for both birth mothers and their children. In longitudinal studies, the listed correlation is based on the latest scores of the children.

SOURCE: After Munsinger, 1975; DeFries and Plomin, 1978.

away children. Because young children's IQs do not correlate well with their own IQ scores later in life, it is unlikely that they will correlate with their birth mothers' IQs. For this reason, the correlations of 0.44 from Skodak and Skeels and 0.31 from the Texas Adoption Project are probably more valid. We shall use the correlation of 0.31 as a conservative estimate of the genetic relationship between birth mothers and their adopted-away children because the Texas study is much larger than the study of Skodak and Skeels and there was less selective placement.

In the earlier chapter on family studies, the best estimate of the single parent-offspring correlation for general cognitive ability was 0.35. Family studies, however, cannot specify the extent to which this familial resemblance is genetic or environmental in origin. The genetic side of adoption studies indicates that it is primarily genetic. We noted earlier that single parent–offspring correlations estimate half of the narrow-sense heritability, although we know that this estimate may be inflated by assortative mating between birth parents. Nonetheless, even allowing for this inflation, the estimate of narrow-sense heritability (0.62) is close to the upper limit estimate of 0.60–0.70 from the family studies. In other words, it suggests that about half of the observed variation in IQ is due to genetic differences.

The opposite side of this coin is that half of the phenotypic variation in IQ is due to environmental differences. The adoption studies suggest that some of this environmental influence is common to parents and their children (between-family environmental influences), but that most operates to make family members different from one another. Table 13.8 summarizes IQ correlations between adoptive parents and their adopted children. The study by Freeman et al. (1928) is inconsistent with the rest, for reasons described earlier (selective and relatively late placement of the adopted children). Excluding the results of that study, the weighted average correlation is 0.14. In other words, about 14 percent of the variance in IQ scores is due to environmental influences that parents and their children have in common, suggesting that the remaining 36 percent may be attributed to within-family environmental influences. The adoption studies also point to a slight, but

TABLE 13.8
IQ correlations between adoptive parents and their adopted children

Study	Adoptive Correlation				Problems
	Mother	N	Father	N	
Burks (1928)	0.19	204	0.07	178	
Freeman et al. (1928)	0.37	255	0.28	180	Selective placement
Leahy (1935)	0.20	186	0.15	178	
Fisch et al. (1976)	0.08	94	—	—	
Scarr and Weinberg (1977)	0.23	109	0.15	111	Selective placement
Texas Adoption Project (Horn et al., 1979)	0.17	459	0.14	462	Slight selective placement
Scarr and Weinberg (1978)	0.09	184	0.16	175	

NOTE: Includes only those adoption studies with IQ data for both adoptive parents.
SOURCE: After Munsinger, 1975; DeFries and Plomin, 1978.

consistent, postnatal maternal effect, in that the correlations for adoptive mothers tend to be higher than the correlations for the adoptive fathers. The only exception is the study by Scarr and Weinberg (1978), which is also the only study to focus on adolescent children.

Although we do not wish to pit heredity against environment, we need to do so briefly in order to make a point about the origin of parent-offspring resemblance for general cognitive abilities. Nine of the ten adoption studies (Freeman's is the exception) that permit a strong test of genetic and environmental influences indicate that heritable influences are greater than between-family environmental influences. This trend is statistically significant, and, as noted earlier, it is found even when selective placement occurs.

Sibling Studies

The adoption study evidence that within-family environment is more important than between-family environment might appear to conflict with our conclusion from the twin studies. Twin studies suggest that between-family environmental influences are more important. However, siblings may well share more environmental influences than parents and their children, and twins probably have a greater share of environmental influences in common than non-twin siblings.

There are some pertinent sibling data from adoption studies. Essentially, the genetic side of the design studies full siblings reared in uncorrelated environments. The environmental side considers genetically unrelated children adopted into the same family. Rather than going into these studies in great detail, we shall summarize them. Table 13.9 presents the results of

TABLE 13.9

IQ correlations between full siblings reared separately

Study	Number of Pairs	Correlation
Hildreth (1925)	78	0.23
Freeman et al. (1928)	125	0.25
Burt (1955)	131	0.46

three studies of full siblings reared in different homes. Because of several problems with Burt's data (see Jensen, 1974, 1978; Hearnshaw, 1979), the most conservative course is to exclude his correlation of 0.46. The other two correlations are quite similar—about 0.24. Because siblings share about half of their segregating genes, we can double this correlation to estimate heritability. The estimate of about 50 percent is in line with the twin study estimates of genetic influence on IQ.

Table 13.10 summarizes correlations between genetically unrelated children reared in the same family. The average correlation, weighted by the size of each sample, is 0.26. In other words, about 26 percent of the IQ variance is caused by environmental influences common to siblings. Thus, for young siblings, this finding agrees with results of twin studies, especially the longitudinal Louisville twin study, in suggesting that between-family environmental influences are quite important.

However, how do these data fit with the results of the Hawaii family study of cognition, which indicated a sibling correlation of 0.31? The genetic side of the adoption design (siblings reared separately) suggests that most of the sibling similarity found in the Hawaii study is due to genetic factors, which does not leave much room for shared environmental influences. The answer may be that the siblings in the Hawaii study are over 14 years of age, whereas almost all adoption studies of unrelated siblings reared together have involved younger children. The notable exception is Scarr and Wein-

TABLE 13.10

IQ correlations between genetically unrelated children reared together

Study	Number of Pairs	Correlation
Freeman et al. (1928)	112	0.36
Leahy (1935)	35	0.08
Scarr and Weinberg (1977)	187	0.33
Texas Adoption Project (1979)	236	0.26
Scarr and Weinberg (1978)	84	−0.03

berg's (1978) study of postadolescent children, which found no IQ correlation at all (−0.03) between unrelated siblings reared in the same homes. Thus, shared environmental influences may contribute more to the similarity of younger than older siblings (as in the Hawaii family study of cognition).

Identical Twins Reared Separately

Much has been written about the relatively few cases of identical twins reared apart, perhaps because this particular adoption design is so easy to understand. If pairs of genetically identical individuals are reared in uncorrelated environments, their correlation directly estimates broad-sense heritability. However, use of this design is severely hampered by the rarity of such twins. The four major studies reported in the literature (Burt, 1966; Juel-Nielsen, 1965; Newman, Freeman, and Holzinger, 1937; Shields, 1962) were reviewed in detail by Jensen (1970). Since no significant differences were found among the samples in the four studies, the IQ data were pooled and reanalyzed. This analysis yielded an intraclass correlation of 0.82 for the 122 twins in the pooled sample.

Cyril Burt reported a correlation of 0.84 for 53 pairs of separated identical twins. However, as mentioned previously, there are several problems with his data. The problems with Burt's data create some doubt that they should be included in the pooled data on identical twins reared separately. The average correlation for the other three studies of 69 pairs was 0.74. If some selective placement occurred in these studies, this estimate of a broad-sense heritability of 0.74 would be somewhat inflated. Correcting this estimate downward would make it fit better with the estimates from the other twin studies and the adoption studies.

SPECIFIC COGNITIVE ABILITIES

Since the advent of adoption studies, the spotlight has clearly been on general cognitive ability. Recently, however, as part of a general shift in behavioral genetics away from the investigation of single traits, adoption studies have begun to consider specific abilities. In 1973, a partial adoption study tested 84 adopted Flemish children at the average age of 6 years on the Primary Mental Abilities test (Claeys, 1973). When the relationship between their scores and the social class of their adoptive parents was analyzed, only 2 of 36 reported correlations were significant. No strong conclusions can be drawn, however, because corresponding data for birth parents and control families were not presented.

TABLE 13.11

Correlations for four WAIS subscales for adoptive and control families

WAIS Subscale	Adoptive Family Correlation			Control Family Correlation		
	Father-Child	Mother-Child	Child-Child	Father-Child	Mother-Child	Child-Child
Arithmetic	0.07	−0.03	−0.03	0.30	0.24	0.24
Vocabulary	0.24	0.23	0.11	0.39	0.33	0.22
Block design	0.02	0.13	0.09	0.32	0.29	0.25
Picture arrangement	−0.04	−0.01	0.04	0.06	0.19	0.16
N	175	184	84	270	270	168

SOURCE: After Scarr and Weinberg, 1978.

As mentioned earlier, another partial adoption study (Scarr and Weinberg, 1978) administered four subscales of the WAIS to adoptive and control families in which children were at least 16 years old. Although scores on the subscales of the WAIS are less reliable than the full-scale WAIS score, the results (Table 13.11) are interesting. For all four subtests, the control correlations are higher than the adoptive correlations, suggesting genetic influence. For vocabulary, significant adoptive family correlations suggest some between-family environmental influence. The results of this study provide a glimpse of the potential usefulness of adoption studies in disentangling genetic and environmental influences on behaviors other than general cognitive ability. Data such as these could permit powerful multivariate quantitative genetic analyses of the cross-correlations among specific cognitive abilities (see Chapters 9–12), but no work along these lines has yet been reported.

PSYCHOPATHOLOGY

During the past decade, there has been an impressive amount of behavioral genetics research on psychopathology. Several books on the genetics of psychopathology (e.g., Fieve, Rosenthal, and Brill, 1975) have been published, and there have been numerous reviews of specific areas of psychopathology, especially schizophrenia (Erlenmeyer-Kimling, 1976; Gottesman and Shields, 1976; Shields, 1977; Tsuang, 1976; Zerbin-Rudin, 1974). Adoption studies have made a major contribution to this burst of research activity. There were no adoption studies of schizophrenia before 1966; since then there have been several that will be summarized in this section.

Schizophrenia

Because of the relative rarity of schizophrenia and the qualitative nature of its diagnosis, adoption studies have not been able to take advantage of quantitative genetics methods that are used to analyze normal variation in a population. Instead, they compare the frequency of schizophrenia in individuals who are genetically related to schizophrenics, but reared in unaffected adoptive families, with the frequency in adopted individuals who have no known cases of schizophrenia among their biological relatives. In order to facilitate our discussion of these studies, we shall use the 2 × 2 design illustrated in Table 13.12 as a paradigm for comparing the effects of genetic relatedness and rearing environment. Most of the studies have used partial adoption designs, in that they test the influence of genetic relatedness by making comparisons between children with normal biological relatives reared by normal adoptive parents (Bn/An) and children with affected biological relatives reared by normal adoptive parents (Ba/An). A few studies have tested the influence of the rearing environment (Bn/An versus Bn/Aa). No study of schizophrenia has included the Ba/Aa cell—individuals with schizophrenic biological relatives reared with schizophrenic adoptive relatives. This is not surprising because of the obvious difficulty of obtaining such a sample.

Within this paradigm, two major strategies have been employed. The most common strategy is called the *adoptees' study method,* which is illustrated in Figure 13.5. In the index families, the affected person is a biological relative (usually a parent) of the adoptee. The adoptees are reared by unaffected adoptive parents. The measure obtained by the study is the incidence of schizophrenia in the adoptees. The control group consists of adoptees whose birth parents and adoptive parents have no known psychopathology. A theory of genetic influence on schizophrenia would predict a greater incidence in the adoptees who have affected birth parents. Environmental theory, on the other hand, would predict that there would be no difference in incidence of psychopathology in the two groups of adoptees.

The other strategy is called the *adoptees' family method.* (See Figure 13.6.) In this case, we begin with adoptees who are affected (probands) and adoptees who are unaffected, and then determine the incidence of schizo-

TABLE 13.12
Paradigm for adoption studies of psychopathology

Biological Relatives	Adoptive Relatives	
	Normal (An)	Affected (Aa)
Normal (Bn)	Bn/An	Bn/Aa
Affected (Ba)	Ba/An	Ba/Aa

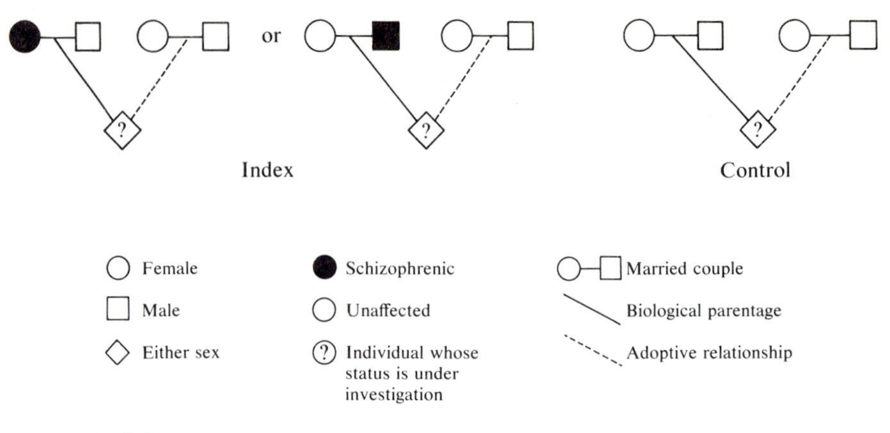

FIGURE 13.5
Research design of adoptees' study method.

phrenia in their biological and adoptive relatives. A theory of genetic influence would predict that more biological relatives of the schizophrenic adoptees would be affected, as compared with biological relatives of the control (unaffected) adoptees. If environmental influences are important, the incidence in adoptive relatives of the schizophrenic adoptees should be greater than that in adoptive relatives of the control adoptees.

The first adoption study of schizophrenia (Heston, 1966) used the adoptees' study method. The study identified hospitalized chronic schizophrenic women, who had been hospitalized while pregnant and whose children had been placed in foundling homes or foster homes during the first two weeks of life. The children of 47 such women were interviewed at the average age of 36, and were compared to 50 adoptees whose birth parents had no known psychopathology. The well-known results, summarized in Table 13.13, indicate significant genetic influence. All five of the affected adoptees had been reared by normal adoptive parents (the Ba/An condition). All these individuals had been hospitalized, and three were chronic schizophrenics hospitalized for several years. Four other Ba/An individuals were regarded as schizophrenic or borderline schizophrenic by one or two of the three psychiatric raters, so that a broader definition of schizophrenia would indicate

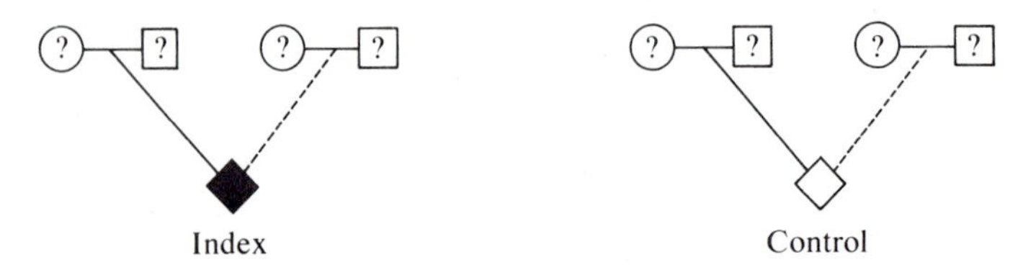

FIGURE 13.6
Research design of adoptees' family method. See Figure 13.5 for key to symbols.

TABLE 13.13

Chronic schizophrenia in adopted offspring of schizophrenic and nonschizophrenic birth mothers

Birth Mothers	Foundling and Foster Homes	
	Normal	Affected
Normal	0% (0/50)	—
Affected	11% (5/47)	—

SOURCE: After Heston, 1966.

occurrence of the disorder in 20 percent (9/47) of the offspring of schizophrenic mothers. Because half of the Ba/An individuals demonstrated some degree of psychosocial disability, Heston suggested that the definition of schizophrenia should be broadened to include schizoid dimensions (an inclusive definition, which we referred to earlier as "soft schizophrenic spectrum").

The major findings of Leonard Heston's study—that schizophrenia is heritable, and that the heritable complex may include schizoid dimensions, as well as borderline schizophrenias—have been confirmed by David Rosenthal and his associates in their adoption studies in Denmark. These investigators used the *Folkeregister* of Copenhagen to find approximately 5,500 individuals who had been adopted between 1924 and 1947. Rosenthal and co-workers (1968, 1971, 1972) employed the same design as that used by Heston, but with the added experimental control provided by systematic assessment of adoptees, the use of blind psychiatric interviews, and the fact that schizophrenic mothers were not in a mental hospital while pregnant. After the birth parents of the adoptees were identified through the *Folkeregister,* their names were traced in a psychiatric register, hospital records were obtained, and a consensus diagnosis was used to select schizophrenic mothers or fathers whose children had been placed in adoptive homes. Forty-four birth parents (32 mothers and 12 fathers) who were diagnosed as certain or uncertain chronic schizophrenics were in this group, and their adopted-away children (Ba/An) were matched to controls whose birth parents had no psychiatric history (Bn/An). The adoptees, 33 years old on the average and thus still at substantial risk, were interviewed for three to five hours by an interviewer blind to the status of their birth parents.

Table 13.14 presents the incidence of schizophrenia (chronic), "hard schizophrenic spectrum," and "soft schizophrenic spectrum" for the two groups of adoptees. For all three classifications, the results suggest substantial genetic influence. There is a lower incidence of schizophrenia in the adoptees than in Heston's study, but it is likely that the birth parents in this study were less severely affected.

The unusually high rates of psychopathology in the control data (Bn/An) shown in Table 13.14 is consistent in the early Danish studies because

TABLE 13.14

Schizophrenia and hard and soft schizophrenic spectrum in adopted offspring of schizophrenic and nonschizophrenic birth parents

Birth Parents	Adoptive Parents	
	Normal	Affected
Normal	0% (0/67)*	—
	4% (3/67)†	—
	18% (12/67)‡	—
Affected	7% (3/44)*	—
	14% (6/44)†	—
	27% (12/44)‡	—

* Chronic schizophrenia.

† Chronic, acute, borderline, and uncertain schizophrenia (hard spectrum).

‡ Hard spectrum plus paranoid and schizoid personality (soft spectrum).

SOURCE: Rosenthal et al., 1968.

these studies relied on hospital records to assess psychiatric status and may have overlooked significant psychopathology in the control birth parents. For example, psychiatric interviews in a later study (Wender et al., 1974) revealed 36 persons who were schizophrenic, but who had not been hospitalized. The birth parents of controls have been interviewed, and it appears that one-third of them fall in the schizophrenic spectrum. Thus, with respect to the early studies, they conclude that "our controls are a poor control group and ... our technique of selection has minimized the differences between the control and index groups" (Wender et al., 1974, p. 127). This bias is conservative in terms of demonstrating genetic influence.

In the early Rosenthal study described above, adoptees were included in the Ba/An group if they had only one schizophrenic parent, and data were obtained only on that parent. In a follow-up study (Rosenthal, 1975), the other birth parent of each adoptee was interviewed. The results of interviewing these "co-parents" suggest considerable assortative mating in terms of soft schizophrenic spectrum. The effects of assortative mating have been discussed previously. In the present context, this co-parents study provides additional support for a genetic hypothesis: The adopted-away offspring of chronic schizophrenic birth mothers or fathers were significantly more often diagnosed as in the schizophrenic spectrum when the co-parent also exhibited some type of schizophrenic disorder.

Other studies in Denmark (Kety et al., 1968, 1971, 1976; Kety et al., 1975) have used the adoptees' family method rather than the adoptees' study method used in the studies described above. In other words, the investigators first identified schizophrenic adoptees, and then assessed the psychiatric status of their biological and adoptive relatives. Thus, they focused on

TABLE 13.15

Design of adoption studies reported by Kety and associates

	Nonschizophrenic Adoptees	Schizophrenic Adoptees
Biological relatives	Control group for genetic test	Test of genetic influence
Adoptive relatives	Control group for environmental test	Test of environmental (rearing) influence

the extent of psychopathology among the relatives of schizophrenic and nonschizophrenic adoptees. The design of these studies is illustrated in Table 13.15.

The pool of adoptees in the studies reported by Rosenthal and associates included 507 individuals who had been admitted to a psychiatric facility. Of these, 33 were diagnosed as falling in the hard schizophrenic spectrum (16 were chronic schizophrenics), and these index adoptees (probands) were matched to controls having no history of psychiatric problems. The *Folkeregister* was searched for the names of parents, siblings, and half-siblings from the biological and adoptive families of the index and control adoptees. Nearly all of the birth parents (N = 126) and adoptive parents (N = 129) were identified. Although the search revealed few full siblings of the index or control adoptees (N = 32), a large number of biological half-siblings (N = 176) was found.

The results of an early study relying on hospital diagnoses were mixed (Kety et al., 1968, 1971), but recent results based on extensive psychiatric interviews (Kety et al., 1975, 1976) are less ambiguous. Approximately 90 percent of the relatives were interviewed, and psychiatric diagnoses were obtained by consensus. Table 13.16 shows the frequency of hard schizophrenic spectrum in first-degree biological relatives, biological half-siblings, and adoptive relatives of schizophrenic and nonschizophrenic adoptees.

TABLE 13.16

Hard schizophrenic spectrum (based on psychiatric interviews) in biological and adoptive relatives of schizophrenic and nonschizophrenic adoptees

	Nonschizophrenic Adoptees	Schizophrenic Adoptees
First-degree biological relatives	4% (3/68)	12% (8/68)
Adoptive parents and adoptive siblings	4% (4/90)	3% (2/73)
Biological half-siblings (total sample)	3% (3/104)	16% (16/101)
Biological half-siblings (paternal only)	3% (2/64)*	18% (11/61)

* Both psychiatric interviews and hospital diagnoses.

SOURCE: Kety et al., 1976.

The data for first-degree biological relatives support a genetic hypothesis, as do the results for the biological half-siblings.

The comparison of biological half-siblings who have the same father (paternal half-siblings) with those who have the same mother (maternal half-siblings) is particularly useful for examining the possibility that results of adoption studies may be affected by prenatal or early maternal care, rather than by genetic transmission. Data on paternal half-siblings are not influenced by these environmental variables. Using results of both hospital diagnoses and psychiatric interviews, the frequency of hard schizophrenic spectrum in paternal half-siblings of schizophrenic and nonschizophrenic adoptees was determined (Table 13.16). These important data again confirm a genetic hypothesis, suggesting that prenatal maternal environmental variables do not play an important role in the results of these adoption studies.

Jon L. Karlsson (1966, 1970) conducted several small adoption studies in Iceland. Different designs were used, but three studies yield 52 first-degree relatives separated from schizophrenic probands. Of these relatives, 12 (23 percent) were schizophrenic, providing strong support for a genetic hypothesis. This percentage is considerably higher than the usual familial incidence in first-degree relatives, possibly due to greater severity of schizophrenia in the probands in these studies.

Investigators in Denmark have also tested the effect of a schizophrenic rearing environment by studying adoptees whose birth parents were not affected, but whose adoptive parents were schizophrenic. Using the same pool of adoptees as in the studies reported by Rosenthal and Seymour Kety, hospital records were searched for those whose adoptive parents had been diagnosed as schizophrenic (Wender et al., 1974). A total of 28 such cases were found. Of these schizophrenic adoptive parents, 9 were chronic, 9 were acute, 4 were borderline, and 6 were diagnosed as falling in the soft schizophrenic spectrum. The adoptees in this group (the Bn/Aa cell in Table 13.12) were compared to those in the two cells of the design that were discussed earlier (Bn/An and Ba/An). Diagnosis of the adoptees was by means of a 3- to 5-hour psychiatric interview and 1½ days of psychological testing (providing a rich data source that will continue to be mined in the future). The essential aspects of the cases were typed on cards and sorted by four raters into twenty categories of severity. Adoptees with scores higher than 15 were considered to be in the hard schizophrenic spectrum.

The 24 adoptees classified in the hard schizophrenic spectrum were found to be distributed in the three groups, as indicated in Table 13.17. These findings suggest that being reared by a parent in the schizophrenic spectrum is not sufficient to produce schizophrenia. The influence of genetic factors is again supported by the comparison between adoptees with no schizophrenic birth parents (Bn/An and Bn/Aa) and those with a birth parent in the schizophrenic spectrum (Ba/An).

These results tend to confirm the findings of an earlier study by the same investigators (Wender, Rosenthal, and Kety, 1968) that was conducted

TABLE 13.17

Hard schizophrenic spectrum as a function of rearing and schizophrenic status of birth parents

Birth Parents	Adoptive Parents	
	Normal	Affected
Normal	10% (8/79)	11% (3/28)
Affected	19% (13/69)	—

SOURCE: Wender et al., 1974.

in the United States using a different design. Natural parents rearing their own schizophrenic offspring were compared to adoptive parents of schizophrenics, and adoptive parents of unaffected children. Significantly higher psychopathology was found among the natural parents of schizophrenics than among the adoptive parents of schizophrenics. This finding supports a genetic hypothesis. However, the adoptive parents of schizophrenics received a significantly higher psychopathology rating than adoptive parents of normal children, suggesting that either these parents contributed to schizophrenia in their children or they responded to the worry caused by having a chronically ill child.

Because this has been the only study to suggest a possible role for schizophrenic rearing, a similar design was used in a subsequent investigation (Wender et al., 1977). The study involved the adoptive status of all patients in fourteen New York metropolitan hospitals. Individuals who had been adopted before 1 year of age and who were currently 15 to 30 years old were diagnosed by consensus for disorders falling in the hard schizophrenic spectrum. The 33 parents of 19 such adoptees were compared to 33 natural parents who had reared schizophrenic children. The parents were diagnosed by psychiatric interviews for hard and soft schizophrenic spectrum, and the results are summarized in Table 13.18. Data for the soft schizophrenic spec-

TABLE 13.18

Schizophrenic spectrum in the birth and adoptive parents of schizophrenics

	Diagnosis by Psychiatric Consensus		Diagnosis by Computer Analysis	
	Hard Spectrum	Soft Spectrum	Hard Spectrum	Soft Spectrum
Birth parents who reared schizophrenics	18% (6/33)	45% (15/33)	18% (6/33)	36% (12/33)
Adoptive parents of schizophrenics	15% (5/33)	15% (5/33)	3% (1/33)	6% (2/33)

SOURCE: After Wender et al., 1977.

trum strongly support a genetic hypothesis and deny the influences of rearing, although results for the hard spectrum are not as clearcut. Where the diagnoses were based on computer analysis rather than psychiatric interviews, the results support a genetic hypothesis for both types of schizophrenic spectrum. They also suggest that rearing by a seriously deviant parent is not a sufficient condition for the development of schizophrenia.

In closing this section on adoption studies of schizophrenia, it is appropriate to mention another type of adoption design involving identical twins reared separately. Gottesman and Shields (1972) have reviewed studies including a total of 14 pairs of identical twins, separated by the age of 2 years, in which at least one co-twin was schizophrenic. Of the 14 pairs, 10 were concordant for schizophrenia, a rate that is even higher than the incidence in identical twins reared together. Although this approach is inherently limited by small sample sizes, its results corroborate other adoption data in suggesting a major role for genetic influences.

Summary of Adoption Studies on Schizophrenia

With respect to schizophrenia, results of the adoption studies clearly point to genetic influence. Including the study by Heston, the Danish studies, and the study by Karlsson (with the results given in Tables 13.13, 13.14, and 13.15, and in the summary of Karlsson's data), there are 211 first-degree biological relatives of schizophrenic adoptees and 185 control individuals. The incidence of schizophrenia among the biological relatives of schizophrenics was 13 percent (28/211), while that in the control group was 1.6 percent (3/185). This summary result is consistent with the findings of each individual study. Thus, an overall view of the adoption study data on schizophrenia clearly allows us to reject the hypothesis of no genetic influence.

The finding of schizophrenia in 13 percent of the biological relatives in these studies deserves more discussion. This is the incidence in biological relatives who share no environment with the schizophrenic individual. It is actually higher than the incidence in biological relatives of schizophrenics sharing the same family environment, as mentioned in our discussion of family studies in Chapter 11. This suggests that all of the familial resemblance for schizophrenia is due to heredity, and that none is due to between-family environmental influences, whose hypothesized role in the etiology of schizophrenia has seemed so reasonable. This conclusion is supported by the cross-fostering research that found no increase in the incidence of schizophrenia in adoptees reared by schizoid adoptive parents.

This does not mean that the environment is unimportant in triggering schizophrenia. However, it does indicate that the environmental culprit that we have traditionally blamed for schizophrenia (between-family influences) is actually blameless. As was true in the family and twin studies, the adoption studies suggest that the environment plays a very substantial role, but that it operates within families, making members of the same family distinct from one another. Unfortunately, we are a long way from tracking down these within-family environmental influences. From the point of view of prevention, we can only hope that Gottesman and Shields are wrong when they suggest that "the 'culprits' may be nonspecific, time-limited in their effectiveness, and idiosyncratic" (1976, p. 379).

Psychopathology Other Than Schizophrenia

One adoption study of psychosis other than schizophrenia has recently been reported. In Belgium, J. Mendlewicz and J. D. Rainer (1977) used the adoptees' family method to study bipolar manic-depressive psychosis. The adoptive parents and birth parents of 22 unaffected adoptees and 29 bipolar manic-depressive adoptees (probands) were included in the study. Table 13.19 shows the results for the affective spectrum, which includes bipolar, unipolar, schizoaffective, and cyclothymic disorders. The results are similar to those that we have just described for schizophrenia. For the adoptive parents, the incidence of affective disorder is similar regardless of whether their adopted child was affected. However, the incidence of affective disorder is much greater in the birth parents of probands (31 percent) than in the birth parents of unaffected adoptees (2 percent). Mendlewicz and Rainer also studied a group of 31 bipolar probands reared by their natural parents, and found the incidence of affective disorder in this group to be 26 percent. Thus, parents who rear manic-depressive children have no greater incidence of affective disorder than birth parents of manic-depressive children who are adopted away. This study also found that the bipolar-unipolar distinction involves considerable overlap. The incidence of bipolar disorder in the birth parents of the bipolar probands was 7 percent, whereas the incidence of unipolar depression in the birth parents was 21 percent. Similarly, among the natural parents of bipolar probands, 3 percent exhibited the bipolar disorder and 18 percent showed unipolar depression. However, the data do provide

TABLE 13.19

An adoption study of bipolar manic-depressive psychosis

	Unaffected Adoptees (N = 22)	Bipolar Manic-Depressive Adoptees (N = 29)
Birth parents in affective spectrum	2%	31%
Adoptive parents in affective spectrum	10%	12%

SOURCE: After Mendlewicz and Rainer, 1977.

strong evidence for heritable influences within the realm of affective disorders. Another adoption study examined adopted offspring of birth parents with heterogeneous psychopathology (Cunningham et al., 1975; Cadoret et al., 1975). Some evidence was found for specific inheritance of psychoses, affective disorders, antisocial personality, and mental retardation, although the sample size for each disorder was small.

Adoption studies have begun to focus on less severe psychopathology. For example, there have been a few investigations of criminal behavior and the related behavioral syndromes of psychopathy and antisocial personality in Denmark (Schulsinger, 1972; Hutchings and Mednick, 1975) and in the United States (Crowe, 1972, 1974). These studies included 321 first-degree biological relatives of adopted criminal or psychopathic probands and 316 controls (biological relatives of adoptees who have shown no criminality). Twenty-five percent (82/321) of the biological relatives of criminal probands either had criminal records or were diagnosed as psychopathic. In the control group, only 13 percent (41/316) of the biological relatives were similarly diagnosed. These studies thus provide significant evidence for the involvement of heredity in criminal behavior.

The possibility of genetic influence on criminal behavior is bound to cause an even more emotional response than the issue of genetic influence on IQ in those individuals who think that this means that crime is destined by DNA. We hope that the meaning of genetic influences is now clear—that there can be no allele for criminal behavior, just as there is no allele for nose length. Possible pleiotropic effects of genes on the very complex behaviors labeled as criminal could include effects on diverse characteristics, such as body build, abnormal EEG patterns, IQ differences, and psychopathology (Mednick and Christiansen, 1977).

Alcoholism is another socially important phenotype that has received the attention of adoption studies (Goodwin, 1979). Such studies (Goodwin et al., 1973, 1974; Schuckit, Goodwin, and Winokur, 1972) have included 77 offspring of alcoholic birth parents reared by nonalcoholic adoptive parents. The control group consisted of 182 individuals whose birth and rearing parents were nonalcoholic. The incidence of alcoholism in the offspring of the alcoholic birth parents was 22 percent (17/77); in the control groups, it was 4.4 percent (8/182). These data also permit us to reject the hypothesis of no genetic influence.

SUMMARY

Adoption studies provide the most convincing demonstration of the genetic influence on complex human behaviors. Issues of representativeness, selective placement, and assortative mating were discussed. Adoption studies, like the twin studies reviewed in Chapter 12, suggest that about half of the observed variation in IQ scores is due to genetic differences. Though the majority of the environmental variance occurs between families for young siblings, such between-family variance has somewhat less influence on the resemblance between parents and offspring and between older siblings. Specific cognitive abilities show about the same level of heritability (0.50) as general cognitive ability. As yet, there have been no multivariate analyses of adoption data.

For psychopathology, we emphasized the more serious psychoses, particularly schizophrenia and manic-depressive disorders. Overall, the data suggest that the familial resemblance long known to occur for psychosis is due to heredity and not to between-family environmental influences. The same data, however, provide strong evidence for the role of within-family environmental factors.

14

Directions in Behavioral Genetics

We shall now attempt a synthesis of some of the material presented in the preceding chapters. Now that the basics of behavioral genetics have been presented, we shall also discuss more fully some of the controversies alluded to in the first chapter.

RECAPITULATION AND SYNTHESIS

Genes and evolution are at the foundation of behavioral genetics. Genes have played a major role throughout the course of evolution, and genetic variability remains the basis of the evolutionary process. We discussed evolution from the perspective of genetic variability, and then introduced Mendelian genetics, which explains the way that genes are transmitted from one generation to the next. There are many examples of single-gene influences on behavior, and we reviewed a few of these. We then discussed the chemical basis for understanding how genes can affect behavior. Genes control the production of proteins which then interact in physiological systems, thus indirectly influencing behavior. The fact that the alleles of each gene are located at a certain position on a particular pair of chromosomes has implications for behavior. We then turned to the topics of population genetics and quantitative genetics. In the last four chapters, we considered the methods of quantitative genetics in some detail and illustrated their behavioral application.

Genetics of General Cognitive Ability

In this section, we shall summarize the data on general cognitive ability presented in the last three chapters. The relative influence of genetic and environmental variables on mental ability is one of the oldest continuously studied questions in the behavioral sciences, and it has been the source of many controversies surrounding behavioral genetics. During the last three years, larger samples of mental ability data have been reported than in the previous fifty years combined. In aggregate, these data suggest that individual differences in general cognitive ability are less heritable than previously believed. The older data are compatible with a high heritability, perhaps 0.70 or greater, whereas the newer data suggest a heritability closer to 0.50.

Table 14.1 lists IQ correlations for the new and old data (Plomin and DeFries, 1980a). The older data are the same as those illustrated in Figure 11.1. Although there are no new adoption data for identical twins or non-twin

TABLE 14.1
Correlation coefficients for old and new IQ data

	Old Data		New Data	
	Correlation	Number of Pairs	Correlation	Number of Pairs
Genetically identical:				
Same individual tested twice	—	—	0.87	456
Identical twins reared together	0.87	1,082	0.86	1,300
Identical twins reared apart	0.75	107	—	—
Genetically related (first-degree):				
Fraternal twins reared together:				
Same sex	0.53	2,052	0.62	864
Opposite sex	0.53	(total)	0.62	358
Non-twin siblings reared together	0.49	8,228	0.34	776
Non-twin siblings reared apart	0.40	125	—	—
Parent-child living together	0.50	371	0.35	3,973
Parent-child separated by adoption	0.45	63	0.31	345
Genetically unrelated:				
Unrelated children reared together	0.23	195	0.25	601
Adoptive parent-adopted child	0.20	not reported	0.15	1,594
Unrelated persons reared apart	−0.01	15,086	—	—

SOURCE: After Plomin and DeFries, 1980a.

siblings reared separately, cognitive data from the last three years include 4,749 pairings using the family design, 3,540 pairings using the adoption design, and 2,164 pairs of twins. As discussed in Chapter 11, data from the Hawaii family study of cognition indicate lower familiality for both parent-offspring and sibling relationships than do the older data. In Chapter 12, we discussed the recent large-scale NMSQT twin study, which yielded a correlation of 0.86 for 1,300 pairs of identical twins and a correlation of 0.62 for 864 pairs of fraternal twins, thus suggesting a lower heritability than the older data. New adoption data, discussed in Chapter 13, also suggest less genetic influence than the older data. Overall, these new data suggest that genetic differences explain about 50 percent of the IQ score variance.

There is likely to be no simple explanation for the differences between the new and old data. Possible explanations include environmental or genetic changes in the population during the past few decades, as well as differences in the nature of the tests employed, in sample size and representativeness, and in methods of test administration and statistical procedures, such as adjustment for age.

The new data, like the older data, provide evidence that environmental influences are important for IQ. However, contrary to the usual assumptions, the data suggest that the environmental factors that influence IQ correlations between parents and their offspring and between older siblings tend to operate within families (making family members different from one another) rather than between families (making family members similar to one another). We have few clues as to what these important within-family environmental influences might be.

Although we have scrupulously avoided the temptation to pit nature against nurture, it is clear that these data implicate genes as a major systematic influence on the development of individual differences in IQ, even though the newer data suggest a heritability of 0.50 rather than 0.70. In fact, we know of no such specific environmental influences that account for as much as 10 percent of the variance in IQ.

The new behavioral genetic data for mental ability are summarized in terms of average absolute differences in Figure 14.1. It is possible to convert correlations into average differences if the standard deviation is known, and if the variable is normally distributed. The relationship between correlations (r) and absolute average differences ($\bar{z}$) is: $\bar{z} = 1.13s\sqrt{1 - r}$. For pairs of randomly selected individuals for whom the IQ correlation is zero, the average IQ difference is 17, given a standard deviation (s) of 15 for IQ measures. In contrast, genetically unrelated individuals sharing family environments differ by an average of 15 IQ points. Genetically related first-degree relatives differ by 13 IQ points on the average, whereas genetically identical individuals (identical twins, as well as the same individuals tested on different occasions) differ by only about 6 IQ points. The influence of both genetic and environmental factors is clearly evident.

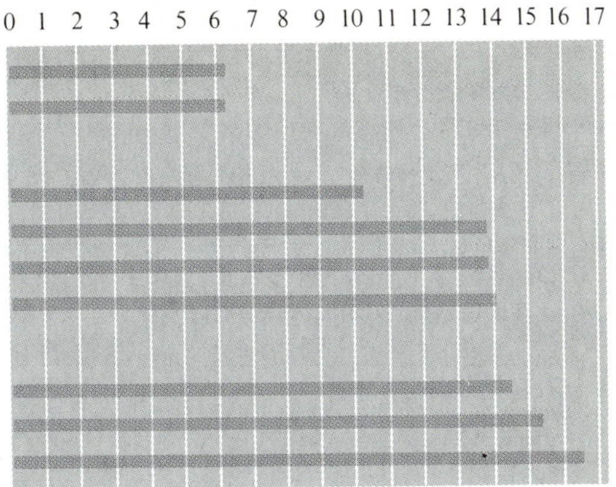

FIGURE 14.1

Average absolute differences in IQ points based on the new data in Table 14.1. (From "Genetics and intelligence: recent data" by R. Plomin and J. C. DeFries. *Intelligence,* 4, 22. Copyright © 1980. Reprinted by permission.)

Fitting Models to IQ Data

In our presentation of the IQ data in the last three chapters, our focus on one method at a time may have made the methods seem unrelated to one another. To the contrary, the most powerful quantitative genetic analyses result from the use of data from various sources and methods. R. B. Cattell (1953, 1960) has long argued for the need for this approach, which he called Multiple Abstract Variance Analysis (MAVA). Human behavioral geneticists have recently attempted to construct models of genetic and environmental influence of IQ data using various quantitative genetics methods (Eaves, 1975; Fulker, 1973; Goldberger, 1978; Jinks and Fulker, 1970; Loehlin, 1978, 1979; Rao and Morton, 1974, 1978; Rao, Morton, and Yee, 1974, 1976).

The major benefits of such models include making one's assumptions explicit and being able to compare competing models with one another. As we have seen, quantitative genetics methods involve assumptions about additive and nonadditive genetic variance, assortative mating, genotype-environment correlation and interaction, and shared environment. The basic idea of model fitting is to construct a reasonable model for genetic and environmental influences and to solve for the components of variance in the model (Goldberger and Duncan, 1973). The fit of the model to the observed familial correlations can be tested by comparing correlations predicted by the model to those that are observed. The details of model fitting are beyond the scope of this introductory text (see, for example, Eaves et al., 1978), and, indeed, the confusion that arose when such models were first applied to IQ data has not yet been completely resolved. Loehlin has attempted to sort out the conflicting results of these recent attempts to fit models to IQ data and made some cogent conclusions concerning model fitting:

> This analysis would seem to point two morals. One is that it is easy to make mistakes in specifying causal models of how heredity and environment affect traits.... The second moral, and I believe the more important one, is that conclusions depend on assumptions, and that therefore when making somewhat arbitrary assumptions (e.g., no selective placement, no dominance, no G × E covariance) in order to solve a complex model, the theorist is well advised not only to offer some justification for the choice he makes but also to offer the reader some information about what the consequences would be if he were to make it differently.... This should not be construed as implying that such model building is pointless. On the contrary, it is in my view extremely valuable. It is only when assumptions are embedded into explicit models that one can see what the consequences of these assumptions *are* (as opposed to what the theorist thinks they might be). (Loehlin, 1978, p. 430)

CONTROVERSIES

Now that you have become acquainted with some of the basic themes of behavioral genetics, its controversial aspects are perhaps more understandable. One issue that continues to be resurrected is the so-called nature-nurture problem. The reason for the repeated resurrection of this supposedly dead issue is a basic misunderstanding of the relationship between genes and environment. We shall try to clarify this issue further in the following discussion of interactionism, genotype-environment correlation, and genotype-environment interaction. A second controversial issue concerns the etiology of group differences in behavior. Racial-ethnic, class, and sex differences are all group differences that present sticky problems. The third and final issue involves relieving some of the frustration we feel when people impute political motives to behavioral genetics research. We have therefore chosen to conclude with a discussion of science and politics.

Nature–Nurture Arguments

As we noted in the first chapter, it is now generally agreed that both heredity and environment (nature *and* nurture) play a role in influencing behavior. However, the nature *or* nurture dichotomy has often been replaced by the equally mistaken view that the relative influences of heredity and environment cannot be analyzed. If we took this view seriously, it would mean that we could not study the effects of genetic factors on behavior because they are hopelessly enmeshed with the effects of environment. It would also mean that we could not isolate environmental effects because they are inseparable from genetic influences.

"Interactionism" and Genotype-Environment Interaction

A well-known discussion of this topic asserted that the "organism is a product of its genes and its past environment" (Anastasi, 1958, p. 197). The point was that there could be no behavior without both environment and genes, a view called "interactionism" (Plomin, DeFries, and Loehlin, 1977). However, behavioral genetics now emphasizes that the influences of genes and environment can be dissociated. By now, we hope that this does not sound paradoxical. In case it does, an analogy might help. A sailboat needs both sails and a hull. The "behavior" of a sailboat (speed, turning, ability to point into the wind, planing ability) depends on the design of its sails and the design of its hull. The aerodynamic shape of the sails, their number and size, and their positioning are important. The depth, width, length, and shape of the hull are also important. Obviously, for sailboats, there can be no behavior without both sails and hull, but this does not restrict us from asking about the independent contributions of sail design and hull design to the behavior of sailboats. These two factors can interact—certain hull designs are useful only with particular sail designs—but this interaction does not prevent us from specifying the independent contributions of sails and hulls. For example, regardless of the design of their sails, flat, broad-beamed boats plane better, while catamarans go faster, but do not turn as easily. In the same way, we can say that behavior requires both genes and environment, that these influences interact, and that we can determine their independent contributions to behavior.

The sailboat analogy is useful in making another general point about interactions and behavioral genetics. How would you determine whether hull design or sail design affected the behavior of sailboats? You would try different designs to see which makes a difference in the boat's behavior. All experiments study differences (variance); an independent variable is manipulated to produce differences between groups. In other words, we study things that make a difference. Behavioral genetics is the study of genetic and environmental influences that make a difference in behavior. Just as hull design is more important than sail design for turning, while sail design is more important for straightaway speed, genetic differences may be more important for some behaviors and environmental differences may be more important for others.

Thus, even though there can be no behavior without both genes and environment, we can ask about the relative contributions of genes or environment to a particular behavior. Behavioral genetics analyses assess individual differences in a population, and ascribe the individual differences to environmental and genetic variability. Environmental differences can occur when genetic differences do not exist (for example, individual differences observed within pairs of identical twins). Also, genetic differences can be expressed in the absence of environmental differences (for example, differences among members of a genetically heterogeneous population reared in the same controlled laboratory environment).

Genotype-environment (GE) interaction refers to the possibility that individuals of different genotypes may respond differently to specific environments. For example, children of greater ability may respond better to a given school environment. GE interaction is due to the nonlinear combination of genetic and environmental effects. It is the extent to which observed variance can be predicted from both influences considered together. In terms of the sailboat analogy, even though we can assess the independent effects of sail design and hull design, there may be some aspects of sailboat behavior that can only be predicted from knowledge of both sail and hull design. In other words, certain hull designs may work well only with particular sail designs.

Interactionism, however, often connotes something more than this statistical interaction, and it has begun to take on an almost mystical aura. One of the more useful connotations of interactionism is that the organism actively interacts with its environment. This view, however, is better conveyed by GE correlation than by GE interaction. As we shall see, GE correlation meshes genes and environment and leads to a consideration of the organism as an active agent.

Positive Genotype-Environment Correlation

Genotype-environment correlation refers to the differential exposure of genotypes to environments. In other words, GE correlation is a function of the frequency with which certain genotypes and certain environments occur together. For example, if talented children are exposed to a special training that enhances their talent, there is a positive correlation between genetic differences in talent and environmental differences. We shall consider three types of GE correlation: passive, reactive, and active.

The most frequently mentioned type of GE correlation occurs when parents give their children genes and an environment that are both favorable (or unfavorable) for the development of a particular trait. For example, given that verbal ability is inherited to some extent, parents who are gifted in verbal ability provide their children with both genes and an environment conducive to the development of verbal ability (Meredith, 1973). We call this *passive GE correlation,* because it occurs independently of the behavior of the individual in question. The child passively receives genes and an environment rich in verbal stimulation.

Passive GE correlation has been discussed much more than other types (for example, Jencks, 1972), but it is not the most common form of GE correlation. A second type can be called *reactive,* in that people may react differently to persons of different genotypes. Reactive GE correlation differs from passive in that it is not limited to environments provided by relatives. Teachers, for example, may recognize cognitive abilities in their students and furnish an enriched environment to maximize their talents. In terms of personality, peers and adults may reciprocate the affection and attention of a highly sociable child. In short, people may respond to genotypic differences among individuals in such a way that they provide an environment that reflects and correlates with those genotypic differences.

Active GE correlation occurs when individuals are not merely passive recipients of their environment. Rather, they may contribute to their own environment and may actively seek one related to their genetic propensities. For example, bright children may seek peers, adults, or inanimate aspects of their environment that foster their cognitive growth. Similar examples exist for personality traits. An active child can create mayhem in the most placid environment, and a sociable child will seek out others and perhaps create imaginary playmates if real ones are not at hand. Genotypic differences among individuals may cause them to seek environments correlated with their genotypes.

The three types of GE correlation are summarized in Table 14.2. Because active or reactive GE correlation occurs as a result of the interaction of individuals with their own environments, these types are particularly relevant to the interactionist orientation.

We should note that the trichotomy of passive, reactive, and active correlation is merely one of conceptual convenience. Many real-life cases of GE correlation are intermediate, ambiguous, or mixed in character. Indeed, even the distinction between correlation and interaction is often difficult to make in practice. Much depends on how one chooses to define genotypes and environment. Once genotypes and environments are specified, the distinction between correlation and interaction is no longer ambiguous.

Negative Genotype-Environment Correlation

GE correlation can be negative as well as positive. For example, children of lower ability may be given special attention to boost their perfor-

TABLE 14.2
Three types of genotype-environment correlations

Type	Description	Pertinent Environment
Passive	Children have genotypes linked to their environment	Parents and siblings
Reactive	People react to children on the basis of the children's genotypes	Anybody
Active	Children seek an environment conducive to their genotype	Anything

SOURCE: After Plomin, DeFries, and Loehlin, 1977.

mance closer to the norm. Cattell suggested that, for personality, negative GE correlation is more common than positive. Using dominance as an example, he piquantly noted that "society likes to 'cut down' individuals naturally too dominant and to help the humble inherit the earth" (1973, p. 145). He described this type of negative genotype-environment correlation as "coercion to the biosocial norm." His examples are of reactive genotype-environment correlation, as the word *coercion* would imply. The other two types may also be negative. An example of negative passive genotype-environment correlation in personality involves anger. Emotionally labile parents, who are easily angered, may have children with a proclivity to be quick-tempered. Yet such parents are likely to assail expressions of anger in their children. Negative active genotype-environment correlation at first thought seems unlikely, because we would not expect individuals to seek environments that rub against the grain of their disposition. On the other hand, it does seem reasonable to suppose, for example, that some emotionally unstable individuals might seek calm environments and stable friends to steady their psyches, thus producing negative genotype-environment correlations.

The Effects of Genotype-Environment Interaction and Correlation in Behavioral Genetics Analyses

In Chapter 9, we described an example of phenotypic variance and its genetic and environmental components. We showed that genetic and environmental variances would remain the same, even if the differences due to genes and environment correlated perfectly; but phenotypic variance would be increased by the GE correlation (Jensen, 1974). Similarly, GE interaction can increase phenotypic variance. GE interaction and correlation can also contribute to the phenotypic resemblance between relatives, and thus have an effect on behavioral genetics analyses. Familial resemblance includes not only genetic and environmental similarities, but also any shared GE correlation or interaction. For example, the unique inherited qualities of one family member can contribute to the environment of the other, creating a GE correlation. Also, GE interactions affecting one member of the family may be similar to those experienced by other members.

We shall briefly summarize the ways in which GE interactions and correlations affect twin and adoption studies. (For details, see Plomin et al., 1977.) In twin studies, we compare the resemblance of identical and fraternal twins. If GE interaction and correlation contributed equally to the similarity of both types of twins, estimates of genetic influence based on the difference between the correlations for identical and fraternal twins would not be affected. In this case, both effects are included in the environmental estimate. However, to the extent that GE correlations and interactions occur more often for identical twins, the twin method will overestimate genetic influence. At the same time, environmental influence will also be overestimated, because only part of the increase in phenotypic variance caused by the GE interaction or correlation is interpreted as genetic variance. Thus, the absolute size of both genetic and environmental variances may be overestimated in a twin study in the presence of significant GE correlation and interaction. But the relative estimates of genetic and environmental influences may not be affected.

One of the major advantages of working with an adoption design is that it is unaffected by most of these complications. GE interaction does not affect either the genetic side or the environmental side of the full adoption design. The adoption design also avoids GE correlation of the passive kind, although the reactive and active types may have some effect. The direction of this effect depends on whether the GE correlation is positive or negative.

Although GE interaction and correlation are not likely to have a great effect on behavioral genetics analyses, they are worthy of study in their own right. Behavioral genetics methods afford a unique opportunity to study these concepts, as discussed in the next section.

Tests of Genotype-Environment Correlation and Interaction

One of the most important, yet only recently discovered, potential uses of the adoption design is in the isolation of specific GE interactions and correlations. For example, adoption studies that include environmental assessments could measure reactive and active GE correlations (Plomin et al., 1977). Because data are not yet available to permit a test of such correlations, we shall concentrate on analyses of GE interaction.

A search for GE interactions in adoption study data can reveal environmental influences that are particularly potent for certain individuals. For example, environmental influences that account for only 1 percent of the total variance in a population may have a powerful effect on a certain segment of that population. Fifty years ago, if someone had decided that restricting intake of phenylalanine of young children would improve later cognitive development, no overall beneficial effect would have been found (although a damaging effect might well have been observed). However, for one in ten thousand children (those children who have both recessive alleles for phenylketonuria), the intervention would have had a dramatic effect.

The adoption study method can provide a way of scanning for interactions between genotypes and environments. In Chapter 10, we discussed, as an example of GE interaction, a study in which rats from different lines were reared in different environments. The two lines responded differently to the environmental conditions. The adoption test of GE interaction is analogous to the method used in this and other animal studies of GE interaction (reviewed by Erlenmeyer-Kimling, 1972). The design is depicted in Table 14.3, which shows two genotypes reared in two different environments. This is a

TABLE 14.3

Illustrative design for testing genotype-environment interaction

Genotype	Environment	
	Low	High
Low	a	b
High	c	d

2 × 2 design, in which one variable is the genotype, the other is some aspect of the environment, and the dependent variable (for which measures are entered in the four cells) is some behavior. A 2 × 2 analysis of variance will reveal the effect of genotype, independent of the environment (called the "main effect" for G); the effect of environment, independent of genotype ("main effect" for E); and the interaction of these effects (G × E). Animal studies using this design often find significant GE interaction.

Adoption studies permit a similar analysis of human behavior. The genotype of adopted children can be estimated from some measure of the behavior of their birth parents. The environment of adopted children can be any measurable aspect of the children's environment, such as some characteristic of the adoptive parents. For example, consider adoptive children whose birth parents are higher or lower than average in IQ. These children are adopted by couples who are higher or lower than average in IQ. The dependent variable is the IQ of the adopted children in the four groups (cells) in Table 14.3. Cell a contains the IQ scores of adopted children whose birth parents are low in IQ, and who are placed with adoptive parents who are also relatively low in IQ, and so on. As in the animal studies, results of an analysis of variance will indicate the effect of genetic influences independent of the environmental measure, the effect of environmental influences independent of genotype, and the GE interaction.

If genes do not make a difference for the particular behavior, there will be no main effect of genotype. If the measured characteristic of the adoptive parents does not make a difference, there will be no main effect of environment. GE interactions will be observed if children of low- and high-IQ birth parents are differentially affected by the rearing environment. The test of GE interaction is not limited to the environment based on the measured characteristic of the adoptive parents. Any aspect of the environment can be studied in a similar manner.

The ideal data to analyze in a search for GE interaction would result from a study including many behavioral measures on both birth parents, both adoptive parents, and the adopted children, as well as extensive assessments of the adopted children's environments. Although no reported adoption studies have collected all these data, our reanalysis of the results obtained by Skodak and Skeels and by Munsinger (discussed in the previous chapter) illustrate the use of the method. The measure of the environment in both studies is limited: educational level of the adoptive parents in Skodak and Skeels' study, and socioeconomic status of the adoptive parents in Munsinger's study. We used midparent educational level or socioeconomic status and split the birth parents and the adoptive parents into groups above or below the mean. Thus, there were high and low groups of both birth parents and adoptive parents. The dependent variable was the IQ of the adopted children in the four groups.

For Skodak and Skeels' study, we reanalyzed the IQ data on 59 children for whom the educational levels of both birth parents and both adoptive parents were known. The average IQ of the children in each of the four groups is shown in Table 14.4. The average IQs of the children indicate that the differences in the children's environment (as indexed by educational level of their adoptive parents) had no effect on their IQs. However, genetic influences (as indexed by the educational level of their birth parents) were considerable. Our analysis reveals a highly significant main effect of genotype, but no significant environmental effect. In addition, the GE interaction is not statistically significant.

Data from Munsinger's adoption study yielded similar results. He recorded socioeconomic status information for the birth parents and the adoptive parents of 41 children. Table 14.5 summarizes IQ scores of the children born to parents below or above the mean for socioeconomic status, and reared by adoptive parents below or above the mean for socioeconomic status. There was no selective placement. The IQ means of the children in Table 14.5 indicate that, as in Skodak and Skeels' study, the genetic variable is highly significant, but the environmental variable is not. Once again, the GE interaction term is not significant.

TABLE 14.4

Test of genotype-environment interaction for IQ: data reanalyzed from Skodak and Skeels (1949)

Genotype	Environment		Row Mean
	Low	High	
Low	111.8 N = 16	106.3 N = 12	109.4
High	122.5 N = 13	121.4 N = 18	121.9
Column mean	116.6	115.4	

NOTE: The children were born to parents above or below the mean for years of education (9.9), and reared by adoptive parents above or below the mean (12.1 years). The data are from tests of 13-year-olds, using the 1937 Stanford-Binet.

SOURCE: After Plomin, DeFries, and Loehlin, 1977.

TABLE 14.5

Test of genotype-environment interaction for IQ: data reanalyzed from Munsinger (1975)

Genotype	Environment		Row Mean
	Low	High	
Low	3.0 N = 10	2.5 N = 14	2.7
High	5.0 N = 17	5.1 N = 10	5.0
Column mean	4.3	3.6	

NOTE: The children were born to parents above or below the mean for socioeconomic status (3.3), and reared by adoptive parents above or below the mean (2.8). The children's IQs are reported as stanine scores, which are essentially percentile scores where 5.0 is equivalent to the fiftieth percentile.

SOURCE: After Plomin, DeFries, and Loehlin, 1977.

This procrustean 2 × 2 design is only meant to be illustrative. Most behaviors, including IQ, are continuously distributed. Rather than arbitrarily dividing individuals into groups such as those above and below the mean, continuous variation should be analyzed in a continuous manner. A method for doing this is available (Plomin et al., 1977). The conclusions were the same when this more adequate method was applied to the data just described.

These analyses were based on indirect measures of birth parents' genotypes and limited measures of the environment of the adoptive home. We reiterate the need for detailed data on the birth and adoptive parents, and on the adopted children, as well as for extensive environmental assessments. More complete information would permit an interesting variety of analyses. For example, one could analyze the effects of adopted children's genotypes, child-rearing practices of adoptive parents, and the interaction between the two, using some relevant aspect of the adopted children's behavior as a dependent variable. Any aspect of the genotype or the environment can be screened in this way for GE interaction, with respect to any trait in the children, provided that all three variables are measurable. This use of adoption data to screen for GE interaction is an unusually promising tool for the more refined analysis of environmental effects in psychology.

Group Differences

For some reason, human beings seem inclined to think in terms of groups and group differences: young-old, experimental-control, male-female, black-white. Although group differences have their place, we need to be careful to avoid slipping into typological traps. It is easy to start thinking in terms of nonoverlapping distributions for two groups, even though we know that differences between groups are usually small in comparison to individual differences within the groups. An extreme example is the relationship between birth order, family size, and IQ. Earlier-born children and children in smaller families tend to have higher IQs. For large data sets, this relationship looks quite orderly. For example, on the average, second-born children have higher IQs than third-born children, and children in two-child families have higher IQs than those in three-child families. Mathematical models can predict these group averages with great accuracy (Zajonc and Marcus, 1975). However, such models account for only about 2 percent of the total variance of IQ in a population (Grotevant, Scarr, and Weinberg, 1977). In other words, if all that you know about someone is birth order and family size, you will make a very poor prediction of that person's IQ.

With respect to cognitive ability, group comparisons involving gender, ethnic, and social class differences have captured the attention of behavioral scientists. All of these group differences are small compared to individual differences within the groups. For example, the average performance of males on spatial ability tests is consistently better than that of females, making it appear that there is an important sex difference in this specific cognitive ability. However, the overlap between the distributions for males and females is at least 80 percent. It is obvious that one would do a very poor job of predicting spatial ability solely on the basis of gender.

It is especially important to distinguish differences *between* groups from differences *within* groups in discussions of ethnic or social class differences in IQ. For both blacks and whites in the United States, social class (independent of race) accounts for about 8 percent of the variance of IQ. Race (independent of social class) accounts for about 14 percent. However, individual differences, independent of race and class, are responsible for 78 percent of the variance. In fact, the average IQ difference between full siblings in a family is twice as great as the average differences between social classes, and as great as the average difference between blacks and whites (Jensen, 1976). It seems clear that each person must be evaluated individually, not on the basis of race or class membership, in situations where IQ is a relevant consideration. Thus, although group differences appear deceptively easy to grasp and to talk about, individual differences within groups make a far more important contribution to total variance.

The IQ difference between blacks and whites is the most widely studied group difference, and also the most controversial, so we shall use it as our example. Most studies indicate that blacks, on the average, score about one standard deviation (about 15 IQ points) lower than whites on conventional tests of general intelligence. The etiology of the difference is very much disputed. Some argue that it is simply caused by cultural bias of IQ tests. However, we cannot conclude that the tests are biased just because two groups differ. Although the hypothesis of test bias is reasonable, research does not bear it out. IQ tests predict school and occupational performance equally well for blacks and whites. Also, reliability, rank order of item difficulties, correlations with age, and the results of factor analysis are essentially the same for both groups (Jensen, 1976, 1979).

If the IQ difference between blacks and whites is not due to test bias, then what is the cause of the difference? A controversy of unprecedented proportions in the behavioral sciences began in 1969 with the publication of a monograph by Arthur Jensen, in which he reviewed the evidence for the heritability of IQ and hypothesized that genetic factors may be implicated in the observed difference between blacks and whites. In the abstract, the possibility is clear. If two populations have been to some extent reproductively isolated for an appreciable period of time, they will likely come to have different allelic frequencies for a number of genes. There is no reason to believe that genes affecting intelligence are different from any others in this respect. Thus, two groups that have been separated reproductively might well have different frequencies of genes relating to intelligence. The difficulty is in determining if the observed difference in IQ is due to genetic differences or to the manifestly unequal environmental opportunities that have been available to the two groups.

The fact that intelligence is heritable within black and white groups does not mean that the difference between the groups is heritable. More generally, the causes of differences within groups are not necessarily related to the causes of differences between groups. This is just as true for environmental factors as for genetic ones. In terms of genetic sources of variance, does the finding of a high within-group heritability imply that an observed difference between groups is also heritable? DeFries (1972) has shown that the answer is no. In fact, even if heritability within each of the two groups were 100 percent, the difference between the groups could be completely environmental in origin. For example, suppose that two groups were equal in means and variance. Then suppose that some environmental variables (such as those related to prejudice, for example) led to lower scores for everyone in one group. If these variables had equal effects on all members of that group, variance within the group would not be changed, even though the mean would be lower. Thus, heritability within both groups would still be 1.0, and the difference between the groups would be entirely environmental. Though this example is extreme, it makes the point that high within-group heritability does not necessarily imply that an observed difference between group means is also highly heritable.

Although we still have not answered the question about the etiology of the IQ difference between blacks and whites, the above discussion does bring us closer to an answer by pointing out that the sources of differences within groups—either genetic or environmental—are not necessarily the sources of differences between groups. No topic is more difficult to review than the etiology of racial differences in IQ, but a book by Loehlin, Gardner Lindzey, and James Spuhler (*Race Differences in Intelligence,* 1975) presents an admirably balanced review and evaluation of studies in this area. The book includes highly readable discussions of general issues, such as the meaning of race, intelligence, and heritability. The more relevant aspect for our present purpose is the evaluation of studies on the etiology of racial differences in IQ.

Their conclusion—and the conclusion reached in other recent reviews, including those by Philip Vernon (1979) and Lee Willerman (1979)—is one that will not please extremists on either side. However, if it is correct, it would in part explain why so much heat has been generated by interpretations of the results of such studies. Loehlin and coauthors conclude that the data are essentially equivocal, and thus the data are compatible with a wide range of views as to the etiology of racial difference in IQ. They suggest that the reason for the ambiguous results may be that the IQ difference is due to both environmental and genetic factors. They also stress that differences among individuals within racial-ethnic and socioeconomic groups greatly exceed differences between such groups. Finally, they emphasize that only these limited conclusions are justified by the evidence: "When we have mentioned our general conclusion to colleagues—that the solid evidence to date is compatible with a relatively broad range of intellectual positions on the 'race-IQ' question—a typical response is, 'yes, but what do you *really* think?' Well, what we really think is just that" (Loehlin, Lindzey, and Spuhler, 1975, p. 257).

Some of the findings that these authors reviewed to reach this conclusion are as follows: The IQ difference between blacks and whites in the United States has not decreased from the time of World War I to World War II to the Vietnam War, despite increasing equality of education. The group difference is apparent by about three years of age (which suggests the need for early developmental studies of IQ). The groups of studies bearing most directly on the question of the etiology of racial differences are studies of racial mixture, but each of these investigations has some important limitations. Studies of blood groups characteristic of Europeans and of African blacks have determined that blacks in the United States, on the average, have inherited a substantial number of "European genes." To what extent do these European genes associate with IQ in blacks? The answer from two small studies (summarized by Loehlin, Vandenberg, and Osborne, 1973) is "not at all." For sixteen blood-group genes, no relationship was found between higher IQ among the blacks and those genes found more frequently among the whites. Unfortunately, the power of this method to find such relationships is likely to be quite limited (Loehlin et al., 1973). An early study in the United States (Witty and Jenkins, 1936) also found no greater than average European heritage among black children with very high IQs.

The most direct study of racial admixture involved illegitimate children of United States servicemen in Germany following World War II (Eyferth, 1961). Comparing the average IQs of the children of white and black fathers provides a test of the role of heredity in the racial difference in IQ. The rearing environments of the children in the two groups were similar in that they were all reared by German mothers and the groups were matched for social status. The results showed no overall IQ difference between the children of black fathers and those of white fathers. Despite the apparent persuasiveness of this finding, it is not conclusive because of the possibility that there were no IQ differences between these two groups of fathers. Unfortunately, this possible bias cannot be tested because no IQ data are available for the fathers.

Thus, we emphasize again that the question of the etiology of racial differences in IQ has simply not been answered. This does not mean it cannot be answered, and Loehlin et al. (1975) propose some possible tests. They also raise the important issue of values. Should we *want* to know the answer? This brings us full circle to some of the issues raised in the first chapter. The misconception that genes are responsible for destiny leads many to suggest that such research will do irreparable harm if genetic differences are found to underlie group differences in IQ. But one could argue in just the opposite direction. The phenylketonuria story suggests that a difference that is largely genetic in origin can permit more powerful environmental interventions than a difference caused by amorphous environmental factors.

The final paragraph in the book by Loehlin and coauthors summarizes their answer to this question of values:

> We do *not* believe that the lack of a definitive answer to the questions with which we began is either disastrous or disappointing. Moral and political questions never have had scientific answers. The factual questions involved, if phrased in limited and specific form, should indeed be answerable, and it is probably worth society's time and money to try to answer a good number of them. It is part of our own fundamental conviction as social scientists that on the whole better and wiser decisions are made with knowledge than without. (Loehlin et al., 1975, p. 258)

Science and Politics

We share this view that better decisions can be made with knowledge than without. But knowledge alone by no means accounts for societal and political decisions. Values are just as important in the decision-making process, and decisions (bad or good) can be made with or without knowledge. The relationship between facts and values is complex. Most scientists view themselves as fact-seekers, ferreting out facts using the objective and verifiable methods of science. Values and politics, in this view, are what other people do with scientific facts. At the opposite extreme is the view that the scientific method is not at all objective, but rather riddled with values and politics. Some even argue that this should be so—that science is politics and should be used as a political tool.

Values undoubtedly do enter the scientific process from the beginning, when we decide what problems to study, and when we interpret the findings and their implications. Scientists need to worry about how their findings will be used, even though this is often frustrating because they have little control over the uses of their data. The frustration is the worst for scientists who work on the most important problems. Some research is so unimportant that no one ever gets excited about it or its implications. Powerful, new findings with important implications for society can create problems as well as solutions. For example, consider the new techniques of prenatal screening. Amniocentesis has obvious benefits in terms of detecting chromosomal and genetic disorders before birth. Combined with therapeutic abortion, it can relieve parents and society of a tremendous burden of birth defects. However, it raises ethical problems concerning abortion and creates the possibility of abuses, such as compulsory screening and mandatory abortion if defects are found. Despite the problems created by scientific findings, we would not want to cut off the flow of knowledge in order to cut down on the problems.

We equate knowledge with the scientific method. We are not so naive as to argue that scientists can or should eschew politics. However, when scientists face a problem, they must be objective if they want to know the answer, rather than merely providing support for preconceived notions. Science is not politics, and there are some telling examples of what happens when politics and science become mixed.

Lysenko

Time and again, science has been interpreted and misinterpreted to advance political causes. Two notorious examples related to genetics include the eugenics movement in the early part of the twentieth century and the Nazi carnage during World War II. Moreover, an attempt to equate science and politics occurred in Russia from the 1930s until the 1960s.

The story has been told many times. An interesting version of the political take-over of science in the USSR was written by the biologist Sir Julian Huxley, who sympathized with Marxism and was in Russia during the rise of T. D. Lysenko. (See also Medvedev, 1969.) Lysenko, like Lamarck, believed that acquired characteristics could be inherited. He attempted to apply this belief to agriculture, obviously with little success, but his belief in his theory overcame his belief in the scientific method. With the help of a Marxist philosopher in the 1930s, Lysenko mounted a political campaign to brand Mendelian scientists as idealistic and antichange. Huxley surmises that the real reasons for the adverse reaction to Mendelism were matters of values:

> One is the dislike of Mendelism because Mendelian heredity, with its self-copying genes and its random undirected mutations, seems to offer too much resistance to man's desire to change nature, and to elude the control we would

> like to impose. Lamarckism, on the other hand, holds out a promise of speedy control. ... This is relevant not only in agriculture, but also in human affairs, for it would be politically very convenient and agreeable if a few generations of life under improved communist conditions would level up the genetic quality of the population of the U.S.S.R. ... The second ideological reason ... is a dislike of Mendelism because it implies human inequality, and because it can be taken to imply human helplessness in the face of genetic predestination. (Huxley, 1949, pp. 182–184)

By 1939, Lysenko had a firm grip on the reins of science, and some of the Stalinist purges were used to eliminate the leading geneticists. In a conference in 1948, Lysenko announced that his views were endorsed by the Central Committee of the Party. The Soviet Academy of Sciences supported Lysenko, as indicated in their letter to Joseph Stalin:

> A pledge is here given by the Praesidium of the U.S.S.R. Academy of Sciences to further [Lysenko's] biology and to root out unpatriotic, idealistic, [Mendelian] ideology. ... [Lysenko's] materialist direction in biology is the only acceptable form of science, because it is based on dialectical materialism and on the revolutionary principle of changing Nature for the benefit of the people. [Mendelian] idealist teaching is pseudo-scientific, because it is founded on the notion of the divine origin of the world and assumes eternal and unalterable scientific laws. The struggle between the two ideas has taken the form of the ideological class-struggle between socialism and capitalism on the international scale, and between the majority of Soviet scientists and few remaining Russian scientists who have retained traces of bourgeois ideology, on a smaller scale. There is no place for compromise, [Lysenkoism] and [Mendelism] cannot be reconciled. (Huxley, 1949, pp. 39–40)

It became impossible to challenge Lysenko's views, and this signaled the end of Mendelian genetics in Russia for two decades. Geneticists who refused to acquiesce left Russia, and geneticists who remained were forced to recant. Many lost their positions, and some lost their lives. By 1964, the crop failures that resulted from Lysenko's policies were too much to be ignored, and may even have contributed to the downfall of Nikita Khrushchev, who had supported him during the post-Stalinist political upheaval. Lysenko was disgraced and Mendelian genetics returned to Russia.

Lysenko's belief in the inheritance of acquired characteristics and many of his experiments have no scientific merit, as Soviet scientists now recognize. However, the scientific issues are less important in the present context than understanding the process of politicization and its disastrous results. Huxley observed that Lysenkoism is not really a scientific issue, but rather an example of "a preconceived idea, which has been imposed on the facts instead of arising out of them; when the facts do not fit the idea, their relevance or even their existence is denied.... It is essentially nonscientific or prescientific doctrine applied to a branch of scientific study, not a branch of science in its own right" (Huxley, 1949, p. 23).

The methods used by Lysenko to force his ideas on science deserve special attention to sensitize us to such politicization. First, the scientific controversy quickly became personal. Attacks were focused on men and their motives, rather than on their ideas (Brill, 1975). Second, jingoism prevailed. Simplistic slogans with popular appeal took the place of honest discussions of complex issues. A third and related tactic was Lysenko's greater use of the popular press than of scientific journals to promulgate his views. Fourth, he disavowed the scientific method. For example, Lysenko told his staff that "to obtain a certain result, one must wish to obtain such a particular result; if you want a particular result you will obtain it" (Lerner and Libby, 1976, p. 396). Control groups and statistical tests of significance were seldom used. Fifth, no compromise or middle ground was allowed. The issue was treated as a class struggle in science, and the revolutionary slogan "who is not with us is against us" was applied.

Political Issues Today

The example of Lysenkoism suggests the problems that can occur when science is controlled by politics. We are sorry to note that there are signs of similar tactics of politicization in relation to human behavioral genetics today. Although scientific criticism is healthy, attempts to politicize the field of behavioral genetics are ill-considered because of the bizarre course that politicization can take once it has begun. Scientific criticism should not be ad hominem, rely on jingoism or the popular press, ignore the role of the scientific method, not attempt to polarize controversies. It is unfortunate that some opponents of human behavioral genetics research (particularly in relation to IQ) have used these polemical tactics. For example, a newborn chromosomal screening program was stopped after the principal investigator was subjected to considerable personal harassment (Culliton, 1975). Although many critics of Arthur Jensen's position dealt with the issues at a scientific level, many did not (Jensen, 1972). Also, E. O. Wilson became the focus of some scurrilous attacks after the publication of his book on sociobiology (Wade, 1976; Wilson, 1978).

It should be obvious that politicization is not the answer, even if individuals are troubled by the value implications of behavioral genetics, rather than the scientific issues. As discussed in the first chapter, many people believe that equality means biological identity and that any differences among individuals must be caused by environmental inequality. The values underlying this belief are no doubt noble. However, it must be pointed out that Lysenko began with the good intention of advancing science by making science more responsive to socialist needs. If we were able to distinguish between concern with values and scientific issues, there would perhaps be less rancor in discussions of both values and scientific issues.

We suggest that the hypotheses of biological identity (that is, lack of hereditary individual differences) and environmental causation of individual differences are scientific questions. Social equality, on the other hand, is a value, and a very important one. If examined closely, there need be no contradiction between the value of equality and the possibility of genetic differences among individuals. Surely we do not treat people equally because there are no differences among them. People differ in morphological characters (such as height, strength, age, sex, and so on), and they also differ behaviorally. However, none of these differences implies that discrimination is justified. Equality of opportunity and equality before the law are independent of the question of individual differences, regardless of their genetic or environmental etiology.

The basic scientific message of behavioral genetics is individuality. Recognition of, and respect for, individual differences is essential to the ethic of individual worth. Proper attention to individual needs, including the provision of the environmental circumstances that will optimize the development of each person, is a utopian ideal and no more attainable than other utopias. Nevertheless, we can approach this ideal more closely if we recognize individuality than if we do not. Although the requisite knowledge will require much research effort and expense, it would seem to warrant a high priority. Human individuality is the fundamental natural resource of our species.

THE IMPORTANCE OF BEHAVIORAL GENETICS

In the first chapter, we indicated that our goal is to communicate the principle that both heredity and environment can be responsible for individual differences in behavior. We hope that the reader is convinced that genetic differences can account for a substantial portion of observed variation in behavior, even for such complex human characters as IQ. However, we have failed in our goal if the reader thinks that behavioral genetics ignores environmental influences. We have emphasized examples in which behavioral genetics studies have provided critical information concerning the role of the environment. One of the most important aspects of behavioral genetics is its ability to study both environmental and genetic influences on behavior, without assuming that one or the other is omnipotent.

Most early behavioral genetics research sought only to demonstrate that genes can affect behavior. More recently, research has attempted to determine the extent to which individual differences for a particular behavior can be attributed to genetic or environmental factors. Although this is just the first step in understanding behavior, it is clearly a reasonable approach. However, there is much more to be learned from behavioral genetics. We shall briefly consider the importance of behavioral genetics in understanding more about behavior, genetic influences on behavior, environmental influences on behavior, and the interface between genes and environment.

Behavior

Behavioral genetics methods can be used to analyze the heterogeneity of behavior. Many phenotypes important to society, and thus of interest to behavioral scientists, seem hopelessly complex. For examples, behavioral genetics analyses have made some progress in differentiating components of both intelligence and psychosis. Behavioral genetics studies have analyzed specific cognitive abilities and their relationship to IQ. It appears that there is a substantial genetic relationship among quite diverse specific cognitive abilities. In the area of psychopathology, behavioral genetics studies have demonstrated that the two major psychoses, schizophrenia and depressive psychosis, are genetically distinct. These studies also suggest the possibility of etiologically different subtypes of depressive psychosis.

These analyses are versions of multivariate genetic-environmental analysis. We have just scratched the surface of possibilities for such analyses. In addition to breaking down the heterogeneity of complex phenotypes such as IQ and psychoses, these analyses may also be useful in constructing etiologically pure behavioral complexes. For instance, multivariate analysis can begin at a basic level of behavior—with the items on a personality questionnaire or individual symptoms of psychopathology—and build scales or syndromes that are purely genetic or purely environmental in origin. In other words, instead of working with complex phenotypes that are undoubtedly composed of a complex mixture of genetic and environmental influences, multivariate genetic-environmental analyses can create new phenotypes that are etiologically simpler.

Genetic Influences on Behavior

We can go beyond the estimation of heritability, and study the influence of chromosomal abnormalities and search for possible major genes. If major genes are found, we can determine the number and their location on chromosomes. We can also begin the difficult task of tracing the physiological intermediaries between the gene's enzyme activity and its ultimate indirect effect on behavior.

Another important, yet relatively unexplored, direction for research may reveal the developmental unfolding of genetic influences on behavior. For example, as described in Chapter 12, there is some indication that spurts and lags in the development of IQ may be influenced by genetic factors.

Environmental Influences on Behavior

Similarly, our analysis of environmental influences need not end with the determination of environmentality. Indeed, it is our belief that behavioral genetics will increase our understanding of environmental, not just genetic, influences on behavior. One example is the use of behavioral genetics designs to identify specific environmental factors. Adoption studies can be used to isolate specific environmental effects free of genetic influence. Identical twins are particularly useful in the search for within-family environmental influences because they are genetically identical. Any differences within pairs must be caused by within-family environmental factors. In terms of behavior in nonhuman animals, we have seen that reciprocal crosses, cross-fostering, and ovary transplants can be used to identify prenatal and postnatal maternal influence.

Behavioral genetics is also useful in studying the role of the environment in analyses of within-family and between-family components of environmental variance. It is generally accepted that environmental influences operate between families, making members of a family similar to one another. However, in some cases, most notably for psychoses, behavioral genetics studies conclude that nearly all relevant environmental variance operates within families, making members of a family different from one another.

Developmental analyses of environmental influences are also needed. Multivariate genetic-environmental analysis can be profitably applied to longitudinal data to determine the extent to which developmental continuity is mediated by environmental continuity or genetic continuity. In other words, do the same environmental factors affect a particular character at two points in development, or are different environmental factors operating during these different developmental stages?

The Interface Between Genes and Environment

The interface between genes and environment is likely to be the ground for very exciting discoveries. Studying environmental influences without consideration of genetic differences among individuals leads to findings that cannot be widely applied. As Scarr and Weinberg have pointed out: "In its baldest form, naive environmentalism has led us into an intervention fallacy. By assuming that all of the variance in behavior was environmentally determined, we have blithely promised a world of change that we have not delivered, at great cost to the participants, the public, and ourselves" (1978, p. 690).

As discussed earlier in this chapter, analyses of genotype-environment interaction are likely to yield important information concerning environmental influences that are particularly powerful for certain individuals. For example, educational methods appropriate for some individuals may not be effective for others. We have only begun to use behavioral genetics methods to study such problems.

Behavioral genetics research can go far beyond simple descriptions of the relative contribution of genetic and environmental influences to individual behavioral differences. For human behavior, we have used the examples of cognitive ability and psychopathology because these areas of research have always been concerned with individual differences. More behavioral genetics research has been conducted in these two areas than in all the other areas of behavioral science combined. However, even in these areas, sophisticated behavioral genetics research has just begun. Only the most rudimentary methods of behavioral genetics have been applied in other important areas of psychology, including perception, learning, memory and information processing, language and communication, personality, social psychology, and psychopathology other than psychoses. Lamentably, there is barely a trace of behavioral genetics in other areas of the social sciences, such as anthropology, sociology, and economics. Thus, although behavioral genetics has convincingly demonstrated that genetic variability can affect behavior, it is still in its infancy in terms of its potential application to the behavioral and social sciences.

SUMMARY

More mental-ability data were published between 1976 and 1979 than in the previous fifty years. Results of recent family, twin, and adoption studies suggest that the heritability of general cognitive ability is less than previously believed—about 0.50 rather than 0.70 or greater.

Three types of controversies recur in behavioral genetics: nature-nurture arguments, issues concerning the etiology of group differences, and problems involving the relationship between science and politics. The nature-nurture issue is related to genotype-environment (GE) interaction and correlation. GE interaction occurs when individuals of different genotypes respond differently to the same environment. Three types of GE correlation are discussed: passive, reactive, and active. Behavioral genetics analyses, particularly adoption studies, are capable of isolating specific GE interactions and correlations.

Variance between groups is usually much less important than variance within groups. For IQ, individual differences, independent of race and class, account for three times as much variance as race and social class. Also, the causes of differences within groups are not necessarily related to the causes of differences between groups. The results of studies of racial differences in IQ are equivocal. Such research also raises questions of values. Although values enter the scientific process, science must remain as independent as possible from politics. Lysenko's politicization of genetics in the USSR serves as a reminder of the disastrous consequences that can occur when science becomes controlled by politics. The value of social equality is not compromised by the finding of genetic differences among individuals.

Behavioral genetics studies can contribute much more than information about heritability and environmentality. Multivariate genetic-environmental analyses are particularly useful in analyzing the heterogeneity of complex behavioral phenotypes. Similarly, behavioral genetics can contribute greatly to the identification of specific environmental factors, the role of within-family and between-family environmental influences, and developmental aspects of the environment as it affects behavior. Genotype-environment interaction is likely to be a powerful tool for isolating environmental influences important for certain individuals. Such analyses have not yet been fully applied, even with regard to such characters as IQ and psychosis. The future of behavioral genetics lies in its application to other areas of behavioral science.

References

Adler, J. 1976. The sensing of chemicals by bacteria. *Scientific American,* 234, 40–47.

Allen, G. 1975. *Life science in the twentieth century.* New York: John Wiley.

Ambrus, C. M.; Ambrus, J. L.; Horvath, C.; Pedersen, H.; Sharma, S.; Kant, C.; Mirand, E.; Guthrie, R.; and Paul, T. 1978. Phenylalanine depletion for the management of phenylketonuria: Use of enzyme reactors with immobilized enzymes. *Science,* 201, 837–839.

Anastasi, A. 1958. Heredity, environment, and the question "How?" *Psychological Review,* 65, 197–208.

Angst, J. 1966. Zur ätiologie und nosologie endogener depressiver psychosen. *Monographie aus dem gesamtgebiete der Neurologie und Psychiatrie* (*Berlin*), 112.

Anisman, H. 1975. Task complexity as a factor in eliciting heterosis in mice: Aversively motivated behaviors. *Journal of Comparative and Physiological Psychology,* 89, 976–984.

Ayala, F. J. 1978. The mechanism of evolution. *Scientific American,* 239, 56–69.

Bailey, D. W. 1971. Recombinant inbred strains. *Transplantation,* 11, 325–327.

Baker, W. K. 1978. A genetic framework for *Drosophila* development. *Annual Review of Genetics,* 12, 451–470.

Barash, D. P. 1975. Ecology of paternal behavior in the hoary marmot (*Marmota caligata*): An evolutionary interpretation. *Journal of Mammalogy,* 56, 612–615.

———. 1977. *Sociobiology and behavior.* New York: Elsevier.

Barnicot, N. A. 1950. Taste deficiency for phenylthiourea in African Negroes and Chinese. *Annals of Eugenics,* 15, 248–254.

Baron, M.; Gershon, E. S.; Rudy, V.; Jonas, W. Z.; and Buchsbaum, M. 1975. Lithium carbonate response in depression. *Archives of General Psychiatry,* 32, 1107–1111.

Barr, M. L., and Bertram, E. G. 1949. A morphological distinction between neurones of the male and female, and the behaviour of the nucleolar satellite during accelerated nucleoprotein synthesis. *Nature,* 163, 676.

Bashi, J. 1977. Effects of inbreeding on cognitive performance. *Nature,* 266, 440–442.

Bateson, W. 1909. *Mendel's principles of heredity.* Cambridge, England: Cambridge University Press.

Bateson, W.; Saunders, E. R.; and Punnett, R. C. 1906. Experimental studies in the physiology of heredity. *Reports to the Evolution Committee of the Royal Society,* 3, 1–53.

Beadle, G. W., and Tatum, E. L. 1941. Experimental control of developmental reaction. *American Naturalist,* 75, 107–116.

Beckwith, L. 1971. Relationships between attributes of mothers and their infants' IQ scores. *Child Development,* 42, 1083–1097.

Belmaker, R.; Pollin, W.; Wyatt, R. J.; and Cohen, S. 1974. A follow-up of monozygotic twins discordant for schizophrenia. *Archives of General Psychiatry,* 30, 219–222.

Benzer, S. 1967. Behavioral mutants of *Drosophila* isolated by countercurrent distribution. *Proceedings of the National Academy of Sciences,* 58, 1112–1119.

Berg, C. J. 1978. Development and evolution of behavior in mollusks, with emphasis on changes in stereotypy. In *The development of behavior: Comparative and evolutionary aspects,* eds. G. M. Burghardt and M. Bekoff, pp. 3–17. New York: Garland STOM Press.

Bertelsen, A.; Harvald, B.; and Hauge, M. 1977. A Danish study of manic-depressive disorders. *British Journal of Psychiatry,* 130, 330–351.

Bessman, S. P.; Williamson, M. L.; and Koch, R. 1978. Diet, genetics, and mental retardation interaction between phenylketonuric heterozygous mother and fetus to produce nonspecific diminution of IQ: Evidence in support of the justification hypothesis. *Proceedings of the National Academy of Sciences,* 78, 1562–1566.

Blixt, S. 1975. Why didn't Gregor Mendel find linkage? *Nature,* 256, 206.

Bock, R. D., and Vandenberg, S. G. 1968. Components of heritable variation in mental test scores. In *Progress in human behavior genetics,* ed. S. G. Vandenberg, pp. 233–260. Baltimore: Johns Hopkins University Press.

Bodmer, W. F., and Cavalli-Sforza, L. L. 1976. *Genetics, evolution, and man.* San Francisco: W. H. Freeman and Company.

Böök, J. A. 1957. Genetical investigation in a north Swedish population: The offspring of first-cousin marriages. *Annals of Human Genetics,* 21, 191–221.

Boolootian, R. A. 1971. *Slide guide for human genetics.* New York: John Wiley.

Borgaonkar, D. S. 1977. *Chromosomal variation in man.* 2d ed. New York: Alan R. Liss.

Boring, E. G. 1950. *A history of experimental psychology.* 2d ed. New York: Appleton.

Bouchard, T. J., and McGee, M. G. 1977. Sex differences in human spatial ability: Not an X-linked recessive gene effect. *Social Biology,* 24, 332–335.

Boué, J. G. 1977. Chromosomal studies in more than 900 spontaneous abortuses. Paper presented at the Teratology Society Meeting, 1974. Cited in D. W. Smith, 1977.

Bovet, D. 1977. Strain differences in learning in the mouse. In *Genetics, environment and intelligence,* ed. A. Oliverio, pp. 79–92. Amsterdam: North-Holland Publishing Company.

Bovet, D.; Bovet-Nitti, F.; and Oliverio, A. 1969. Genetic aspects of learning and memory in mice. *Science,* 163, 139–149.

Brill, H. 1975. Presidential address: Nature and nurture as political issues. In *Genetic research in psychiatry,* eds. R. R. Fieve, D. Rosenthal, and H. Brill, pp. 283–288. Baltimore: Johns Hopkins University Press.

Broadhurst, P. L. 1960. Experiments in psychogenetics. Applications of biometrical genetics to the inheritance of behaviour. In *Experiments in personality.* Psychogenetics and psychopharmacology, vol. 1, ed. H. J. Eysenck, pp. 1–102. London: Routledge & Kegan Paul.

———. 1975. The Maudsley reactive and nonreactive strains of rats: A survey. *Behavior Genetics,* 5, 299–319.

———. 1976. The Maudsley reactive and nonreactive strains of rats: A clarification. *Behavior Genetics,* 6, 363–365.

———. 1978. *Drugs and the inheritance of behavior.* New York: Plenum Press.

———. 1979, in press. The experimental approach to behavioral evolution. In *Theoretical advances in behavior genetics,* ed. J. R. Royce. Alphen aan den Rijn, Netherlands: Sijthoff Noordhoff International.

Bruce, E. J., and Ayala, F. J. 1978. Humans and apes are genetically very similar. *Nature,* 276, 264–265.

Bruell, J. H. 1964*a*. Heterotic inheritance of wheelrunning in mice. *Journal of Comparative and Physiological Psychology,* 58, 159–163.

———. 1964*b*. Inheritance of behavioral and physiological characters of mice and the problem of heterosis. *American Zoologist,* 4, 125–138.

———. 1967. Behavioral heterosis. In *Behavior-genetic analysis,* ed. J. Hirsch, pp. 270–274. New York: McGraw-Hill.

Bruun, K.; Markkanen, T.; and Partanen, J. 1966. *Inheritance of drinking behavior, a study of adult twins.* Helsinki: The Finnish Foundation for Alcohol Research.

Bulmer, M. G. 1970. *The biology of twinning in man.* Oxford: Clarendon Press.

Burks, B. S. 1928. The relative influence of nature and nurture upon mental development: A comparative study of foster parent–foster child resemblance and true parent–true child resemblance. In *Nature and nurture: Their influence upon intelligence. Twenty-Seventh Yearbook of the National Society for the Study of Education,* part 1, 219–316.

Burt, C. 1955. The evidence for the concept of intelligence. *British Journal of Educational Psychology,* 25, 158–177.

———. 1966. The genetic determination of differences in intelligence: A study of monozygotic twins reared together and apart. *British Journal of Psychology,* 57, 137–153.

Cadoret, R. J.; Cunningham, L.; Loftus, R.; and Edwards, J. 1975. Studies of adoptees from psychiatrically disturbed biological parents. II. Temperament, hyperactive, antisocial, and developmental variables. *Journal of Pediatrics,* 87, 301–306.

Campbell, D. T. 1975. On the conflicts between biological and social evolution and between psychology and moral tradition. *American Psychologist,* 30, 1103–1126.

Caplan, A. I., and Ordahl, C. P. 1978. Irreversible gene repression model for control of development. *Science,* 201, 120–201.

Carroll, J. B., and Maxwell, S. E. 1979. Individual differences in cognitive abilities. *Annual Review of Psychology,* 30, 603–640.

Carter-Saltzman, L., and Scarr, S. 1975. Blood group, behavioral and morphological differences among dizygotic twins. *Social Biology,* 22, 373–374.

———. 1977. MZ or DZ? Only your blood grouping laboratory knows for sure. *Behavior Genetics,* 7, 273–280.

Casler, L. 1976. Maternal intelligence and institutionalized children's developmental quotients: A correlational study. *Developmental Psychology,* 12, 64–67.

Caspersson, T.; Farber, S.; Foley, G. E.; Kudynowski, J.; Modest, E. J.; Simonsson, E.; Wagh, U.; and Zech, L. 1968. Chemical differentiation along metaphase chromosomes. *Experimental Cell Research,* 49, 219–222.

Caspersson, T., Lomakka, G., and Møller, A. 1971. Computerized chromosome identification by aid of the quinacrine mustard fluorescence technique. *Hereditas,* 67, 103–109.

Castle, W. E. 1903. The law of heredity of Galton and Mendel and some laws governing race improvement by selection. *Proceedings of the American Academy of Sciences,* 39, 233–242.

Cattell, R. B. 1953. Research designs in psychological genetics with special reference to the multiple variance analysis method. *American Journal of Human Genetics,* 5, 76–93.

———. 1960. The multiple abstract variance analysis equations and solutions: For nature-nurture research on continuous variables. *Psychological Review,* 67, 353–372.

———. 1973. *Personality and mood by questionnaire.* San Francisco: Jossey-Bass.

Caviness, V. S., Jr., and Rakic, P. 1979, in press. Mechanisms of cortical development: A view from mutations in mice. *Annual Review of the Neurosciences.*

Cederlöf, R.; Friberg, L.; Jonsson, E.; and Kaij, L. 1961. Studies on similarity diagnosis with the aid of mailed questionnaires. *Acta Genetica et Statistica Medica,* 11, 338–362.

Cederlöf, R., and Lorich, U. 1978. The Swedish twin registry. In *Twin research, Part B, Biology and Epidemiology,* ed. W. E. Nance, pp. 189–196. New York: Alan R. Liss.

Changeux, J.-P. 1965. The control of biochemical reactions. *Scientific American,* 212, 36–45.

Chen, C.-S., and Fuller, J. L. 1976. Selection for spontaneous or priming-induced audiogenic seizure susceptibility in mice. *Journal of Comparative and Physiological Psychology,* 90, 765–772.

Claeys, W. 1973. Primary abilities and field-independence of adopted children. *Behavior Genetics,* 3, 323–338.

Clarke, B. 1975. The causes of biological diversity. *Scientific American,* 233, 50–60.

Cloninger, C. R.; Reich, T.; and Guze, S. B. 1975*a*. The multifactorial model of disease transmission: II. Sex differences in the familial transmission of sociopathy (antisocial personality). *British Journal of Psychiatry,* 127, 11–22.

———. 1975*b*. The multifactorial model of disease transmission: III. Familial relationship between sociopathy and hysteria (Briquet's syndrome). *British Journal of Psychiatry,* 127, 23–32.

Cohen, R.; Bloch, N.; Flum, Y.; Kadar, M.; and Goldschmidt, E. 1963. School attainment in an immigrant village. In *The genetics of migrant and isolate populations,* ed. E. Goldschmidt, pp. 350–351. Baltimore: Williams & Wilkins.

Cohen, D. J.; Dibble, E.; Grawe, J. M.; and Pollin, W. 1973. Separating identical from fraternal twins. *Archives of General Psychiatry,* 29, 465–469.

———. 1975. Reliably separating identical from fraternal twins. *Archives of General Psychiatry,* 32, 1371–1375.

Collins, R. L. 1970. A new genetic locus mapped from behavioral variation in mice: Audiogenic seizure prone (*asp*). *Behavior Genetics,* 1, 99–109.

Collins, R. L., and Fuller, J. L. 1968. Audiogenic seizure prone (*asp*): A gene affecting behavior in linkage group VIII of the mouse. *Science,* 162, 1137–1139.

Connolly, J. A. 1978. Intelligence levels of Down's syndrome children. *American Journal of Mental Deficiency,* 83, 193–196.

Connolly, K. 1968. Report on a behavioural selection experiment. *Psychological Reports,* 23, 625–627.

Cooper, R. M., and Zubek, J. P. 1958. Effects of enriched and restricted early environments on the learning ability of bright and dull rats. *Canadian Journal of Psychology,* 12, 159–164.

Corey, L. A., and Nance, W. E. 1978. The monozygotic half-sib model: A tool for epidemiologic research. In *Twin research, Part A, Psychology and methodology,* ed. W. E. Nance, pp. 201–210. New York: Alan R. Liss.

Corney, G. 1978. Twin placentation and some effects on twins of known zygosity. In *Twin research, Part B, Biology and epidemiology,* ed. W. E. Nance, pp. 9–16. New York: Alan R. Liss.

Coyne, J. A. 1976. Lack of genic similarity between two sibling species of *Drosophila* as revealed by varied techniques. *Genetics,* 84, 593–607.

Crick, F. 1979. Split genes and RNA splicing. *Science,* 204, 264–271.

Crow, J. F., and Kimura, M. 1970. *An introduction to population genetics theory.* New York: Harper & Row.

Crowe, R. R. 1972. The adopted offspring of women criminal offenders: A study of their arrest records. *Archives of General Psychiatry,* 27, 600–603.

———. 1974. An adoption study of antisocial personality. *Archives of General Psychiatry,* 31, 785–791.

Culliton, B. J. 1975*a*. Amniocentesis: HEW backs test for prenatal diagnosis of disease. *Science,* 190, 537–540.

———. 1975*b*. XYY: Harvard researcher under fire stops newborn screening. *Science,* 188, 1284–1285.

———. 1976. Genetic screening: States may be writing the wrong kinds of laws. *Science,* 191, 926–929.

Cunningham, L.; Cadoret, R. J.; Loftus, R.; and Edwards, J. E. 1975. Studies of adoptees from psychiatrically disturbed biological parents: Psychiatric conditions in childhood and adolescence. *British Journal of Psychiatry,* 126, 534–549.

Darwin, C. 1859. *On the origin of species by means of natural selection, or the preservation of favoured races in the struggle for life.* London: John Murray. (New York: Modern Library, 1967.)

———. 1868. *The variation of animals and plants under domestication.* New York: Orange Judd.

———. 1871. *The descent of man and selection in relation to sex.* London: John Murray. (New York: Modern Library, 1967.)

———. 1872. *The expression of the emotions in man and animals.* London: John Murray.

———. 1888. Letter from C. Darwin to C. Lyell, 1858. In *The life and letters of Charles Darwin,* vol. 2, ed. F. Darwin, p. 117. London: John Murray.

———. 1896. *Journal of researches into the natural history and geology of the countries visited during the voyage of H. M. S. Beagle round the world, under the command of Capt. Fitz Roy, T. N.* New York: Appleton.

Davis, H. 1849. *The works of Plato,* vol. 2. London: Bohn.

Dawkins, R. 1976. *The selfish gene.* New York: Oxford University Press.

Deckard, B. S.; Lieff, B.; Schlesinger, K.; and DeFries, J. C. 1976. Developmental patterns of seizure susceptibility in inbred strains of mice. *Developmental Psychobiology,* 9, 17–24.

Deckard, B. S.; Tepper, J. M.; and Schlesinger, K. 1976. Selective breeding for acoustic priming. *Behavior Genetics,* 6, 375–383.

DeFries, J. C. 1969. Pleiotropic effects of albinism on open field behaviour in mice. *Nature,* 221, 65–66.

———. 1972. Quantitative aspects of genetics and environment in the determination of behavior. In *Genetics, environment, and behavior: Implications for educational policy,* eds. L. Ehrman, G. S. Omenn, and E. Caspari, pp. 6–16. New York: Academic Press.

DeFries, J. C.; Ashton, G. C.; Johnson, R. C.; Kuse, A. R.; McClearn, G. E.; Mi, M. P., Rashad, M. N.; Vandenberg, S. G.; and Wilson, J. R. 1976. Parent-offspring resemblance for specific cognitive abilities in two ethnic groups. *Nature,* 261, 131–133.

DeFries, J. C.; Gervais, M. C.; and Thomas, E. A. 1978. Response to 30 generations of selection for open-field activity in laboratory mice. *Behavior Genetics,* 8, 3–13.

DeFries, J. C., and Hegmann, J. P. 1970. Genetic analysis of open-field behavior. In *Contributions to behavior-genetic analysis: The mouse as a prototype,* eds. G. Lindzey and D. D. Thiessen, pp. 23–56. New York: Appleton-Century-Crofts.

DeFries, J. C.; Hegmann, J. P.; and Weir, M. W. 1966. Open-field behavior in mice: Evidence for a major gene effect mediated by the visual system. *Science,* 154, 1577–1579.

DeFries, J. C.; Johnson, R. C.; Kuse, A. R.; McClearn, G. E.; Polovina, J.; Vandenberg, S. G.; and Wilson, J. R. 1979. Familial resemblance for specific cognitive abilities. *Behavior Genetics,* 9, 23–43.

DeFries, J. C.; Kuse, A. R.; and Vandenberg, S. G. 1979, in press. Genetic correlations, environmental correlations and behavior. In *Theoretical advances in behavior genetics,* ed. J. R. Royce. Alphen aan den Rijn, Netherlands: Sijthoff Noordhoff International.

DeFries, J. C., and Plomin, R. 1978. Behavioral genetics. *Annual Review of Psychology,* 29, 473–515.

DeFries, J. C.; Thomas, E. A.; Hegmann, J. P.; and Weir, M. W. 1967. Open-field behavior in mice: Analysis of maternal effects by means of ovarian transplantation. *Psychonomic Science,* 8, 207–208.

DeFries, J. C.; Vandenberg, S. G.; and McClearn, G. E. 1976. The genetics of specific cognitive abilities. *Annual Review of Genetics,* 10, 179–207.

deGrouchy, J.; Turleau, C.; and Finaz, C. 1978. Chromosomal phylogeny of the primates. *Annual Review of Genetics,* 12, 289–328.

Dickerson, R. E. 1978. Chemical evolution and the origin of life. *Scientific American,* 239, 70–86.

Dobzhansky, Th. 1964. *Heredity and the nature of man.* New York: Harcourt, Brace & World.

Dobzhansky, Th., and Pavlovsky, O. 1971. Experimentally created incipient species of *Drosophila*. *Nature,* 230, 289–292.

Dudai, Y.; Jan, Y. N.; Byers, D.; Quinn, W. G.; and Benzer, S. 1976. Dunce, a mutant of *Drosophila* deficient in learning. *Proceedings of the National Academy of Sciences,* 73, 1684–1688.

Dunner, D. L., and Fieve, R. R. 1975. Psychiatric illness in fathers of men with bipolar primary affective disorder. *Archives of General Psychiatry,* 32, 1134–1137.

Dunner, D. L.; Fleiss, J. L.; Addonizio, G.; and Fieve, R. R. 1976. Assortative mating in primary affective disorder. *Biological Psychiatry,* 11, 43–51.

Dunner, D. L.; Gershon, E. S.; and Goodwin, F. K. 1976. Heritable factors in the severity of affective illness. *Biological Psychiatry,* 11, 31–42.

Dworkin, R., and Haber, S. 1979, in press. *Handbook of behavior genetics.* Hillsdale, N.J.: Lawrence Erlbaum Associates.

East, E. M., and Hayes, H. K. 1911. Inheritance in maize. *Bulletin of the Connecticut Agricultural Experiment Station,* 167, 1–142.

Eaves, L. J. 1975. Testing models for variation in intelligence. *Heredity,* 34, 132–136.

Eaves, L. J., and Eysenck, H. J. 1974. Genetics and the development of social attitudes. *Nature,* 249, 288–289.

Eaves, L. J., and Gale, J. S. 1974. A method for analyzing the genetic basis of covariation. *Behavior Genetics,* 4, 253–267.

Eaves, L. J.; Last, K. A.; Young, P. A.; and Martin, N. G. 1978. Model-fitting approaches to the analysis of human behaviour. *Heredity,* 41, 249–320.

Ebert, P. D., and Hyde, J. S. 1976. Selection for agonistic behavior in wild female *Mus musculus*. *Behavior Genetics,* 6, 291–304.

Ehrman, L. 1970. The mating advantage of rare males in *Drosophila*. *Proceedings of the National Academy of Sciences,* 65, 345–348.

Ehrman, L., and Parsons, P. A. 1976. *The genetics of behavior.* Sunderland, Mass.: Sinauer Associates.

Ehrman, L., and Probber, J. 1978. Rare *Drosophila* males: The mysterious matter of choice. *American Scientist,* 66, 216–222.

Eiseley, L. 1959. Charles Darwin, Edward Blyth, and the theory of natural selection. *Proceedings of the American Philosophical Society,* 103, 94–158.

Elardo, R.; Bradley, R.; and Caldwell, B. 1975. The relation of infants' home environments to mental test performance from six to thirty-six months: A longitudinal analysis. *Child Development,* 46, 71–76.

Eleftheriou, B. E., ed. 1975. *Psychopharmacogenetics.* New York: Plenum Press.

Eleftheriou, B. E., and Elias, P. K. 1975. Recombinant inbred strains: A novel genetic approach for psychopharmacogeneticists. In *Psychopharmacogenetics,* ed. B. E. Eleftheriou, pp. 43–71. New York: Plenum Press.

Elston, R. C., and Stewart, J. 1971. A general model for the genetic analysis of pedigree data. *Human Heredity,* 21, 523–542.

Emerson, R. A., and East, E. M. 1913. The inheritance of quantitative characters in maize. *University of Nebraska Research Bulletin,* 2, 5–120.

Epstein, C. J., and Golbus, M. S. 1977. Prenatal diagnosis of genetic diseases. *American Scientist,* 65, 703–711.

Erlenmeyer-Kimling, L. 1972. Gene-environment interactions and the variability of behavior. In *Genetics, environment and behavior: Implications for educational policy,* eds. L. Ehrman, G. Omenn, and E. Caspari, pp. 181–218. New York: Academic Press.

———. 1975. A prospective study of children at risk for schizophrenia: Methodological considerations and some preliminary findings. In *Life history research on psychopathology,* eds. R. D. Wirt, G. Winokur, and M. Roff, pp. 217–252. Minneapolis: University of Minnesota Press.

———. 1976. Schizophrenia: A bag of dilemmas. Papers from a workshop on "Genetic Counseling in Psychiatric Disorders: State of the Science and the Art." *Social Biology,* 23, 123–134.

Erlenmeyer-Kimling, L.; Cornblatt, B.; and Fleiss, J. 1979. High-risk research in schizophrenia, *Psychiatric Annals,* 9, 1–13.

Erlenmeyer-Kimling, L., and Jarvik, L. F. 1963. Genetics and intelligence: A review. *Science,* 142, 1477–1479.

Essen-Möller, E. 1941. Empirische Ahnlichkeitsdiagnose bei Zwillingen. *Hereditas,* 27, 1.

Eyferth, K. 1961. Leistungen verschiedener Grup en von Besatzungskindern in Hamburg-Wechsler Intelligenztest für Kinder (HAWIK). *Archiv für die gesamte Psychologie,* 113, 222–241.

Eysenck, H. J., and Broadhurst, P. L. 1964. Experiments with animals: Introduction. In *Experiments in motivation,* ed. H. J. Eysenck, pp. 285–291. New York: Macmillan.

Falconer, D. S. 1960. *Introduction to quantitative genetics.* New York: Ronald Press.

———. 1965. The inheritance of liability to certain diseases estimated from the incidence among relatives. *Annals of Human Genetics,* 29, 51–76.

Fantino, E., and Logan, C. A. 1979. *The experimental analysis of behavior.* San Francisco: W. H. Freeman and Company.

Fieve, R. R.; Rosenthal, D.; and Brill, H., eds. 1975. *Genetics research in psychiatry.* Baltimore: Johns Hopkins University Press.

Fisch, R. O.; Bilek, M. K.; Deinard, A. S.; and Chang, P.-N. 1976. Growth, behavioral, and psychologic measurements of adopted children: The influences of genetic and socioeconomic factors in a prospective study. *Behavioral Pediatrics,* 89, 494–500.

Fischer, M.; Harvald, B.; and Hauge, M. 1969. A Danish twin study of schizophrenia. *British Journal of Psychiatry,* 115, 981–990.

Fisher, R. A. 1918. The correlation between relatives on the supposition of Mendelian inheritance. *Transactions of the Royal Society of Edinburgh,* 52, 399–433.

———. 1930. *The genetical theory of natural selection.* Oxford: Clarendon Press.

Flanagan, J. R. 1977. A method for fate mapping the foci of lethal and behavioral mutants in *Drosophila melanogaster. Genetics,* 85, 587–607.

Følling, A.; Mohr, O. L.; and Ruud, L. 1945. *Oligophrenia phenylpyrouvica,* a recessive syndrome in man. *Norske Videnskaps/Akademi I Oslo, Matematisk-Naturvidenskapelig Klasse,* 13, 1–44.

Fowler, W. A. 1965. A study of process and method in three-year-old twins and triplets learning to read. *Genetic Psychology Monographs,* 72, 3–90.

Fox, J. L. 1975. Evolution and education. *Science,* 187, 389–390.

Freeman, F. N.; Holzinger, K. J.; and Mitchell, B. C. 1928. The influence of environment on the intelligence, school achievement, and conduct of foster children. *Nature and nurture: Their influence upon intelligence. Twenty-Seventh Yearbook of the National Society for the Study of Education,* part 1, 27, 103–217.

Friedman, E.; Shopsin, B.; Sathananthan, G.; and Gershon, E. 1974. Blood platelet monoamine oxidase activity in psychiatric patients. *American Journal of Psychiatry,* 131, 1392–1394.

Friedman, G. D., and Lewis, A. M. 1978. The Kaiser-Permanente twin registry. In *Twin research, Part B, Biology and epidemiology*, ed. W. E. Nance, pp. 173–178. New York: Alan R. Liss.

Friedmann, T. 1971. Prenatal diagnosis of genetic disease. *Scientific American*, 225, 34–42.

Fulker, D. W. 1973. A biometrical genetic approach to intelligence and schizophrenia. *Social Biology*, 20, 266–275.

———. 1978. Multivariate extensions of a biometrical model of twin data. In *Twin research, Part A, Psychology and methodology*, ed. W. E. Nance, pp. 217–236. New York: Alan R. Liss.

———. 1979, in press. Some implications of biometrical genetical analysis for psychological research. In *Theoretical advances in behavior genetics*, ed. J. R. Royce. Alphen aan den Rijn, Netherlands: Sijthoff Noordhoff International.

Fuller, J. L.; Easler, C.; and Smith, M. E. 1950. Inheritance of audiogenic seizure susceptibility in the mouse. *Genetics*, 35, 622–632.

Fuller, J. L., and Thompson, W. R. 1960. *Behavior genetics*. New York: John Wiley.

———. 1978. *Foundations of behavior genetics*. St. Louis: C. V. Mosby.

Galton, F. 1865. Hereditary talent and character. *Macmillan's Magazine*, 12, 157–166, 318–327.

———. 1869. *Hereditary genius: An inquiry into its laws and consequences*. London: Macmillan. (Cleveland: World Publishing Co., 1962.)

———. 1871. Gregariousness in cattle and in men. *Macmillan's Magazine*, 23, 353–357.

———. 1876. The history of twins as a criterion of the relative powers of nature and nurture. *Royal Anthropological Institute of Great Britain and Ireland Journal*, 6, 391–406.

———. 1883. *Inquiries into human faculty and its development*. London: Macmillan.

Garrod, A. E. 1908. The Croonian lectures on inborn errors of metabolism, I, II, III, IV. *Lancet*, 2, 1–7, 73–79, 142–148, 214–220.

Gehring, W. J. 1976. Developmental genetics of *Drosophila*. *Annual Review of Genetics*, 10, 209–252.

Gesell, A. L., and Thompson, H. 1929. Learning and growth in identical infant twins. *Genetic Psychology Monographs*, 6, 5–120.

Giblett, F. R. 1969. *Genetic markers in human blood*. London: Blackwell.

Ginsburg, B. E., and Miller, D. S. 1963. Genetic factors in audiogenic seizures. *Colloques Nationaux du Centre National de la Recherche Scientifique, Paris*, 112, 217–225.

Goldberger, A. S. 1978. Pitfalls in the resolution of IQ inheritance. In *Genetic epidemiology*, eds. N. E. Morton and C. S. Chung, pp. 195–215. New York: Academic Press.

Goldberger, A. S., and Duncan, O. D., eds. 1973. *Structural equation models in the social sciences*. New York: Seminar Press.

Goodman, M.; Koen, A. L.; Barnabas, J.; and Moore, G. W. 1971. Evolving primate genes and proteins. In *Comparative genetics in monkeys, apes, and man*, ed. A. B. Chiarelli, pp. 150–212. New York: Academic Press.

Goodwin, D. W. 1979. Alcoholism and heredity. *Archives of General Psychiatry*, 36, 57–61.

Goodwin, D. W.; Schulsinger, F.; Hermansen, L.; Guze, S. B.; and Winokur, G. 1973. Alcohol problems in adoptees raised apart from alcoholic biological parents. *Archives of General Psychiatry*, 28, 238–243.

Goodwin, D. W.; Schulsinger, F.; Moller, N.; Hermansen, L.; Winokur, G.; and Guze, S. B. 1974. Drinking problems in adopted and nonadopted sons of alcoholics. *Archives of General Psychiatry,* 31, 164–169.

Gorlin, R. J. 1977. Classical chromosome disorders. In *New chromosomal syndromes,* ed. J. J. Yunis, pp. 59–118. New York: Academic Press.

Gottesman, I. I., and Shields, J. 1972. *Schizophrenia and genetics: A twin study vantage point.* New York: Academic Press.

———. 1976. A critical review of recent adoption, twin, and family studies of schizophrenia: Behavioral genetics perspectives. *Schizophrenia Bulletin,* 2, 360–401.

———. 1977. Twin studies and schizophrenia a decade later. In *Contributions to the psychopathology of schizophrenia,* ed. B. A. Maher. New York: Academic Press.

Grabiner, J. V., and Miller, P. D. 1974. Effects of the Scopes trial. *Science,* 185, 832–837.

Green, E. L. 1966. Breeding systems. In *Biology of the laboratory mouse.* 2d ed., ed. E. L. Green. New York: McGraw-Hill.

Gregory, M. S.; Silvers, A.; and Sutch, D., eds. 1978. *Sociobiology and human nature.* San Francisco: Jossey-Bass.

Griffiths, M. I., and Phillips, C J. 1976. *Twin research* (Birmingham, 1968–1972). Institute of Child Health, University of Birmingham.

Grossfield, J. 1976. Behavioral mutants of *Drosophila.* In *Handbook of genetics,* ed. R. C. King. New York: Plenum Press.

Grotevant, H. D.; Scarr, S.; and Weinberg, R. A. 1977. Intellectual development in family constellations with adopted and natural children: A test of the Zajonc and Markus model. *Child Development,* 48, 1699–1703.

Gruneberg, H. 1952. *The genetics of the mouse.* The Hague: Martinus Nijhoff.

Guilford, J. P. 1967. *The nature of human intelligence.* New York: McGraw-Hill.

Guttman, R. 1974. Genetic analysis of analytical spatial ability: Raven's Progressive Matrices. *Behavior Genetics,* 4, 273–284.

Haggard, E. A. 1958. *Intraclass correlation and the analysis of variance.* New York: Holt, Rinehart & Winston.

Haldane, J. B. S. 1936. A search for incomplete sex-linkage in man. *Annals of Eugenics,* 7, 28–57.

Hall, C. S. 1934. Emotional behavior in the rat. I. Defecation and urination as measures of individual differences in emotionality. *Journal of Comparative Psychology,* 18, 385–403.

Hall, J. C. 1977*a*. Behavioral analysis in *Drosophila* mosaics. In *Genetic mosaics and cell differentiation,* ed. W. J. Gehring, pp. 1–78. New York: Springer.

———. 1977*b*. Portions of the central nervous system controlling reproductive behavior in *Drosophila melanogaster. Behavior Genetics,* 7, 291–312.

Halstead, L. B. 1978. New light on Piltdown hoax? *Nature,* 276, 11–13.

Hamilton, W. D. 1964. The genetical theory of social behaviour (I and II). *Journal of Theoretical Biology,* 7, 1–52.

Hardy, G. H. 1908. Mendelian proportions in a mixed population. *Science,* 28, 49–50.

Harris, H. 1967. Enzyme variation in man: Some general aspects. In *Proceedings of the Third International Congress of Human Genetics,* eds. J. F. Crow and J. V. Neel, pp. 207–214. Baltimore: Johns Hopkins University Press.

Hay, D. A. 1975. Strain differences in maze-learning ability of *Drosophila melanogaster. Nature,* 257, 44–46.

———. 1976. The behavioral phenotype and mating behavior of two inbred strains of *Drosophila melanogaster*. *Behavior Genetics,* 6, 161–170.

Hazel, L. N. 1943. The genetic basis for constructing selection indexes. *Genetics,* 28, 476–490.

Hearnshaw, L. S. 1979. *Cyril Burt, psychologist.* Ithaca, N.Y.: Cornell University Press.

Hegmann, J. P. 1975. The response to selection for altered conduction velocity in mice. *Behavioral Biology,* 13, 413–423.

Hegmann, J. P., and DeFries, J. C. 1970. Are genetic correlations and environmental correlations correlated? *Nature,* 226, 284–286.

Henderson, N. D. 1967. Prior treatment effects on open field behavior of mice—A genetic analysis. *Animal Behaviour,* 15, 365–376.

———. 1972. Relative effects of early rearing environment on discrimination learning in housemice. *Journal of Comparative and Physiological Psychology,* 79, 243–253.

Henry, K. R. 1967. Audiogenic seizure susceptibility induced in C57BL/6J mice by prior auditory exposure. *Science,* 158, 938–940.

Henry, K. R., and Schlesinger, K. 1967. Effects of the albino and dilute loci on mouse behavior. *Journal of Comparative and Physiological Psychology,* 63, 320–323.

Heston, L. L. 1966. Psychiatric disorders in foster home reared children of schizophrenic mothers. *British Journal of Psychiatry,* 112, 819–825.

Hildreth, G. H. 1925. *The resemblance of siblings in intelligence and achievement.* New York: Columbia University Teachers College.

Hilgard, J. R. 1933. The effect of early and delayed practice on memory and motor performance studied by the method of cotwin control. *Genetic Psychology Monographs,* 14, 493–567.

Hirsch, J. 1963. Behavior genetics and individuality understood. *Science,* 142, 1436–1442.

Hockett, C. F. 1973. *Man's place in nature.* New York: McGraw-Hill.

Homyk, T., Jr. 1977. Behavioral mutants of *Drosophila melanogaster.* II. Behavioral analysis and focus mapping. *Genetics,* 87, 105–128.

Homyk, T., Jr., and Sheppard, D. E. 1977. Behavioral mutants of *Drosophila melanogaster.* I. Isolation and mapping of mutations which decrease flight ability. *Genetics,* 87, 95–104.

Honzik, M. P. 1957. Developmental studies of parent-child resemblance in intelligence. *Child Development,* 28, 215–228.

Hook, E. B. 1973. Behavioral implications of the human XYY genotype. *Science,* 179, 139–150.

Hopkinson, D. A., and Harris, H. 1971. Recent work on isozymes in man. *Annual Review of Genetics,* 5, 5–32.

Horn, J. M.; Loehlin, J. C.; and Willerman, L. 1979. Intellectual resemblance among adoptive and biological relatives: The Texas adoption project. *Behavior Genetics,* 9, 177–208.

Hotta, Y., and Benzer, S. 1970. Genetic dissection of the *Drosophila* nervous system by means of mosaics. *Proceedings of the National Academy of Sciences,* 67, 1156–1163.

Hrubec, Z., and Neel, J. V. 1978. The National Academy of Sciences-National Research Council twin registry: Ten years of operation. In *Twin research, Part B, Biology and epidemiology,* ed. W. E. Nance, pp. 153–172. New York: Alan R. Liss.

Hsia, D. Y.-Y. 1968. *Human developmental genetics.* Chicago: Year Book Medical Publishers.

———. 1970. Phenylketonuria and its variants. In *Progress in medical genetics,* vol. 7, eds. A. G. Steinberg and A. G. Bearn, pp. 29–68. New York: Grune & Stratton.

Hsia, D. Y.-Y.; Knox, W. E.; Quinn, K. V.; and Paine, R. S. 1958. A one-year controlled study of the effect of low-phenylalanine diet on phenylketonuria. *Pediatrics,* 21, 178–202.

Husen, T. 1959. *Psychological twin research.* Stockholm: Almqvist & Wiksell.

Hutchings, B., and Mednick, S. A. 1975. Registered criminality in the adoptive and biological parents of registered male criminal adoptees. In *Genetic research in psychiatry,* eds. R. R. Fieve, D. Rosenthal, and H. Brill, pp. 105–116. Baltimore: Johns Hopkins University Press.

Hutchins, R. M., and Adler, M. J. 1968. The idea of equality. In *The great ideas today,* eds. R. M. Hutchins and M. J. Adler, pp. 302–350. Chicago: Encyclopaedia Britannica, 1968.

Huxley, J. 1949. *Heredity east and west: Lysenko and world science.* New York: Schuman.

Hyde, J. S. 1974. Inheritance of learning ability in mice: A diallel-environmental analysis. *Journal of Comparative and Physiological Psychology,* 86, 116–123.

Itakura, K.; Hirose, T.; Crea, R.; Riggs, A. D.; Heyneker, H. L.; Bolivar, F.; and Boyer, H. W. 1977. Expression in *Escherichia coli* of a chemically synthesized gene for the hormone somatostatin. *Science,* 198, 1056–1063.

Jacob, F., and Monod, J. 1961. On the regulation of gene activity. *Cold Spring Harbor Symposia on Quantitative Biology,* 26, 193–209.

Jacobs, P. A.; Brunton, M.; Melville, M. M.; Brittain, R. P.; and McClemont, W. F. 1965. Aggressive behaviour, mental sub-normality, and the XYY male. *Nature,* 208, 1351–1352.

Jarvik, L. F. 1975. Thoughts on the psychobiology of aging. *American Psychologist,* 30, 576–583.

Jencks, C. 1972. *Inequality: A reassessment of the effect of family and schooling in America.* New York: Basic Books.

Jensen, A. R. 1967. Estimation of the limits of heritability of traits by comparison of monozygotic and dizygotic twins. *Proceedings of the National Academy of Sciences,* 58, 149–156.

———. 1969. How much can we boost IQ and scholastic achievement? *Harvard Educational Review,* 39(1), 1–123.

———. 1970. IQ's of identical twins reared apart. *Behavior Genetics,* 1, 133–148.

———. 1972. *Genetics and education.* New York: Harper & Row.

———. 1974*a*. Kinship correlations reported by Sir Cyril Burt. *Behavior Genetics,* 4, 1–28.

———. 1974*b*. The problem of genotype-environment correlation in the estimation of heritability from monozygotic and dizygotic twins. *Acta Geneticae Medicae et Gemellogiae,* 23, 21.

———. 1976. Test bias and construct validity. *Phi Delta Kappan,* December, 340–346.

———. 1978*a*. Genetic and behavioral effects of nonrandom mating. In *Human variation: The biopsychology of age, race and sex,* eds. R. T. Osborne, C. E. Noble, and N. Weyl, pp. 51–105. New York: Academic Press.

———. 1978*b*. Sir Cyril Burt in perspective. *American Psychologist,* 33, 499–503.

———. 1979. *Bias in mental tests*. New York: Free Press.

Jinks, J. L., and Fulker, D. W. 1970. Comparison of the biometrical genetical, MAVA, and classical approaches to the analysis of human behavior. *Psychological Bulletin,* 75, 311–349.

Johnson, R. C.; DeFries, J. C.; Wilson, J. R.; McClearn, G. E.; Vandenberg, S. G.; Ashton, G. C.; Mi, M. P.; and Rashad, M. N. 1976. Assortative marriage for specific cognitive abilities in two ethnic groups. *Human Biology,* 48, 343–352.

Juel-Nielsen, N. 1965. Individual and environment: A psychiatric-psychological investigation of monozygous twins reared apart. *Acta Psychiatrica et Neurologica Scandinavica,* supplement 183.

Kallmann, F. J. 1946. The genetic theory of schizophrenia: An analysis of 691 schizophrenic twin index families. *American Journal of Psychiatry,* 103, 309–322.

Kamin, L. J. 1974. *The science and politics of I.Q.* Potomac, Md.: Lawrence Erlbaum Associates.

Kaprio, J.; Sarna, S.; Koskenvuo, M.; and Rantasalo, I. 1978. The Finnish twin registry: Formation and compilation, questionnaire study, zygosity determination procedures, and research program. In *Twin research, Part B, Biology and epidemiology,* ed. W. E. Nance, pp. 179–184. New York: Alan R. Liss.

Karlsson, J. L. 1966. *The biologic basis of schizophrenia*. Springfield, Ill.: Thomas.

———. 1970. The rate of schizophrenia in foster-reared close relatives of schizophrenic index cases. *Biological Psychiatry,* 2, 285–290.

Keeler, C. 1968. Some oddities in the delayed appreciation of "Castle's law." *Journal of Heredity,* 59, 110–112.

Kekic, V., and Marinkovic, D. 1974. Multiple-choice selection for light preference in *Drosophila subobscura*. *Behavior Genetics,* 4, 285–300.

Kessler, S. 1975. Extra chromosomes and criminality. In *Genetic research in psychiatry,* eds. R. R. Fieve, D. Rosenthal, and H. Brill, pp. 65–73. Baltimore: Johns Hopkins University Press.

Kessler, S., and Moos, R. H. 1970. The XYY karyotype and criminality: A review. *Journal of Psychiatric Research,* 7, 153–170.

Kety, S. S. 1975. The Paul H. Hoch award lecture: Progress toward an understanding of the biological substrates of schizophrenia. In *Genetic research in psychiatry,* eds. R. R. Fieve, D. Rosenthal, and H. Brill, pp. 15–26. Baltimore: Johns Hopkins University Press.

Kety, S. S.; Rosenthal, D.; Wender, P. H.; and Schulsinger, F. 1968. The types and prevalence of mental illness in the biological and adoptive families of adopted schizophrenics. *Journal of Psychiatric Research,* 6, 345–362.

———. 1971. Mental illness in the biological and adoptive families of adopted schizophrenics. *American Journal of Psychiatry,* 128, 302–306.

———. 1976. Studies based on a total sample of adopted individuals and their relatives: Why they were necessary, what they demonstrated and failed to demonstrate. *Schizophrenia Bulletin,* 2, 413–428.

Kety, S. S.; Rosenthal, D.; Wender, P. H.; Schulsinger, F.; and Jacobsen, B. 1975. Mental illness in the biological and adoptive families of adopted individuals who have become schizophrenic: A preliminary report based on psychiatric interviews. In *Genetic research in psychiatry,* eds. R. R. Fieve, D. Rosenthal, and H. Brill, pp. 147–166. Baltimore: Johns Hopkins University Press.

Khorana, H. G. 1979. Total synthesis of a gene. *Science,* 203, 614–625.

Kidd, K. K., and Matthysse, S. 1978. Research designs for the study of gene-environment interactions in psychiatric disorders. *Archives of General Psychiatry,* 35, 925–932.

King, M. C., and Wilson, A. C. 1975. Evolution at two levels in humans and chimpanzees. *Science,* 188, 107–115.

Klein, T. W.; DeFries, J. C.; and Finkbeiner, C. T. 1973. Heritability and genetic correlations: Standard errors of estimates and sample size. *Behavior Genetics,* 4, 355–364.

Konner, M. J. 1977. Evolution of human behavior development. In *Cross-cultural and social class influences in infancy,* eds. H. Leiderman and S. Tulkin. New York: Academic Press.

Kringlen, E. 1966. Schizophrenia in twins, an epidemiological-clinical study. *Psychiatry,* 29, 172–184.

———. 1978. Norwegian twin registers. In *Twin research, Part B, Biology and epidemiology,* ed. W. F. Nance, pp. 185–188. New York: Alan R. Liss.

Kung, C.; Chang, S. Y.; Satow, Y.; Van Houten, J.; and Hansma, H. 1975. Genetic dissection of behavior in *Paramecium. Science,* 188, 898–904.

Kupfer, D. J.; Pickar, D.; Himmelhoch, J. M.; and Detre, T. P. 1975. Are there two types of unipolar depression? *Archives of General Psychiatry,* 32, 866–871.

Lack, D. 1953. Darwin's finches. *Scientific American,* 188(4), 66–72.

Lawrence, E. M. 1931. An investigation into the relation between intelligence and inheritance. *British Journal of Psychology,* supplement 16, 1–80.

Leahy, A. M. 1935. A study of certain selective factors influencing prediction of the mental status of adopted children. *Journal of Genetic Psychology,* 41, 294–329.

Lee, R. B., and DeVore, I., eds. 1968. *Man the hunter.* Chicago: Aldine.

Lehtovaara, A. 1938. *Psychologische Zwillingsuntersuchungen.* Helsinki: Finnish Academy of Science.

Lerner, I. M. 1968. *Heredity, evolution, and society.* San Francisco: W. H. Freeman and Company.

Lerner, I. M., and Libby, W. J. 1976. *Heredity, evolution, and society.* San Francisco: W. H. Freeman and Company.

Lewandowski, R. C., and Yunis, J. J. 1977. Phenotypic mapping in man. In *New chromosomal syndromes,* ed. J. J. Yunis, pp. 369–394. New York: Academic Press.

Li, C. C. 1975. *Path analysis: A primer.* Pacific Grove, Calif.: Boxwood Press.

Loehlin, J. C. 1965. A heredity-environment analysis of personality inventory data. In *Methods and goals in human behavior genetics,* ed. S. G. Vandenberg, pp. 163–170. New York: Academic Press.

———. 1978. Heredity-environment analyses of Jencks's IQ correlations. *Behavior Genetics,* 8, 415–436.

———. 1979, in press. Combining data from different groups in human behavior genetics. In *Theoretical advances in behavior genetics.* ed. J. R. Royce. Alphen aan den Rijn, Netherlands; Sijthoff Noordhoff International.

Loehlin, J. C.; Lindzey, G.; and Spuhler, J. N. 1975. *Race differences in intelligence.* San Francisco: W. H. Freeman and Company.

Loehlin, J. C., and Nichols, R. C. 1976. *Heredity, environment and personality.* Austin: University of Texas Press.

Loehlin, J. C.; Sharan, S.; and Jacoby, R. 1978. In pursuit of the "spatial gene": A family study. *Behavior Genetics,* 8, 227–241.

Loehlin, J. C., and Vandenberg, S. G. 1968. Genetic and environmental components in the covariation of cognitive abilities: An additive model. In *Progress in human behavior genetics,* ed. S. G. Vandenberg, pp. 261–285. Baltimore: Johns Hopkins University Press.

Loehlin, J. C.; Vandenberg, S. G.; and Osborne, R. T. 1973. Blood group genes and Negro-white ability differences. *Behavior Genetics,* 3, 263–270.

Loehlin, J. C.; Willerman, L.; and Vandenberg, S. G. 1974. Blood group and behavioral differences among dizygotic twins: A failure to replicate. *Social Biology,* 21, 205–206.

Lubs, H. A., and de la Cruz, F., eds. 1977. *Genetic counseling: A monograph of the National Institute of Child Health and Human Development.* New York: Raven Press.

Lush, J. L. 1949. Heritability of quantitative characters in farm animals. *Hereditas,* suppl. vol., 356–375.

———. 1951. Genetics and animal breeding. In *Genetics in the twentieth century,* ed. L. C. Dunn, pp. 493–525. New York: Macmillan.

Lyon, M. F. 1958. Twirler: A mutant affecting the inner ear of the house mouse. *Journal of Embryology and Experimental Morphology,* part 1, 105–116.

McAskie, M., and Clarke, A. M. 1976. Parent-offspring resemblance in intelligence: Theories and evidence. *British Journal of Psychology,* 67, 243–273.

McCabe, M. S. 1976. Reactive psychoses and schizophrenia with good prognosis. *Archives of General Psychiatry,* 33, 571–576.

McClearn, G. E. 1960. Strain differences in activity of mice: Influence of illumination. *Journal of Comparative and Physiological Psychology,* 53, 142–143.

———. 1963. The inheritance of behavior. In *Psychology in the making,* ed. L. J. Postman, pp. 144–152. New York: Alfred A. Knopf.

———. 1976. Experimental behavioural genetics. In *Aspects of genetics in paediatrics,* ed. D. Barltrop, pp. 31–39. London: Fellowship of Postdoctorate Medicine, 1976.

McClearn, G. E., and DeFries, J. C. 1973. *Introduction to behavioral genetics.* San Francisco: W. H. Freeman and Company.

McClearn, G. E.; Wilson, J. R.; and Meredith, W. 1970. The use of isogenic and heterogenic mouse stocks in behavioral research. In *Contributions to behavior-genetic analysis: The mouse as a prototype,* eds. G. Lindzey and D. D. Thiessen, pp. 3–22. New York: Appleton-Century-Crofts.

McDougall, W. 1908. *An introduction to social psychology.* London: Methuen.

McKusick, V. A. 1975. *Mendelian inheritance in man.* 4th ed. Baltimore: Johns Hopkins University Press.

Martin, N. G. 1975. The inheritance of scholastic abilities in a sample of twins. *Annals of Human Genetics,* 39, 219–229.

Martin, N. G., and Eaves, L. J. 1977. The genetical analysis of covariance structure. *Heredity,* 38, 79–95.

Martin, N. G.; Eaves, L. J.; and Eysenck, H. J. 1977. Genetical, environmental and personality factors influencing the age of first sexual intercourse in twins. *Journal of Biosocial Science,* 9, 91–97.

Mason, J. J., and Price, E. O. 1973. Escape conditioning in wild and domestic Norway rats. *Journal of Comparative and Physiological Psychology,* 84, 403–407.

Matheny, A. P., and Brown, A. M. 1974. A twin study of genetic influences in reading achievement. *Journal of Learning Disabilities,* 7, 99–102.

Matheny, A. P., and Bruggemann, C. E. 1973. Children's speech: Hereditary components and sex differences. *Folia Phoniatrica,* 25, 442–449.

Matheny, A. P.; Wilson, R. S.; and Dolan, A. B. 1976. Relations between twins' similarity of appearance and behavioral similarity: Testing an assumption. *Behavior Genetics,* 6, 343–352.

Maxson, S. C., and Cowen, J. S. 1976. Electroencephalographic correlates of the audiogenic seizure response of inbred mice. *Physiology and Behavior,* 16, 623–629.

Mayr, E. 1965. *Animal species and evolution.* Cambridge, Mass.: The Belknap Press of Harvard University.

———. 1974. Behavior programs and evolutionary strategies. *American Scientist,* 62, 650–659.

———. 1978. Evolution. *Scientific American,* 239, 47–55.

Mednick, S. A., and Christiansen, K. O., eds. 1977. *Biosocial bases of criminal behavior.* New York: Gardner Press.

Medvedev, Z. A. 1969. *The rise and fall of T. D. Lysenko.* New York: Columbia University Press.

Mendel, G. J. 1866. Versuche Ueber Pflanzenhybriden. *Verhandlungen des Naturforschunden Vereines in Bruenn,* 4, 3–47. (Translated by Royal Horticultural Society of London.) Available in: Dodson, E. O. 1956. *Genetics.* Philadelphia: W. B. Saunders. Also available in: Sinnott, E. W.; Dunn, L. C.; and Dobzhansky, Th. 1950. *Principles of genetics.* New York: McGraw-Hill.

Mendlewicz, J., and Fleiss, J. L. 1974. Linkage studies with X-chromosome markers in bipolar (manic-depressive) and unipolar (depressive) illnesses. *Biological Psychiatry,* 9, 261–294.

Mendlewicz, J., and Rainer, J. D. 1977. Adoption study supporting genetic transmission in manic-depressive illness. *Nature,* 268, 327–329.

Meredith, W. 1973. A model for analyzing heritability in the presence of correlated genetic and environmental effects. *Behavior Genetics,* 3, 271–277.

Merrell, D. J. 1962. *Evolution and genetics: The modern theory of evolution.* New York: Holt, Rinehart & Winston.

Merriman, C. 1924. The intellectual resemblance of twins. *Psychological Monographs,* 33, 1–58.

Mirenva, A. N. 1935. Psychomotor education and the general development of preschool children. *Pedagogical Seminary and Journal of Genetic Psychology,* 46, 433–454.

Mittler, P. 1971. *The study of twins.* Harmondsworth, England: Penguin Books.

———. 1974. *Language development in young twins: Biological, genetic and social aspects.* Paper presented at 1st International Congress on Twin Studies, Rome.

Mohr, J. 1954. *A study of linkage in man.* Copenhagen: Munksgaard.

Money, J. 1964. Two cytogenetic syndromes: Psychologic comparisons. I. Intelligence and specific-factor quotients. *Journal of Psychiatric Research,* 2, 223–231.

———. 1968. Cognitive deficits in Turner's syndrome. In *Progress in human behavior genetics,* ed. S. G. Vandenberg, pp. 27–30. Baltimore: Johns Hopkins University Press.

Money, J.; Annecillo, C.; Van Orman, B.; and Borgaonkar, D. S. 1974. Cytogenetics, hormones and behavior disability: Comparison of XYY and XXY syndromes. *Clinical Genetics,* 6, 370–382.

Montagu, A. 1959. The reception of Darwin's theory. Introduction to *Man's place in nature* by T. H. Huxley. Ann Arbor: University of Michigan Press.

Morgan, T. H.; Sturtevant, A. H.; Muller, H. J.; and Bridges, C. B. 1915. *The mechanism of Mendelian heredity*. New York: Henry Holt.

Morton, N. E. 1958. Segregation analysis in human genetics. *Science,* 127, 79–80.

Motulsky, A. G., and Omenn, G. S. 1975. Special award lecture: Biochemical genetics and psychiatry. In *Genetic research in psychiatry,* eds. R. R. Fieve, D. Rosenthal, and H. Brill, pp. 3–14. Baltimore: Johns Hopkins University Press.

Muller, H. J. 1927. Artificial transmutation of the gene. *Science,* 66, 84–87.

Munsinger, H. 1975*a*. The adopted child's IQ: A critical review. *Psychological Bulletin,* 82, 623–659.

———. 1975*b*. Children's resemblance to their biological and adopting parents in two ethnic groups. *Behavior Genetics,* 5, 239–254.

———. 1977. The identical twin transfusion syndrome: A source of error in estimating IQ resemblance and heritability. *Annals of Human Genetics,* 40, 307–321.

Munsinger, H., and Douglass, A. 1976. The syntactic abilities of identical twins, fraternal twins, and their siblings. *Child Development,* 47, 40–50.

Nadler, C. F., and Borges, W. H. 1966. Chromosomal structure and behavior. In *Lectures in medical genetics,* ed. D. Y.-Y. Hsia, pp. 25–58. Chicago: Year Book Medical Publishers.

Nance, W. E. 1976. Genetic studies of the offspring of identical twins. *Acta Genetica Medicae et Gemellologiae,* 25, 103–113.

Nance, W. E., and Corey, L. A. 1976. Genetic models for the analysis of data from families of identical twins. *Genetics,* 83, 811–826.

Nance, W. E.; Winter, P. M.; Segreti, W. O.; Corey, L. A.; Parisi-Prinzi, G.; and Parisi, P. 1978. A search for evidence of hereditary superfetation in man. In *Twin research, Part B, Biology and epidemiology,* ed. W. E. Nance, pp. 65–70. New York: Alan R. Liss.

Newman, J.; Freeman, F.; and Holzinger, K. 1937. *Twins: A study of heredity and environment.* Chicago: University of Chicago Press.

Nichols, R. C. 1965. The National Merit twin study. In *Methods and goals in human behavior genetics,* ed. S. G. Vandenberg, pp. 231–243. New York: Academic Press.

Nichols, R. C., and Bilbro, W. C. 1966. The diagnosis of twin zygosity. *Acta Genetica,* 16, 265–275.

Nilsson-Ehle, H. 1908–1909. Einige Ergebnisse von Kruezungen bei Hafer und Weisen. *Botanische Notiser,* 257–294.

Noël, B.; Duport, J. P.; Revil, D.; Sussuyer, I.; and Quack, B. 1974. The XYY syndrome: Reality or myth? *Clinical Genetics,* 5, 387–394.

Nuttall, G. H. F. 1904. *Blood immunity and blood relationship*. Cambridge, England: Cambridge University Press.

Nylander, P. P. S. 1978. Causes of high twinning frequencies in Nigeria. In *Twin research, Part B, Biology and epidemiology,* ed. W. E. Nance, pp. 35–44. New York: Alan R. Liss.

Oliverio, A.; Eleftheriou, B. E.; and Bailey, D. W. 1973. A gene influencing active avoidance performance in mice. *Physiology and Behavior,* 11, 497–501.

Omenn, G. S. 1978. Prenatal diagnosis of genetic disorders. *Science,* 200, 952–958.

Packard, A. S. 1901. *Lamarck, the founder of evolution*. New York: Longmans, Green.

Padeh, B.; Wahlsten, D.; and DeFries, J. C. 1974. Operant discrimination learning and operant bar-pressing rates in inbred and heterogeneous laboratory mice. *Behavior Genetics,* 4, 383–393.

Palm, J. 1961. Transplantation of ovarian tissue. In *Transplantation of tissues and cells,* eds. R. E. Billingham and W. K. Silvers, pp. 49–56. Philadelphia: The Wistar Institute Press.

Parkinson, J. S. 1977. Behavioral genetics in bacteria. *Annual Review of Genetics,* 11, 397–414.

Pearson, K. 1924. *The life, letters and labours of Francis Galton,* vol. 1. London: Cambridge University Press.

Penrose, L. S., and Smith, G. F. 1966. *Down's anomaly.* London: J. & A. Churchill.

Perris, C. 1966. A study of bipolar (manic-depressive) and unipolar recurrent depressive psychoses. *Acta Psychiatrica Scandinavica,* 42, supplement 194.

———. 1976. Frequency and hereditary aspects of depression. In *Depression: Behavioral, biochemical, diagnostic and treatment concepts,* eds. D. M. Gallant and G. M. Simpson. New York: Spectrum.

Petit, C. 1951. Le rôle de l'isolement sexuel dans l'évolution des populations de *Drosophila melanogaster. Bulletin Biologique de la France et de la Belgique,* 85, 392–418.

Petit, C., and Ehrman, L. 1969. Sexual selection in *Drosophila.* In *Evolutionary biology,* vol. 3, eds. Th. Dobzhansky, M. K. Hecht, and W. C. Steere, pp. 177–217. New York: Appleton-Century-Crofts.

Plomin, R., and DeFries, J. C. 1980*a*. Genetics and intelligence: Recent data. *Intelligence,* 4, 15–24.

———. 1980*b*, in press. Multivariate behavioral genetic analysis of twin data on scholastic abilities. *Behavior Genetics.*

Plomin, R.; DeFries, J. C.; and Loehlin, J. C. 1977. Genotype-environment interaction and correlation in the analysis of human behavior. *Psychological Bulletin,* 84, 309–322.

Plomin, R.; DeFries, J. C.; and Roberts, M. K. 1977. Assortative mating by unwed biological parents of adopted children. *Science,* 196, 449–450.

Plomin, R., and Kuse, A. R. 1979. Comment in response to S. L. Washburn: Human behavior and the behavior of other organisms. *American Psychologist,* 34, 188–190.

Plomin, R., and Willerman, L. 1975. A cotwin control study and a twin study of reflection-impulsivity in children. *Journal of Educational Psychology,* 67, 537–543.

Plomin, R.; Willerman, L.; and Loehlin, J. C. 1976. Resemblance in appearance and the equal environments assumption in twin studies of personality traits. *Behavior Genetics,* 6, 43–52.

Polani, P. E.; Briggs, J. H.; Ford, C. E.; Clarke, C. M.; and Berg, J. M. 1960. A mongoloid girl with 46 chromosomes. *Lancet,* 1, 721–724.

Polivanov, S. 1975. Response of *Drosophila persimilis* to phototactic and geotactic selection. *Behavior Genetics,* 5, 255–267.

Porter, I. H. 1977. Evolution of genetic counseling in America. In *Genetic counseling: A monograph of the National Institute of Child Health and Human Development,* eds. H. A. Lubs and F. de la Cruz, pp. 17–34. New York: Raven Press.

Price, W. H., and Jacobs, P. A. 1970. The 47,XYY male with special reference to behavior. *Seminars in Psychiatry,* 2, 30–39.

Quinn, W. G.; Harris, W. A.; and Benzer, S. 1974. Conditioned behavior in

Drosophila melanogaster. *Proceedings of the National Academy of Sciences*, 71, 708–712.
Rao, D. C., and Morton, N. E. 1978. IQ as a paradigm in genetic epidemiology. In *Genetic epidemiology*, eds. N. E. Morton and C. S. Chung, pp. 145–181. New York: Academic Press.
Rao, D. C.; Morton, N. E.; and Yee, S. 1974. Analysis of family resemblance. II. A linear model for familial correlation. *American Journal of Human Genetics*, 26, 331–359.
———. 1976. Resolution of cultural and biological inheritance by path analysis. *American Journal of Human Genetics*, 28, 228–242.
Rao, T. V. 1978. Maternal age, parity, and twin pregnancies. In *Twin research, Part B, Biology and epidemiology*, ed. W. E. Nance, pp. 99–104. New York: Alan R. Liss.
Reading, A. J. 1966. Effect of maternal environment on the behavior of inbred mice. *Journal of Comparative and Physiological Psychology*, 62, 437–440.
Ready, D. F.; Hanson, T. E.; and Benzer, S. 1976. Development of the *Drosophila* retina, a neurocrystalline lattice. *Developmental Biology*, 53, 217–240.
Regier, D. A.; Goldberg, I. D.; and Taube, C. A. 1978. The de facto US mental health services system. *Archives of General Psychiatry*, 35, 685–693.
Reich, T. 1976. The schizophrenia spectrum: A genetic concept. *Journal of Nervous and Mental Disease*, 162, 3–12.
Reich, T.; Cloninger, C. R.; and Guze, S. B. 1975. The multifactorial model of disease transmission: I. Description of the model and its use in psychiatry. *British Journal of Psychiatry*, 127, 1–10.
Resnick, L. B., ed. 1976. *The nature of intelligence*. Hillsdale, N.J.: Lawrence Erlbaum Associates.
Reznikoff, M.; Domino, G.; Bridges, C.; and Honeyman, M. 1973. Creative abilities in identical and fraternal twins. *Behavior Genetics*, 3, 365–378.
Ritchie-Calder, P. R. 1970. *Leonardo and the age of the eye*. New York: Simon & Schuster.
Roberts, C. A., and Johansson, C. B. 1974. The inheritance of cognitive interest styles among twins. *Journal of Vocational Behavior*, 4, 237–243.
Roberts, R. C. 1967. Some concepts and methods in quantitative genetics. In *Behavior-genetic analysis*, ed. J. Hirsch, pp. 214–257. New York: McGraw-Hill.
Roper, A. G. 1913. *Ancient eugenics*. Oxford, England: Blackwell.
Rose, R. J.; Harris, E. L.; Christian, J. C.; and Nance, W. E. 1979*a*. Genetics variance in nonverbal intelligence: Data from the kinships of identical twins. *Science*, 205, 1153–1155.
Rose, R. J.; Miller, J. Z.; Dumont-Driscoll, M.; and Evans, M. M. 1979*b*. Twin-family studies of perceptual speed ability. *Behavior Genetics*, 9, 71–86.
Rosenthal, D. 1970. *Genetic theory and abnormal behavior*. New York: McGraw-Hill.
———. 1972. Three adoption studies of heredity in the schizophrenic disorders. *International Journal of Mental Health*, 1, 63–75.
———. 1975*a*. Discussion: The concept of subschizophrenic disorders. In *Genetic research in psychiatry*, eds. R. R. Fieve, D. Rosenthal, and H. Brill, pp. 199–208. Baltimore: Johns Hopkins University Press.
———. 1975*b*. The spectrum concept in schizophrenic and manic-depressive disorders. In *Biology of the major psychoses*, ed. D. X. Freedman. New York: Raven Press.

Rosenthal, D.; Wender, P. H.; Kety, S. S.; Schulsinger, F.; Welner, J.; and Ostergaard, L. 1968. Schizophrenics' offspring reared in adoptive homes. *Journal of Psychiatric Research,* 6, 377–391.

Rosenthal, D.; Wender, P. H.; Kety, S. S.; Welner, J.; and Schulsinger, F. 1971. The adopted-away offspring of schizophrenics. *American Journal of Psychiatry,* 128, 307–311.

Ruddle, F. H., and Kucherlapati, R. S. 1974. Hybrid cells and human genes. *Scientific American,* 232, 37–44.

Russell, W. L., and Douglas, P. M. 1945. Offspring from unborn mothers. *Proceedings of the National Academy of Sciences,* 31, 402–405.

Sanchez, O., and Yunis, J. J. 1977. New chromosome techniques and their medical applications. In *New chromosomal syndromes,* ed. J. J. Yunis, pp. 1–54. New York: Academic Press.

Scarr, S. 1968. Environmental bias in twin studies. *Eugenics Quarterly,* 15, 34–40.

———. 1976. An evolutionary perspective of infant intelligence: Species patterns and individual variations. In *Origins of intelligence,* ed. M. Lewis, pp. 165–198. New York: Plenum Press.

———. 1977. *Genetic effects on human behavior: Recent family studies.* Lecture, Annual Meeting of the American Psychological Association, San Francisco.

Scarr, S., and Carter-Saltzman, L. 1979, in press. Twin method: Defense of a critical assumption. *Behavior Genetics.*

Scarr, S., and Weinberg, R. A. 1977. Intellectual similarities within families of both adopted and biological children. *Intelligence,* 1, 170–191.

———. 1978. The influence of "family background" on intellectual attainment. *American Sociological Review,* 43, 674–692.

Schiff, M.; Duyme, M.; Dumaret, A.; Stewart, J.; Tomkiewicz, S.; and Feingold, J. 1978. Intellectual status of working-class children adopted early into upper-middle-class families. *Science,* 200, 1503–1504.

Schmidt, J. 1919. La valeur de l'individu à titre de generateur appreciée suivant la méthode du croisement diallele. *Comptes Rendus des Travaux du Laboratoire Carlsberg,* 14, 1–33.

Schoenfeldt, L. F. 1968. The hereditary components of the Project TALENT two-day test battery. *Measurement and Evaluation in Guidance,* 1, 130–140.

Schopf, J. W. 1978. The evolution of the earliest cells. *Scientific American,* 239, 110–139.

Schuckit, M. A., Goodwin, D. W., and Winokur, G. 1972. A study of alcoholism in half siblings. *American Journal of Psychiatry,* 128, 1132–1135.

Schull, W. J., and Neel, J. V. 1965. *The effects of inbreeding on Japanese children.* New York: Harper & Row.

Schulsinger, F. 1972. Psychopathy: Heredity and environment. *International Journal of Mental Health,* 1, 190–206.

Scott, J. P. 1958. *Animal behavior.* Chicago: University of Chicago Press.

Scott, J. P., and Fuller, J. L. 1965. *Genetics and the social behavior of the dog.* Chicago: University of Chicago Press.

Scriver, C. R. 1977. Screening, counseling, and treatment for phenylketonuria: Lessons learned—A précis. In *Genetic counseling: A monograph of the National Institute of Child Health and Human Development,* eds. H. A. Lubs and F. de la Cruz, pp. 253–268. New York: Raven Press.

Searle, L. V. 1949. The organization of hereditary maze-brightness and maze-dullness. *Genetic Psychology Monographs,* 39, 279–325.

Sefton, A. E., and Siegel, P. B. 1975. Selection for mating ability in Japanese quail. *Poultry Science,* 54, 788–794.

Segreti, W. O.; Winter, P. M.; and Nance, W. E. 1978. Familial studies of monozygotic twinning. In *Twin research, Part B, Biology and epidemiology,* ed. W. E. Nance, pp. 55–60. New York: Alan R. Liss.

Sergovich, F.; Uilenberg, C.; and Pozsonyi, J. 1971. The 49,XXXXX chromosome constitution: Similarities to the 49,XXXXY condition. *Journal of Pediatrics,* 78, 285–290.

Seyfried, T. N.; Yu, R. K.; and Glaser, G. H. 1979. Genetic study of audiogenic seizure susceptibility in B6XD2 recombinant inbred strains of mice. *Genetics,* 91, 114.

Shaffer, J. W. 1962. A specific cognitive deficit observed in gonadal aplasia (Turner's syndrome). *Journal of Clinical Psychology,* 18, 403–406.

Shah, S. A. 1970. *Report on the XYY chromosomal abnormality.* U.S., Public Health Service pub. no. 2103. Washington, D.C.: U.S. Government Printing Office.

Sherman, P. W. 1977. Nepotism and the evolution of alarm calls. *Science,* 197, 1246–1253.

Shields, J. 1962. *Monozygotic twins brought up apart and brought up together.* London: Oxford University Press.

———. 1968. Summary of the genetic evidence. In *The transmission of schizophrenia,* eds. D. Rosenthal and S. S. Kety, pp. 95–126. Oxford, England: Pergamon Press.

———. 1977. Genetics in schizophrenia. In *Schizophrenia today.* eds. D. Kemali, G. Bertholini, and D. Richter, pp. 57–70. Oxford, England: Pergamon Press.

———. 1978. MZA twins: Their use and abuse. In, *Twin research, Part A, Psychology and methodology,* ed. W. E. Nance, pp. 79–93. New York: Alan R. Liss.

Shields, J.; Heston, L. L.; and Gottesman, I. I. 1975. Schizophrenia and the schizoid: The problem for genetic analysis. In *Genetic research in psychiatry,* eds. R. R. Fieve, D. Rosenthal, and H. Brill, pp. 167–197. Baltimore: Johns Hopkins University Press.

Siemens, H. W. 1927. The diagnosis of identity in twins. *Journal of Heredity,* 18, 201–209.

Singh, R. S.; Lewontin, R. C.; and Felton, A. A. 1976. Genetic heterogeneity within electrophoretic "alleles" of xanthine dehydrogenase in *Drosophila pseudoobscura. Genetics,* 84, 609–629.

Skinner, B. F. 1938. *The behavior of organisms: An experimental analysis.* New York: Appleton-Century-Crofts.

Skodak, M., and Skeels, H. M. 1949. A final follow-up of one hundred adopted children. *Journal of Genetic Psychology,* 75, 85–125.

Slater, E., and Cowie, V. 1971. *The genetics of mental disorders.* London: Oxford University Press.

Smith, C. 1970. Heritability of liability and concordance in monozygous twins. *Annals of Human Genetics,* 34, 85–91.

Smith, D. W. 1977. Clinical diagnosis and nature of chromosomal abnormalities. In *New chromosomal syndromes,* ed. J. J. Yunis, pp. 55–117. New York: Academic Press.

Smith, J. W. 1978. The evolution of behavior. *Scientific American,* 239, 176–192.

Smith, R. T. 1965. A comparison of socio-environmental factors in monozygotic and dizygotic twins, testing an assumption. In *Methods and goals in human behavior genetics,* ed. S. G. Vandenberg, pp. 45–62. New York: Academic Press.

Smith, S. M., and Penrose, L. S. 1955. Monozygotic and dizygotic twin diagnosis. *Annals of Human Genetics,* 19, 273–389.

Snygg, D. 1938. The relation between the intelligence of mothers and of their children living in foster homes. *Journal of Genetic Psychology,* 52, 401–406.

Sprott, R. L., and Staats, J. 1975. Behavioral studies using genetically defined mice—A bibliography. *Behavior Genetics,* 5, 27–82.

———. 1978. Behavioral studies using genetically defined mice—A bibliography (July 1973–July 1976). *Behavior Genetics,* 8, 183–206.

———. 1979. Behavioral studies using genetically defined mice—A bibliography (July 1976–August 1978). *Behavior Genetics,* 9, 87–102.

Sprott, R. L., and Stavnes, K. 1975. Effects of situational variables on performance of inbred mice in active- and passive-avoidance situations. *Psychological Reports,* 37, 683–692.

Srb, A. M.; Owen, R. D.; and Edgar, R. S. 1965. *General genetics.* 2d ed. San Francisco: W. H. Freeman and Company.

Stent, G. S. 1963. *Molecular biology of bacterial viruses.* San Francisco: W. H. Freeman and Company.

Stern, C. 1973. *Principles of human genetics.* 3rd ed. San Francisco: W. H. Freeman and Company.

Strayer, L. C. 1930. The relative efficacy of early and deferred vocabulary training studied by the method of cotwin control. *Genetic Psychology Monographs,* 8, 209–319.

Sturtevant, A. H. 1915. Experiments on sex recognition and the problem of sexual selection in *Drosophila. Journal of Animal Behavior,* 5, 351–366.

———. 1965. *A history of genetics.* New York: Harper & Row.

Sumner, A. T.; Robinson, J. A.; and Evans, H. J. 1971. Distinguishing between X, Y, and YY-bearing human spermatozoa by fluorescence and DNA content. *Nature, New Biology,* 229, 231–233.

Taylor, M. A., and Abrams, R. 1975. Acute mania. *Archives of General Psychiatry,* 32, 863–865.

———. 1978. The prevalence of schizophrenia: A reassessment using modern diagnostic criteria. *American Journal of Psychiatry,* 135, 945–948.

Tennes, K.; Puck, M.; Bryant, K.; Frankenburg, W.; and Robinson, A. 1975. A developmental study of girls with trisomy X. *American Journal of Human Genetics,* 27, 71–86.

Theis, S. V. S. 1924. *How foster children turn out.* New York: State Charities Aid Association, publication No. 165.

Thompson, W. R., and Bindra, D. 1952. Motivational and emotional characters of "bright" and "dull" rats. *Canadian Journal of Psychology,* 6, 116–122.

Thorndike, E. L. 1905. Measurement of twins. *Archives of Philosophy, Psychology, and Scientific Methods,* 1, 1–64.

Tienari, P. 1963. Psychiatric illnesses in identical twins. *Acta Psychiatrica Scandinavica,* 39, supplement 171.

Tjio, J. H., and Levan, A. 1956. The chromosome number of man. *Hereditas,* 42, 1–6.

Tobach, E.; Bellin, J. S.; and Das, D. K. 1974. Differences in bitter taste perception in three strains of rats. *Behavior Genetics,* 4, 405–410.

Tolman, E. C. 1924. The inheritance of maze-learning ability in rats. *Journal of Comparative Psychology,* 4, 1–18.

Trivers, R. L. 1974. Parent-offspring conflict. *American Zoologist*, 14, 249–264.

Tsuang, M. T. 1975. Heterogeneity of schizophrenia. *Biological Psychiatry*, 10, 465–474.

———. 1976. Genetic factors in schizophrenia. In *Biological foundations of psychiatry*, eds. R. G. Grenell and S. Gabay. New York: Raven Press.

Tsuang, M. T.; Crowe, R. R.; Winokur, G.; and Clancy, J. 1977. Relatives of schizophrenics, manics, depressives and controls. Paper presented at the 2nd International Conference on Schizophrenia, Rochester, New York.

Tyler, P. A. 1969. "A quantitative genetic analysis of runway learning in mice." Ph.D. dissertation, University of Colorado.

Valentine, J. W. 1978. The evolution of multicellular plants and animals. *Scientific American*, 239, 176–192.

van Abeelen, J. H. F., and van der Kroon, P. H. W. 1967. *Nijmegen waltzer*—a new neurological mutant in the mouse. *Genetical Research*, 10, 117–118.

van Abeelen, J. H. F., ed. 1974. *The genetics of behaviour*. Amsterdam: North-Holland Publishing Co.

Vandenberg, S. G. 1965. Multivariate analysis of twin differences. In *Methods and goals in human behavior genetics*, ed. S. G. Vandenberg, pp. 29–44. New York: Academic Press.

———. 1968. The contribution of twin research to psychology. *Psychological Bulletin*, 66, 327–352.

———. 1971. What do we know today about the inheritance of intelligence and how do we know it? In *Intelligence: Genetic and environmental influences*, ed. R. Cancro, pp. 182–218. New York: Grune & Stratton.

———. 1972. Assortative mating, or who marries whom? *Behavior Genetics*, 2, 127–157.

———. 1976. Twin studies. In *Human behavior genetics*, ed. A. R. Kaplan, pp. 90–150. Springfield, Ill.: Thomas.

Vandenberg, S. G.; Stafford, R. E.; and Brown, A. M. 1968. The Louisville twin study. In *Progress in human behavior genetics*, ed. S. G. Vandenberg, pp. 153–204. Baltimore: Johns Hopkins University Press.

Vernon, P. E. 1979. *Intelligence: Heredity and environment*. San Francisco: W. H. Freeman and Company.

von Linné, Karl. 1735. *Systema Naturea*, Regnum Animale, L. Salvii, Holminae.

Wade, N. 1976*a*. Recombinant DNA: Chimeras set free under guard. *Science*, 193, 215–217.

———. 1976*b*. Sociobiology: Troubled birth for new discipline. *Science*, 191, 1151–1155.

Walker, A., and Leakey, R. E. F. 1978. The hominids of east Turkana. *Scientific American*, 239, 54–66.

Wallace, B. 1968. *Topics in population genetics*. New York: W. W. Norton.

Ward, S. 1977. Invertebrate neurogenetics. *Annual Review of Genetics*, 11, 415–450.

Washburn, S. L. 1960. Tools and human evolution. *Scientific American*, 203, 62–75.

———. 1978*a*. The evolution of man. *Scientific American*, 239, 194–208.

———. 1978*b*. Human behavior and the behavior of other organisms. *American Psychologist*, 33, 405–418.

Washburn, S. L., and Lancaster, C. S. 1968. The evolution of hunting. In *Man the hunter*, eds. R. B. Lee and I. DeVore, pp. 293–303. Chicago: Aldine.

Watanabe, T. K., and Anderson, W. W. 1976. Selection for geotaxis in *Drosophila melanogaster:* Heritability, degree of dominance, and correlated responses to selection. *Behavior Genetics,* 6, 71–86.

Watson, J. B. 1930. *Behaviorism.* New York: W. W. Norton.

Watson, J. D., and Crick, F. H. C. 1953*a*. Genetical implications of the structure of deoxyribonucleic acid. *Nature,* 171, 964–967.

———. 1953*b*. Molecular structure of nucleic acids. A structure for deoxyribose nucleic acids. *Nature,* 171, 737–738.

Weiss, M. C., and Green, H. 1967. Human-mouse hybrid cell lines containing partial complements of human chromosomes and functioning human genes. *Proceedings of the National Academy of Sciences,* 58, 1104–1111.

Wender, P. H.; Rosenthal, D.; and Kety, S. 1968. A psychiatric assessment of the adoptive parents of schizophrenics. In *The transmission of schizophrenia,* eds. D. Rosenthal and S. Kety, pp. 235–250. Oxford, England: Pergamon Press.

Wender, P. H.; Rosenthal, D.; Kety, S. S.; Schulsinger, F.; and Welner, J. 1974. Crossfostering: A research strategy for clarifying the role of genetic and experimental factors in the etiology of schizophrenia. *Archives of General Psychiatry,* 30, 121–128.

Wender, P. H.; Rosenthal, D.; Rainer, J. D.; Greenhill, L.; and Sarlin, M. B. 1977. Schizophrenics' adopting parents: Psychiatric status. *Archives of General Psychiatry,* 34, 777–784.

West-Eberhard, M. J. 1975. The evolution of social behavior by kin selection. *Quarterly Review of Biology,* 50, 1–33.

Whitney, G. D. 1969. Vocalization of mice: A single genetic unit effect. *Journal of Heredity,* 60, 337–340.

Whitney, G.; McClearn, G. E.; and DeFries, J. C. 1970. Heritability of alcohol preference in laboratory mice and rats. *Journal of Heredity,* 61, 165–169.

Wilder, H. H. 1904. Duplicate twins and double monsters. *American Journal of Anatomy,* 3, 387–472.

Willerman, L. 1979*a*, in press. Effects of families on intellectual development. *American Psychologist.*

———. 1979*b*. *The psychology of individual and group differences.* San Francisco: W. H. Freeman and Company.

Willham, R. L., Cox, D. F.; and Karas, G. G. 1963. Genetic variation in a measure of avoidance learning in swine. *Journal of Comparative and Physiological Psychology,* 56, 294–297.

Wilson, E. O. 1975. *Sociobiology, the new synthesis.* Cambridge, Mass.: Harvard University Press.

———. 1978*a*. The attempt to suppress human behavioral genetics. *Journal of General Education,* 29, 277–287.

———. 1978*b*. *On human nature.* Cambridge, Mass.: Harvard University Press.

Wilson, P. T. 1934. A study of twins with special reference to heredity as a factor determining differences in environment. *Human Biology,* 6, 324–354.

Wilson, R. S. 1972. Twins: Early mental development. *Science,* 175, 914–917.

———. 1977*a*. Mental development in twins. In *Genetics, environment and intelligence,* ed. A. Oliverio, pp. 305–336. Alphen aan den Rijn, Netherlands: Elsevier.

———. 1977*b*. Twins and siblings: Concordance for school-age mental development. *Child Development,* 48, 211–223.

———. 1978. Synchronies in mental development: An epigenetic perspective. *Science,* 202, 939–948.

Winokur, G. 1975. The Iowa 500: Heterogeneity and course in manic-depressive illness (bipolar). *Comprehensive Psychiatry,* 16, 125–131.

Winokur, G.; Cadoret, R.; Baker, M.; Dorzab, J. 1975. Depression spectrum disease versus pure depressive disease: Some further data. *British Journal of Psychiatry,* 127, 75–77.

Wispé, L. G., and Thompson, J. N., Jr. 1976. The war between the worlds: Biological versus social evolution and some related issues. *American Psychologist,* 31, 341–347.

Witkin, H. A.; Mednick, S. A.; Schulsinger, F.; Bakkestrom, E.; Christiansen, K. O.; Goodenough, D. R.; Hirschorn, K.; Lundsteen, C.; Owen, D. R.; Philip, J.; Rubin, D. B.; and Stocking, M. 1976. Criminality in XYY and XXY men. *Science,* 193, 547–555.

Witt, G., and Hall, C. S. 1949. The genetics of audiogenic seizures in the house mouse. *Journal of Comparative and Physiological Psychology,* 42, 58–63.

Witty, P. A., and Jenkins, M. D. 1936. Intra-race testing and Negro intelligence. *Journal of Psychology,* 1, 179–192.

Woodworth, R. S. 1948. *Contemporary schools of psychology.* Rev. ed. New York: Ronald Press.

Wright, S. 1921. Systems of mating. *Genetics,* 6, 111–178.

———. 1977. *Evolution and the genetics of populations.* Experimental results and evolutionary deductions, vol. 3. Chicago: University of Chicago Press.

Wyatt, R. J.; Murphy, D. L.; Belmaker, R.; Cohen, S.; Donnelly, C. H.; and Pollin, W. 1973. Reduced monoamine oxidase activity in platelets: A possible genetic marker for vulnerability to schizophrenia. *Science,* 179, 916–918.

Wynne-Edwards, V. C. 1962. *Animal dispersion in relation to social behaviour.* New York: Hafner.

Wyshak, G. 1978. Statistical findings on the effects of fertility drugs on plural births. In *Twin research, Part B, Biology and epidemiology,* ed. W. E. Nance, pp. 17–34. New York: Alan R. Liss.

Wysocki, C. J.; Whitney, G.; and Tucker, D. 1977. Specific anosmia in the laboratory mouse. *Behavior Genetics,* 7, 171–188.

Zajonc, R. B., and Markus, G. B. 1975. Birth order and intellectual development. *Psychological Review,* 82, 74–88.

Zazzo, R. 1960. *Les jumeaux, le couple et la personne.* Paris: Presses Universitaires de France.

Zerbin-Rüdin, E. 1969. Zur genetik der depressiven erkrankungen. In *Das depressive syndrom,* ed. H. Hippius, and H. Selback. Berlin: Verlag Urban Schwarzenberg.

———. 1974. Genetic aspects of schizophrenia (a survey). In *Biological mechanisms of schizophrenia and schizophrenia-like psychoses,* eds. H. Mitsuda and T. Fukuda. Tokyo: Igaku Shoin.

Zirkle, C. 1951. The knowledge of heredity before 1900. In *Genetics in the 20th century,* ed. L. C. Dunn, pp. 35–57. New York: Macmillan.

Index of Names

Abrams, R., 284, 287, 400
Addonizio, G., 385
Adler, J., 119, 120, 379
Adler, M. J., 11, 390
Agassiz, L., 23
Allen, G., 34, 379
Ambrus, C. M., 111, 379
Ambrus, J. L., 379
Anastasi, A., 359, 379
Anderson, W. W., 271, 401
Angst, J., 287, 379
Anisman, H., 241, 379
Annecillo, C., 394
Aristotle, 14
Ashton, G. C., 384, 391
Ayala, F. J., 53, 54, 107, 379, 381

Bagg, H. J., 241
Bailey, D. W., 86, 87, 379 395
Baker, M., 403
Baker, W. K., 127, 379
Bakkestrom, E., 403
Barash, D. P., 60, 379
Barnabas, J. 387
Barnicot, N. A., 76, 379
Baron, M., 287, 380
Barr, M. L., 150, 380
Bashi, J., 196, 197, 380
Bateson, W., 36, 68, 73, 103, 141, 380
Beadle, G. W., 103, 118, 380
Bearn, A. G., 112
Beckwith, L., 332, 380
Bellin, J. S., 241, 400
Belmaker, R., 322, 380, 403
Benzer, S., 118, 123, 124, 126, 127, 128, 129, 130, 380, 385, 389, 396
Berg, C. J., 49, 380
Berg, J. M., 396
Bertelsen, A., 287, 323, 380
Bertram, E. G., 150, 380
Bessman, S. P., 113, 380
Bilbro, W. C., 294, 395
Bilek, M. K., 386
Bindra, D., 264, 400
Blixt, S., 134, 380
Bloch, N., 382
Bock, R. D., 316, 380
Bodmer, W. F., 164, 108, 380
Bolivar, F., 390
Böök, J. A., 196, 380
Boolootian, R. A., 158, 165, 166, 169, 170, 380
Borgaonkar, D. S., 155, 380, 394
Borges, W. H., 151, 395
Boring, E. G., 32, 380
Bouchard, T. J., 100, 380
Boué, J. G., 149, 380
Bovet, D., 243, 245, 246, 247, 381
Bovet-Nitti, F., 246, 381
Boyer, H. W., 390
Bradley, R., 325, 385
Bridges, C., 397
Bridges, C. B., 395
Briggs, J. H., 396
Brill, H., 342, 373, 381, 386
Brittain, R. P., 390
Broadhurst, P. L., 63, 83, 88, 270, 381, 386
Brown, A. M., 298, 306, 294, 401
Bruce, E. J., 54, 381
Bruell, J. H., 261, 381
Bruggemann, C. E., 307, 394
Brunton, M., 390
Bruun, J., 314, 315, 381
Bryan, W. J., 24
Bryant, K., 400
Buchsbaum, M., 380
Bulmer, M. G., 290, 292, 381
Burks, B. S., 335, 339, 381
Burt, C., 340, 341, 381
Byers, D., 385

Cadoret, R. J., 352, 381, 383, 403
Caldwell, B., 325, 385
Campbell, D. T., 58, 381
Caplan, A. I., 128, 381
Carroll, J. B., 276, 382
Carter-Saltzman, L., 293, 295, 296, 306, 382, 398
Casler, L., 334, 338, 382
Caspersson, T., 143, 172, 382
Castle, W. E., 178, 382
Cattell, J. McK., 32
Cattell, R. B., 298, 357, 362, 382
Cavalli-Sforza, L. L., 64, 108, 380
Caviness, V. S., 127, 382
Cederlöf, R., 294, 307, 382
Chang, P.-N., 386

Chang, S. Y., 392
Changeux, J. P., 115, 382
Chen, C.-S., 271, 382
Christian, J. C., 397
Christiansen, K. O., 352, 394, 403
Claeys, W., 335, 341, 382
Clancy, J., 401
Clarke, A. M., 277, 393
Clarke, B., 188, 190, 382
Clarke, C. M., 396
Cloninger, C. R., 284, 382, 397
Cohen, D. J., 294, 298, 383
Cohen, R., 196, 382
Cohen, S., 380, 403
Collins, R. L., 79, 81, 383
Connolly, J. A., 158, 383
Connolly, K., 237, 383
Cooper, R. M., 247, 248, 383
Corey, L. A., 304, 383, 395
Cornblatt, B., 285, 386
Corney, G., 290, 291, 383
Correns, C., 36
Cowen, J. S., 241, 394
Cowie, V., 284
Coyne, J. A., 37, 383
Crea, R., 390
Crick, F. C. H., 104, 108, 110, 383, 402
Crow, J. F., 57, 383
Crowe, R. R., 352, 383, 401
Culliton, B. J., 91, 162, 373, 383
Cunningham, L., 352, 381, 383

Darrow, C., 24
Darwin, C., 3, 13, 16, 17, 19, 20, 21, 22, 23, 24, 25, 32, 33, 34, 37, 38, 41, 56, 58, 64, 74, 118, 383
Darwin, E., 15, 16, 25
Das, D. K., 241, 400
Davis, H., 14, 384
Dawkins, R., 39, 59, 384
Deckard, B. S., 241, 271, 384
DeFries, J. C., 2, 6, 83, 84, 85, 86, 100, 218, 219, 223, 228, 229, 230, 237, 238, 239, 241, 243, 245, 249, 251, 252, 253, 254, 265, 266, 267, 268, 269, 277, 279, 280, 281, 282, 316, 317, 327, 328, 330, 338, 339, 355, 357, 359, 361, 365, 366, 368, 384, 389, 391, 392, 393, 395, 396, 402
deGrouchy, J., 148, 149, 384
Deinard, A. S., 386
de la Cruz, F., 91, 393
Detre, T. P., 392
DeVore, I., 49
de Vries, H., 36
Dibble, E., 383
Dickerson, R. E., 38, 39, 384
Dobzhansky, Th., 24, 226, 384, 385
Dolan, A. B., 394
Domino, G., 397
Donnelly, C. H., 403
Dorzab, J., 403
Douglas, J. A., 44
Douglas, P. M., 250, 398
Douglass, A., 307, 395
Down, L., 160
Dudai, Y., 127, 385
Dumaret, A., 398
Dumont-Driscoll, M., 397
Duncan, O. D., 357, 387
Dunner, D. L., 286, 287, 328, 385
Duport, J. P., 395
Duyme, M., 398
Dworkin, R., xi, 385

Easler, C., 79, 387
East, E. M., 211, 385
Eaves, L. J., 229, 307, 316, 357, 385, 393
Ebert, P. D., 271, 385
Edgar, R. S., 117, 400
Edwards, J., 381, 383
Ehrman, L., xi, 47, 189, 190, 192, 193, 385, 396
Eiseley, L., 16, 385
Elardo, R., 325, 385
Eleftheriou, B. E., 86, 87, 88, 385, 395
Elias, P. K., 88, 385
Elston, R. C., 77, 385
Emerson, R. A., 211, 385
Epstein, C. J., 91, 162, 385
Erlenmeyer-Kimling, L., 277, 278, 285, 363, 385, 386
Essen-Möller, E., 293, 386
Evans, H. J., 400
Eyferth, K., 369, 386
Eysenck, H. J., 83, 307, 385, 386, 393

Falconer, D. S., 212, 224, 236, 251, 261, 265, 284, 301, 302, 386
Fantino, E., 244
Farber, S., 382
Feingold, J., 398
Felton, A. A., 37, 399
Fieve, R. R., 286, 342, 385, 386
Finaz, C., 148, 149, 384
Finkbeiner, C. T., 392
Fisch, R. O., 336, 337, 339, 386
Fischer, M., 320, 321, 386
Fisher, R. A., 211, 260, 386
Fitz-Roy, R., 17, 23
Flanagan, J. R., 129, 386
Fleiss, J. L., 285, 385, 386, 394
Flum, Y., 382
Foley, G. E., 382
Følling, A., 111, 386
Ford, C. E., 396
Fowler, W. A., 306, 386
Fox, J. L., 24, 386
Frankenberg, W., 400
Freeman, F. N., 306, 338, 339, 340, 341, 386, 395
Friberg, L., 382
Friedman, E., 322, 386
Friedman, G. D., 307, 387
Friedmann, T., 163, 387
Fulker, D. W., 229, 316, 357, 387, 391
Fuller, J. L., xi, 2, 4, 79, 81, 225, 256, 257, 258, 271, 306, 383, 387, 398

Gale, J. S., 316, 385
Galton, F., 1, 3, 13, 25, 26, 27, 28, 29, 30, 31, 32, 33, 37, 57, 104, 210, 211, 289, 310, 387
Garrod, A. E., 103, 387
Gehring, W. J., 128, 387
Gershon, E. S., 287, 380, 385, 386
Gervais, M. C., 243, 266, 267, 268, 269, 384
Gesell, A. L., 306, 387
Giblett, F. R., 293, 387
Ginsburg, B. E., 79, 387
Glaser, G. H., 88, 399
Golbus, M. S., 91, 162, 385
Goldberg, I. D., 397
Goldberger, A. S., 357, 387
Goldschmidt, E., 382
Goodenough, D. R., 403
Goodman, M., 52, 55, 387
Goodwin, D. W., 352, 387, 388, 398
Goodwin, F. K., 287, 385
Gorlin, R. J., 154, 166, 168, 388
Gottesman, I. I., 284, 285, 288, 318, 319, 320, 321, 322, 342, 350, 351, 388, 399
Grabiner, J. V., 24, 388
Grawe, J. M., 383
Gray, A., 23
Green, E. L., 85, 388

Green, H., 141, 402
Greenhill, L., 402
Gregory, M. S., 58, 388
Griffiths, M. I., 310, 388
Grossfield, J., 122, 388
Grotevant, H. D., 367, 388
Gruneberg, H., 77, 388
Guilford, J. P., 280, 388
Guthrie, R., 379
Guttman, R., 100, 388
Guze, S. B., 284, 287, 382, 387, 388, 397

Haber, S., xi, 385
Haggard, E. A., 236, 388
Haldane, J. B. S., 59, 141, 388
Hall, C. S., 79, 83, 388, 403
Hall, J. C., 122, 124, 127, 388
Halstead, L. B., 44, 388
Hamilton, W. D., 59, 388
Hansma, H., 392
Hanson, T. E., 124, 397
Hardy, G. H., 177, 388
Harris, E. L., 397
Harris, H., 37, 130, 388, 389
Harris, W. A., 127, 396
Harvald, B., 287, 323, 380, 386
Harvey, W., 15
Hauge, M., 287, 323, 380, 386
Hay, D. A., 241, 388
Hayes, H. K., 211, 385
Hazel, L. N., 389
Hearnshaw, L. S., 340, 389
Heder, A., 192
Hegmann, J. P., 237, 238, 239, 249, 254, 271, 384, 389
Henderson, N. D., 248, 255, 389
Henry, K. R., 79, 85, 389
Hermansen, L., 387, 388
Heston, L. L., 288, 320, 321, 344, 345, 350, 389, 399
Heyneker, H. L., 390
Hildreth, G. H., 389
Hilgard, J. R., 306, 389
Himmelhoch, J. M., 392
Hirose, T., 390
Hirsch, J., 10, 389
Hirschorn, K., 403
Hockett, C. F., 56, 389
Holzinger, K. J., 302, 306, 336, 341, 386, 395
Homyk, T., 124, 129, 389
Honeyman, M., 397
Honzik, M. P., 332, 333, 389
Hook, E. B., 171, 389
Hooker, J. D., 19, 23
Hopkinson, D. A., 37, 389
Horn, J. M., xii, 334, 338, 339, 389
Horvath, C., 379
Hotta, Y., 124, 126, 128, 389
Hrubec, Z., 307, 389
Hsia, D. Y.-Y., 111, 112, 113, 144, 152, 390
Husén, T., 293, 294, 390
Hutchings, B., 352, 390
Hutchins, R. M., 11, 390
Huxley, J., 371, 372, 390
Huxley, T. H., 19, 23
Hyde, J. S., 255, 271, 385, 390

Itakura, K., 109, 390

Jacob, F., 114, 390
Jacobs, P. A., 170, 171, 390, 396
Jacobsen, B., 391
Jacoby, R., 100, 392
Jan, Y. N., 385
Jarvik, L. F., 277, 278, 307, 386, 390
Jencks, C., 360, 390
Jenkins, M. D., 369, 403
Jensen, A. R., 9, 218, 223, 276, 301, 340, 341, 362, 367, 368, 373, 390, 391
Jinks, J. L., 357, 391
Johannsen, W., 36
Johansson, C. B., 307, 397
Johnson, R. C., 198, 328, 384, 391
Jonas, W. Z., 380
Jonsson, E., 382
Juel-Nielsen, N., 341

Kadar, M., 382
Kaij, L., 382
Kallmann, F. J., 320, 391
Kamin, L. J., 1, 391
Kant, C., 379
Kaprio, J., 307, 391
Karlsson, J. L., 348, 350, 391
Keeler, C., 178, 391
Kekic, V., 271, 391
Kessler, S., 171, 391
Kety, S. S., 284, 346, 347, 348, 391, 398, 402
Khorana, H. G., 109, 391
Khrushchev, N., 372
Kidd, K. K., 287, 392
Kimura, M., 57, 383
King, M. C., 54, 392
Klein, T. W., 226, 392
Knox, W. E., 390
Koch, R., 113, 380
Koen, A. L., 387
Konner, M. J., 50, 392
Koskenvuo, M., 391
Kraepelin, E., 283
Kringlen, E., 307, 318, 392
Kucherlapati, R. S., 146, 398
Kudynowski, J., 382
Kung, C., 121, 122, 392
Kupfer, D. J., 287, 392
Kuse, A. R., 54, 229, 239, 282, 384, 396

Lack, D., 18, 392
Lamarck, J. B., 15, 16, 33
Lancaster, C. S., 48, 401
Last, K. A., 385
Lawrence, E. M., 335, 392
Leahy, A. M., 336, 339, 340, 392
Leakey, L., 42, 44
Leakey, R. E. F., 42, 43, 44, 45, 401
Lee, R. B., 49, 392
Lehtovaara, A., 296, 392
Lepp, G. D., 62
Lerner, I. M., 104, 105, 259, 373, 392
Levan, A., 142, 400
Lewandowski, R. C., 155, 392
Lewis, A. M., 307, 387
Lewontin, R. C., 37, 399
Li, C. C., 228, 392
Libby, W. J., 259, 373, 392
Lieff, B., 384
Lindzey, G., 368, 369, 370, 393
Linnaeus (K. von Linné), 15, 56, 401
Locke, J., 32
Loehlin, J. C., 6, 100, 218, 219, 229, 296, 297, 298, 306, 307, 308, 312, 313, 314, 316, 318, 334, 357, 359, 361, 365, 366, 368, 369, 370, 389, 392, 393, 396
Loftus, R., 381, 383
Logan, C. E., 244
Lomakka, G., 172, 382
Lorich, U., 307, 382
Lubs, H. A., 91, 393
Lundsteen, C., 403
Lush, J. L., 33, 226, 227, 393
Lyell, C., 16, 19, 23
Lyon, M. F., 78, 151, 393
Lysenko, T. D., 371, 372, 373

McAskie, M., 277, 393
McCabe, M. S., 287, 393
McClearn, G. E., 2, 85, 86, 139, 264, 270, 271, 272, 280, 384, 391, 393, 402
McClemont, W. F., 390
McDougall, W., 4, 393
McGee, M. G., 100, 380
McKusick, V. A., 101, 393

Marcus, G. B., 367, 403
Marinkovic, D., 271, 391
Markkanen, T., 314, 381
Martin, N. G., 229, 307, 316, 385, 393
Mason, J. J., 393
Matheny, A. P., 298, 307, 393, 394
Matthysse, S., 287, 392
Maxson, S. C., 241, 394
Maxwell, S. E., 276, 382
Mayr, E., 8, 10, 47, 394
Mednick, S. A., 171, 352, 390, 394, 403
Medvedev, Z. A., 371, 394
Melville, M. M., 390
Mendel, G. J., 3, 34, 35, 36, 37, 57, 61, 64, 66, 67, 68, 70, 72, 73, 74, 75, 101, 102, 130, 133, 134, 152, 176, 200, 211, 253, 394
Mendlewicz, J., 287, 351, 394
Meredith, W., 85, 360, 393, 394
Merrell, D. J., 48, 394
Merriman, C., 289, 290, 307, 310, 394
Mi, M. P., 384, 391
Miller, D. S., 79, 387
Miller, J. Z., 397
Miller, P. D., 24, 388
Mirand, E., 379
Mirenva, A. N., 306, 394
Mitchell, B. C., 336, 386
Mittler, P., 299, 307, 394
Modest, E. J., 382
Mohr, J., 394
Mohr, O. L., 141, 386
Møller, A., 172, 382
Moller, N., 388
Money, J., 164, 165, 168, 172, 394
Monod, J., 114, 390
Montagu, A., 23, 395
Moore, G. W., 387
Moos, R. H., 171, 391
Morgan, T. H., 141, 395
Morton, N. E., 77, 357, 395, 397
Motulsky, A. G., 284, 395
Muller, H. J., 117, 395
Munsinger, H., 307, 331, 333, 338, 339, 364, 365, 366, 395
Murphy, D. L., 403

Nadler, C. F., 151, 395
Nance, W. E., 290, 291, 292, 304, 305, 383, 395, 397, 399
Neel, J. V., 196, 307, 389, 398
Newman, J., 306, 341, 395
Newton, I., 20
Nichols, R. C., 229, 294, 296, 297, 298, 302, 307, 308, 312, 313, 314, 316, 318, 392, 395
Nilsson-Ehle, H., 211, 395
Nöel, B., 395
Nuttall, G. H. F., 52, 395
Nylander, P. P. S., 292, 395

Oliverio, A., 87, 88, 245, 246, 381, 395
Omenn, G. S., 91, 284, 395
Ordahl, C. P., 128, 381
Orel, V., 35
Osborne, R. T., 369, 393
Ostergaard, L., 398
Owen, D. R., 403
Owen, R. D., 117, 400

Packard, A. S., 16, 395
Padeh, B., 241, 245, 395
Paine, R. S., 390
Palm, J., 250, 396
Parisi, P., 395
Parisi-Prinzi, G., 395
Parkinson, J. S., 119, 120, 396
Parsons, P. A., xi, 47, 190, 385
Partanen, J., 314, 381
Paul, T., 379
Pavlovsky, O., 24, 385
Pearson, K., 25, 396
Pedersen, H., 379
Penrose, L. S., 161, 293, 396, 400
Perris, C., 286, 396
Petit, C., 189, 192, 396
Philip, J., 403
Phillips, C. J., 310, 388
Piaget, J., 276
Pickar, D., 392
Plato, 14
Plomin, R., 2, 6, 54, 218, 219, 223, 228, 230, 298, 306, 316, 317, 327, 328, 330, 338, 339, 355, 357, 359, 361, 362, 363, 365, 366, 384, 396
Polani, P. E., 160, 396
Polivanov, S., 271, 396
Pollin, W., 380, 383, 403
Polovina, J., 384
Porter, I. H., 91, 396
Postman, L. G., 139, 264
Pozsonyi, J., 399
Price, E. O., 393
Price, W. H., 171, 396
Probber, J., 193, 385
Puck, M., 400
Punnett, R. C., 380

Quack, B., 395
Quinn, K. V., 390
Quinn, W. G., 127, 385, 396

Rainer, J. D., 287, 351, 394, 402
Rakic, P., 127, 382
Rantasalo, I., 391
Rao, D. C., 357, 397
Rao, T. V., 292, 397
Rashad, M. N., 384, 391
Reading, A. J., 250, 251, 397
Ready, D. F., 124, 397
Regier, D. A., 282, 397
Reich, T., 284, 287, 288, 382, 397
Resnick, L. B., 276, 397
Revil, D., 395
Reznikoff, M., 307, 397
Riggs, A. D., 390
Ritchie-Calder, P. R., 15, 397
Roberts, C. A., 307, 397
Roberts, M. K., 223, 328, 396
Roberts, R. C., 226, 397
Robinson, A., 157, 400
Robinson, J. A., 400
Roper, A. G., 14, 397
Rose, R. J., 305, 397
Rosenthal, D., 284, 286, 288, 319, 323, 342, 345, 346, 347, 348, 386, 391, 397, 398, 402
Rubin, D. B., 403
Ruddle, F. H., 146, 398
Rudy, V., 380
Russell, W. L., 250, 398
Ruud, L., 386

Sanchez, O., 144, 145, 149, 398
Sarlin, M. B., 402
Sarna, S., 391
Sathananthan, G., 386
Satow, Y., 392
Saunders, E. R., 380
Scarr, S., 51, 293, 295, 296, 306, 326, 333, 337, 339, 340, 342, 367, 376, 382, 388, 398
Schiff, M., 331, 398
Schlesinger, K., 85, 384, 389
Schmidt, J., 255, 398
Schoenfeldt, L. F., 314, 315, 398
Schopf, J., 39, 398
Schuckit, M. A., 352, 398
Schull, W. J., 196, 398

Schulsinger, F., 352, 387, 388, 391, 402, 403
Scopes, J. T., 23, 24
Scott, J. P., 244, 256, 257, 258, 398
Scriver, C. R., 91, 398
Searle, L. V., 264, 398
Sefton, A. E., 271, 399
Segreti, W. O., 292, 395, 399
Sergovich, F., 167, 399
Seyfried, T. N., 88, 399
Shaffer, J. W., 164, 399
Shah, S. A., 171, 399
Sharan, S., 100, 392
Sharma, S., 379
Shaw, M., 143
Sheppard, D. E., 124, 389
Sherman, P. W., 61, 62, 399
Shields, J., 284, 285, 288, 318, 319, 320, 321, 322, 341, 342, 350, 351, 388, 399
Shopsin, B., 386
Siegel, P. B., 271, 399
Siemens, H. W., 293, 399
Silvers, A., 58, 388
Simonsson, E., 382
Singh, R. S., 37, 399
Skeels, H. M., 331, 332, 333, 338, 364, 365, 399
Skinner, B. F., 10, 399
Skodak, M., 331, 332, 333, 338, 364, 365, 399
Slater, E., 284, 399
Smith, C., 284, 399
Smith, C. A. B., 30, 31
Smith, D. W., 155, 399
Smith, G. F., 155, 157, 159, 161, 396
Smith, J. W., 60, 399
Smith, M. E., 79, 387
Smith, R. T., 296, 399
Smith, S. M., 293, 401
Snygg, D., 334, 338, 400
Sollas, W. J., 44
Spencer, H., 32
Sprott, R. L., 240, 241, 400
Spuhler, J. N., 198, 368, 369, 370, 392
Srb, A. M., 117, 400
Staats, J., 240, 400
Stafford, R. E., 306, 401
Stalin, J., 372
Stavnes, K., 241, 400
Steinberg, A. G., 112
Stent, G. S., 106, 400
Stern, C., 37, 94, 96, 97, 99, 100, 180
Stewart, J., 77, 385, 398
Stocking, M., 403
Strayer, L. C., 306, 401
Sturtevant, A. H., 83, 103, 127, 128, 395, 400
Summer, A. T., 170, 400
Sussuyer, I., 395
Sutch, D., 58, 388

Tatum, E. L., 103, 118, 380
Taube, C. A., 397
Taylor, M. A., 284, 287, 400
Tennes, K., 166, 400
Tepper, J. M., 384
Theis, S. V. S., 331, 400
Thomas, E. A., 84, 242, 243, 266, 267, 268, 269, 270, 384
Thompson, H., 306, 387
Thompson, J. N., 58, 403
Thompson, W. R., xi, 2, 4, 264, 306, 387, 400
Thorndike, E. L., 289, 310, 400
Thurstone, L. L., 280
Tienari, P., 322, 400
Tjio, J. H., 142, 400
Tobach, E., 400
Tolman, E. C., 262, 263, 400
Tomkiewicz, S., 398
Trivers, R. L., 60, 401
Tryon, R., 263, 264
Tsuang, M. T., 287, 342, 401
Tucker, D., 241, 403
Turleau, C., 148, 149, 384
Tyler, P. A., 263, 401

Uilenberg, C., 399

Valentine, J. W., 39, 401
van Abeelen, J. H. F., xi, 77, 78, 401
Vandenberg, S. G., 2, 196, 198, 229, 239, 280, 282, 296, 307, 316, 369, 380, 384, 391, 393, 401
van der Kroon, P. H. W., 77, 78, 401
Van Houten, J., 392
Vernon, P. E., 276, 369, 401
Vesalius, A., 15
Vinci, da, L., 14
von Linné, K. (Linnaeus), 15, 56, 401
von Tschermak, E., 36

Wade, N., 109, 373, 401
Wagh, U., 382
Wahlsten, D., 241, 245, 395
Walker, A., 42, 43, 44, 45, 401
Wallace, A., 19, 23
Wallace, B., 182, 401
Ward, S., 122, 127, 401
Washburn, S. L., 46, 48, 54, 55, 401
Watanabe, T. K., 271, 401
Watson, J. B., 5, 6, 402
Watson, J. D., 104, 116, 137, 402
Weinberg, R. A., 333, 337, 339, 340, 342, 367, 376, 388, 398
Weir, M. W., 384
Weiss, M. C., 141, 402
Welner, J., 398, 402
Wender, P. H., 346, 348, 349, 391, 398, 402
West-Eberhard, M. J., 60, 402
Whitney, G. D., xii, 81, 82, 237, 241, 402, 403
Wilder, H. H., 289, 402
Willerman, L., 276, 298, 306, 331, 334, 369, 389, 393, 396, 402
Willham, R. L., 237, 402
Williamson, M. L., 113, 380
Wilson, A. C., 54, 392
Wilson, E. O., 51, 57, 58, 59, 63, 373, 402
Wilson, J. R., 85, 384, 391, 393
Wilson, P. T., 296, 402
Wilson, R. S., 298, 310, 311, 312, 334, 402, 403
Winokur, G., 287, 352, 387, 388, 398, 401, 403
Winter, P. M., 292, 395, 399
Wispé, L. G., 58, 403
Witkin, H. A., 171, 172, 403
Witt, G., 79, 403
Witty, P. A., 369, 403
Woodworth, R. S., 6, 403
Wright, S., 194, 195, 211, 228, 262, 403
Wundt, W., 32
Wyatt, R. J., 322, 380, 403
Wynne-Edwards, V. C., 59, 403
Wyshak, G., 292, 403
Wysocki, C. J., 241, 403

Yee, S., 357, 397
Young, P. A., 385
Yu, R. K., 88, 399
Yunis, J. J., 144, 145, 149, 155, 392, 398

Zajonc, R. B., 367, 403
Zazzo, R., 296, 403
Zech, L., 382
Zerbin-Rüdin, E., 323, 342, 403
Zirkle, C., 14, 403
Zubek, J. P., 247, 248, 383

Index of Topics

Abilities. *See* Creativity; IQ; Memory; Perceptual speed; Scholastic abilities; Spatial ability; Specific cognitive abilities; Verbal ability
ABO blood group, 35, 293
Activity:
  in *Drosophila,* 237, 271
  in mouse, 83ff, 237ff, 249ff, 265ff
  in rat, 271
Adaptations, evolutionary, 41
Additive genetic value, 212f
Additive genetic variance, 215ff
Adoption studies, 325ff
  adoptees' family method, 343f
  adoptees' study method, 343f
  alcoholism, 352
  assortative mating, 328, 346
  criminality, 352
  identical twins and IQ, 341
  IQ, 331ff
  manic-depressive psychosis, 351f
  parent–offspring and IQ, 331ff
  prenatal environment, 329
  representativeness, 326
  schizophrenia, 343ff
  selective placement, 327f
  sibling studies and IQ, 339ff
Aggressive behavior, 57, 169ff, 271
Aging, 307
Alarm calls, 61f
Albinism, 83ff, 267f
Alcohol:
  preference in mice, 237, 271
  sensitivity in mice, 270ff
Alcoholism, 352
Alkaptonuria, 103
Allele, 35, 68
Allelic frequencies, 175ff
Allopatric species, 47
Altruism, 58ff
Amaurotic idiocy, 91, 113f
Amino acids, 107f
Amino acid sequencing, 52ff
Amniocentesis, 91, 162f
Amphibians, 40f, 56
Aneuploidy, 149
Anthropometric Laboratory of Galton, 28ff
Antigen, 52
Ants, 59
Assortative mating, 197f, 223, 301, 328, 346
Assortment:
  independent, 35f, 67, 72ff
  linkage as exception to, 74, 133f
"At-risk" studies, 285
Attitudes, 307
Audiogenic seizures, 79ff, 88, 271
Australopithecus, 44
  africanus, 43ff
  boisei, 43f
  gracilis, 44
  robustus, 42ff
  zinjanthropus, 42
Autogamy, 121
Avoidance behavior, 121
Avoidance learning. *See* Learning

Backcross, 80
Bacteria, 39, 56, 115f
  behavioral mutants, 119ff
  chemotaxes, 119
Bacteriophage, 109
Balanced polymorphisms, 187ff

Banding, chromosome, 143ff
Barr body, 150ff
Bees, 59
Behaviorism, 4ff
β-galactosidase, 115f
Biometricians, 211
Biopsychology, 8
Bipolar manic-depressive psychosis. *See* Manic-depressive psychosis
Birth order, 367
Blastoderm, 127ff
Blood serum, 52
  See also ABO blood group; Rh-negative blood factor; Zygosity determination

Castle's law, 178
Catastrophism, 16
Chemotaxes, 119
Chickens, 271
Child rearing, 50f, 60f, 63f, 325f
Chimpanzees, 51ff, 148f
Chromatin, 150f
Chromosome(s), 132ff
  abnormalities, 154ff
  acrocentric, 142
  autosomes, 89
  banding patterns, 143ff
  classification, 142ff
  deletion, 149f
  duplication, 149f
  human, 142ff
  inversion, 149f
  metacentric, 142
  monosomy, 149f
  nondisjunction, 149, 151
  sex, 143, 151ff
  translocation, 149f
  trisomy, 149f
Classical analysis, 253ff
Coercion to biosocial norm, 362
Cognitive abilities. *See* IQ; Specific cognitive abilities
Coisogenic strains, 85
Color blindness, 99f, 141
Concordance, 318
Congenic strains, 88
Correlated characters. *See* Multivariate genetic-environmental analysis
Correlation:
  cross-, 232f
  environmental, 228ff, 238ff
  and estimates of heritability, 231ff, 235
  full-sibling, 236
  genetic, 228ff, 238ff
  half-sibling, 236
  intraclass, 236
  twins, 231, 236, 299ff
Co-twin control studies, 305f, 322
Covariance:
  cross-, 230f
  full-sibling, 223ff
  half-sibling, 223
  offspring and parent, 222ff
  of relatives, 219ff
  twins, 223, 300
Creativity, 307
Cri du chat syndrome, 154, 156
Criminality:
  adoption studies, 352
  XYY, 169ff
Cross-fostering, 250
Crossing over, 137ff

Defecation, 238ff, 252, 266
Deoxyribonucleic acid (DNA), 35, 104ff
  hybridization, 53
  recombinant, 53
  as replicator, 38f
  synthetic, 109
Descent of Man, The, 22, 34
Design, argument from, 17, 19
Development:
  and evolution, 59f
  fate mapping, 127ff
  genetic dissection, 127ff
  IQ, 309ff, 332f
Diallel design, 255f
Dihybrid, 67, 72ff, 133f
Diploid, 60
DNA (deoxyribonucleic acid), 35, 104ff
Dogs, 132, 256ff
Dominance, 70, 89ff, 213ff
Down's syndrome, 157ff
  frequency and reproductive age, 160ff
  IQ, 158
  translocation type, 158ff
Drosophila:
  activity, 237, 271
  attached X, 123ff
  bar-eyed, 189
  chromosome number, 132
  and evolution, 47
  fate mapping, 127ff
  frequency-dependent sexual selection, 188ff
  geotaxis, 125f, 271
  gynandromorphs, 124
  learning, 128, 241
  linkage, 141
  mosaics, 123ff
  mutations, 122f
  olfaction, 241
  phototaxis, 123ff, 271
  rare-male advantage, 188ff
  sexual behavior, 47, 83, 122, 124, 271
  unstable X, 124f

Electrophoresis, 52ff
Emotional behavior, 83ff, 264, 271
  See also Open-field behavior

Endophenotype, 322
Environment:
  between-family (shared), 220f, 275f, 303, 309, 329f, 350, 376
  maternal, 249ff, 348
  postnatal, 249
  prenatal, 249, 329, 348
  within-family (independent), 220ff, 276, 303, 309, 329f, 350, 376
  See also Environmentality
Environmentality, 225ff, 302f, 329ff
Enzymes, 110
Epistasis, 215f
Equality, 10ff, 373f
Equilibrium law, 177ff
Ethanol. *See* Alcohol
Eukaryote, 39
Evolution:
  and behavior, 47ff
  Darwin's theory, 20ff, 38ff
  Lamarckian theory, 15f
  of life, 39ff
  opposition to theory, 22ff
  sociobiology, 22, 58ff
Expressivity, 76

F_1, F_2, 67ff, 175ff
Factor analysis, 280, 317
Family studies:
  animal behavior, 234ff
  human behavior, 274ff
Fate mapping, 127ff
Fitness characters, 260f
Følling's disease. *See* Phenylketonuria

Galactosemia, 113
Galápagos Islands, 17f
Gametes, 64, 71ff
Gemmules, 33
Gene(s):
  and behavior, 7ff
  chemical nature, 104ff
  defined, 35f
  dosage, 212f
  human, number of, 37
  major or single, 66ff
  operator, 114f
  and protein synthesis, 105ff
  regulator, 114ff
  structural, 114ff
  synthetic, 109
Genetic counseling, 91
Genetic drift, 182
Genetic engineering, 109
Genetics:
  Mendelian theory, 34ff
  quantitative theory, 200ff
Genius, 26f
Genotype, 36
Genotype–environment correlation, 218, 222, 360ff
Genotype–environment interaction, 216, 222, 247, 359ff, 376
Genotype frequencies, 175ff
Genotypic value, 212
Geotaxis, 125f, 271
Gregariousness, 57

Haploid, 60
Hardy–Weinberg–Castle equilibrium, 177ff
Hemoglobin, 54
Hereditary Genius, 1, 26f
Heredity:
  blending theory, 21, 34, 57, 211
  Greek concepts, 11
  Mendelian theory, 34ff
  pangenesis theory, 33
Heritability:
  and adoption studies, 329ff
  broad-sense, 227, 235, 254, 301
  Falconer's, 301f
  of fitness characters, 260f
  Holzinger's, 301f
  narrow-sense, 227, 235
  and path analysis, 227f
  and selection response, 258ff
  and twin correlations, 231f, 299ff
  and variances of F_1 and F_2, 254
Heterogeneity, 287f, 375
Heterosis, 195
Heterozygote, 70, 179ff
Heterozygote advantage, 57, 186f
Hominid phylogeny. *See Australopithecus* and *Homo*
Homo:
  erectus, 42, 44ff
  habilis, 43ff
  Java man, 42
  Peking man, 42
  Piltdown hoax, 44
  sapiens. *See* Man
  sapiens neanderthalensis, 44, 46f
Homogentisic acid, 103
Homologous chromosomes, 135
Homozygosity and inbreeding, 194f
Homozygote, 70
Hormones, 109
Hunting:
  and human evolution, 48f
  hunter-gatherers, 40f
Huntington's chorea, 90, 93
Hybrid vigor (heterosis), 195
Hysteria, 287

Identity by descent, 194
Imagery, 28
Immunological techniques, 52ff
Inborn errors of metabolism, 103

Inbred strains, 195, 241f
  See also Mouse strains; Strain studies
Inbreeding, 96, 193ff
  and average genotypic value, 195f
  coefficient, 194
  depression, 195, 241
  and homozygosity, 194f, 241
  and IQ, 196f
  in selection experiments, 263ff
Inclusive fitness, 59f
Index case, 90
Instinct, 5, 8
Insulin, 109
Intelligence. *See* IQ; Specific cognitive abilities
Interactionism, 6, 359
Invertebrates, 41, 56
IQ, 1f
  adoption studies, 331ff, 355ff
  assortative mating, 198
  in chromosome anomalies. *See* separate listings
  development in twins, 309ff
  family studies, 26ff, 277ff, 355ff
  fitting models, 357f
  genotype–environmental interaction, 364ff
  inbreeding, 196f
  race, 367f
  twin studies, 277, 296ff, 307ff, 355ff
Isogenic individuals, 253

Karyotype, 142ff
Kinship selection, 22, 59ff
Klinefelter's syndrome, 167ff

Lamarckism, 15f, 33f
Language, 51, 307
Learning:
  in *Drosophila,* 127, 241
  in mice, 87f, 241ff, 248, 271
  in rats, 247f, 262ff
  in swine, 237
Liability, 283f
Linkage, 84, 138, 141
  map distance, 140f
  sex, 97ff, 123ff, 138, 141, 165
  somatic cell hybridization, 144, 146f
Locus, defined, 35, 74
Lyon hypothesis, 150ff

Mammals and evolution, 40, 42, 56
Man:
  adoption studies. *See* separate listing
  chromosomes, 142ff
  evolution, 42ff
  family studies, 274ff
  galactosemia, 113
  Huntington's chorea, 90, 93
  IQ. *See* separate listing
  phenylketonuria. *See* separate listing
  psychopathology. *See* separate listing
  PTC (phenylthiocarbamide) tasting, 75f, 179ff
  schizophrenia, *See* separate listing
  sickle-cell anemia, 54, 91, 109
  Tay-Sachs disease, 91, 113f
  twin studies. *See* separate listing
Manic-depressive psychosis:
  adoption studies, 351f
  bipolar, 286f, 323, 351f
  family studies, 285ff
  twin studies, 323
  unipolar, 286f, 323
Map distance, 140f
Marmots, 60
Maternal effect, 249ff, 305
Mating, systems of, 193ff
Mating behavior. *See* Sexual behavior
Maze learning. *See* Learning
Meiosis, 135ff
Memory, 28, 280ff, 314f
Mendelians, 211
Mental retardation:
  cri du chat syndrome, 154
  Down's syndrome, 158
  and inbreeding, 196
  Klinefelter's syndrome, 167f
  single-gene effects, 101
  triple-X syndrome, 165ff
Metazoan, 56
Midparental value, 223, 235, 277, 329
Migration, 181f
Mitosis, 135ff
MN blood group, 278f, 293
Model fitting, 357f
Mongolism. *See* Down's syndrome
Monkeys:
  New World, 46, 55
  Old World, 42, 46, 55
Monohybrid, 67ff
Monozygotic half-siblings method, 304f
Motoric behavior, 306
Mouse:
  activity, 83ff, 237ff, 249ff, 265ff
  agitans, 78
  albinism, 83ff, 267f
  alcohol preference, 237, 271
  alcohol sensitivity, 270f
  audiogenic seizures, 79ff, 88, 271
  c locus, 83ff
  defecation, 238f, 252, 266
  emotionality, 83ff, 264, 271
  learning, 87f, 241ff
  nerve conduction, 271
  Nijmegen waltzer, 77
  olfaction, 241
  open-field behavior, 83ff, 237ff, 249ff, 265ff
  quaking, 78
  reeler, 78

seizure susceptibility, 241
sleeping time to alcohol, 270ff
twirler, 78
vocalization, 81ff
waltzer, 77f
Mouse strains:
A, 82, 203ff, 244f
BALB/c, 82, 84, 87f, 237ff, 240ff, 244f, 249ff, 255f, 265
C3H, 82, 241f, 255
C57BL, 79ff, 84, 87f, 203ff, 237ff, 240ff, 244f, 249ff, 255f, 265
C57BR, 241
C57L, 245
CBA, 82, 241
DBA, 79, 82, 241f, 244f, 255
HS, 85
I^s, 82f
JK, 81ff
recombinant inbred, 81
SEC, 245
Swiss, 245
Multifactorial model, 4, 211
See also Quantitative genetics
Multivariate genetic-environmental analysis, 228ff, 232f, 238ff, 281f, 286, 375
in family studies, 238ff, 281f
in twin studies, 314ff
Mutation, 103, 117f, 183
dissection of behavior, 118ff

Natural selection:
Darwin's formulation, 21ff
fundamental theorem, 260f
Nature–nurture issue, 4ff, 358ff
Nematode, 121f
Neurospora, 103
Neutralists, 58, 188
Nucleotide bases of DNA, 104ff
Nucleus, 35, 39

Olfaction, 193, 241
One-gene, one-enzyme hypothesis, 103
Open-field behavior, 83ff, 237ff, 249ff, 265ff, 271
Operon model, 114f
Origin of Species, On the, 19ff
Ovary transplantation, 250ff
Overdominance, 71

Pairwise twin concordance, 318
Pangenesis, 21, 33
Paramecia, 121
Paternal behavior, 60, 51
See also Child rearing
Path analysis, 227ff
Pedigree, 90
Penetrance, 76
Perceptual speed, 280ff, 305, 314f
Personality, 198, 296ff, 306
See also Aggressive behavior; Altruism; Criminality; Emotional behavior; Psychopathology
Phenotype, 3, 36
Phenylalanine hydroxylase, 111
Phenylketonuria (PKU):
classical, 8f
dietary therapy, 111ff
discovery of biochemical defect, 110f
heterozygotes, 113
infant screening, 91
IQ, 111
pedigree, 96
variants, 113
Phenylpyruvic acid, 111
Phenylthiocarbamide (PTC) tasting, 75f, 179ff
Photophobia, 86
Photosynthesis, 39f
Phototaxis, in *Drosophila,* 124f, 271
PKU. *See* Phenylketonuria
Pleiotropy, 83ff
Politicization of science, 370ff
Polygenes, 88, 200ff
Polymorphisms:
balanced, 187ff
defined, 57f
heterozygote advantage, 57
frequency-dependent selection, 188ff
frequency-dependent sexual selection, 188f
rare-male advantage, 188f
Polypeptide, 54, 110
Population genetics, 176ff
Predatory behavior, 49, 188
Prenatal environment, 249, 329, 348
Primates:
classification, 42
evolution, 42
genetic similarity to man, 51ff, 149
Proband, 90
Probandwise twin concordance, 318
Prokaryote, 39
Propositus, 90
Prosimians, 55
Psychopathology:
hysteria, 287
incidence, 283
liability, 283f
manic-depressive psychosis, 285ff, 351f
prevalence, 283
risk estimates, 283
schizophrenia, 284f, 319ff
Psychopharmacogenetics, 271
PTC (phenylthiocarbamide) tasting, 75f, 179ff

Quail, 271
Quantitative genetics, 200ff

Race and IQ, 279ff, 367ff
Random drift, 182
Random mating, 177ff
Rare-male advantage, 188ff
Rats:
  activity, 271
  learning, 247f, 262ff
  reactive and nonreactive, 271
  saccharine preference, 271
  taste perception, 241
Reaction time, 28
Reading, 306
Recessiveness, 70, 93ff
Reciprocal cross, 249f
Recombinant inbred strains, 81, 86ff
Regression:
  and estimates of heritability, 231ff, 235
  offspring and parent, 235f, 279
Regulator genes, 114ff
Reproductive fitness, 21ff
Reproductive strategies, 51
  See also Sexual behavior
Reptiles, 40f, 56
Rh-negative blood factor, 179
Ribonucleic acid (RNA), 106ff, 106f

Schizophrenia:
  adoption studies, 343ff
  at-risk studies, 285
  discordant identical twins, 322
  family studies, 284f
  spectrum, 320ff, 345ff
  twin studies, 319ff
Scholastic abilities, 307, 316ff
Segregation:
  analysis, 77
  law of, 35f, 67ff
Seizure susceptibility, 241, 271
  See also Audiogenic seizures
Selection:
  for activity, 265ff
  for aggressiveness, 57
  for alcohol susceptibility, 270ff
  for audiogenic seizures, 271
  balancing, 47, 183ff
  coefficient, 184f
  correlated response, 239, 266
  differential, 259f
  directional, 57
  disruptive, 58
  against dominant allele, 184f
  for fitness characters, 260f
  frequency-dependent, 188
  geotaxis, 271
  group, 58f
  and inbreeding, 263ff
  kinship, 22, 59ff
  for learning ability, 247, 262ff
  nerve conduction velocity, 271
  for reactivity, 271
  and realized heritability, 258ff
  against recessive allele, 184ff
  response, 259f
  for sexual behavior, 57
  stabilizing, 57
Selectionists, 58, 188
Selective placement, 327f
Sex linkage, 97ff, 123ff, 305
  bipolar manic-depressive psychosis, 287
  color blindness, 99f
  hemophilia, 141
  spatial ability, 99f, 165
Sexual behavior, 57
  in chickens, 271
  in *Drosophila*, 47, 83, 122, 124, 188ff, 271
  in man, 307
  in quail, 271
Shuttle box, 87f, 243ff
Sickle-cell anemia, 54, 91, 109
Snails, 49, 190f
Sociobiology, 22, 58ff, 373
  and behavioral genetics, 62ff
Somatic cell hybridization, 144, 146f
Somatostatin, 109
Spatial ability, 99f, 165, 280ff, 305, 314f, 342
Speciation:
  initiated by behavioral change, 47ff
  Linnaean theory, 15, 56
Specific cognitive abilities, 198, 229f
  adoption studies, 341f
  family studies, 280ff
  twin studies, 312ff
Speech, 307
Squirrels, 61f
Stabilizing selection, 57
Statistics:
  correlation, 205ff
  covariance, 204ff
  least-squares, 208
  mean, 202f
  regression, 205ff
  standard deviation, 204ff
  variance, 203ff
  See also Variance components
Strain studies, 240ff
Survival of fittest, 21
Sympatric species, 47

Taste. *See* Phenylthiocarbamide (PTC) tasting
Tay-Sachs disease, 91, 113f
Triple-X syndrome, 155ff
Trisomy-13, 154f
Trisomy-18, 154f
Trisomy-21. *See* Down's syndrome
Truncate selection, 96
Turner's syndrome, 162ff
Twin studies:
  aging, 307
  co-twin control studies, 305f, 322

creative abilities, 307
embryology, 290ff
equal environments assumption, 295ff
families of identical twins, 304f
genetic similarity within pairs of fraternal twins, 306
heritability estimation, 299ff
IQ, 307ff
language, 307
manic-depressive psychosis, 323
methods, 28–32, 389ff, 296ff
schizophrenia, 319ff
scholastic abilities, 307, 316ff
sexual behavior, 307
social attitudes, 307
speech, 307
twin registers, 307
variance components, 223
vocational interests, 307
zygosity determination, 292ff

Uniformitarianism, 16
Unipolar manic-depressive psychosis. *See* Manic-depressive psychosis
Urea, 103
Use and disuse, 15f

Variability:
and evolution, 21ff, 33f, 56ff
sources of, 36, 135, 177ff, 198f
Variance components:
additive, 215ff, 222ff
and adoption studies, 329f
dominance, 215ff, 222ff
environmental, 225ff, 254
epistatic, 215
of F_1, F_2, 254
and family studies, 234ff, 275f
genetic, 222ff, 231ff
genotype–environment correlation, 218, 222, 360ff
genotype–environment interaction, 216, 222, 247, 359ff, 376
and heritability, 224ff, 254
and selection studies, 258ff
and strain studies, 253f
and twin studies, 299ff
Verbal ability, 164, 198, 280ff, 314f, 342
Vertebrates, 56
Violent behavior. *See* Aggressive behavior
Vision, 28
See also Phototaxis
Vocational interests, 298, 307

Wasps, 59, 190
Wildebeests, 60

X linkage. *See* Sex linkage
XO, 162ff
XXX, 155ff
XXY, 167ff
XYY, 169ff

Zebras, 60
Zygosity determination, 292ff